Seaforth WORLD NAVAL REVIEW 2018

Seaforth

WORLD NAVAL REVIEW
2018

Editor
CONRAD WATERS

Seaforth
PUBLISHING

Frontispiece: The new British aircraft carrier *Queen Elizabeth* pictured on 30 June 2017, shortly after her 26 June departure from Rosyth on initial sea trials. She will be the largest warship ever to serve with the Royal Navy when delivered towards the end of 2017. The US Navy also took delivery of its own largest-ever warship, the aircraft carrier *Gerald R. Ford* (CVN-78) in 2017. *(Crown Copyright 2017)*

First published in Great Britain in 2017 by
Seaforth Publishing
An imprint of Pen & Sword Books Ltd
47 Church Street, Barnsley
S Yorkshire S70 2AS

www.seaforthpublishing.com
Email info@seaforthpublishing.com

British Library Cataloguing in Publication Data
A CIP data record for this book is available from the British Library

ISBN 978-1-5267-2009-2 (Hardback)

ISBN 978-1-5267-2010-8 (Kindle)

ISBN 978-1-5267-2011-5 (ePub)

Typeset and designed by Stephen Dent
Printed in China by 1010 Printing International Ltd.

CONTENTS

Section 1: Overview

Section 2: World Fleet Reviews

Section 3: Significant Ships

Section 4: Technological Reviews

Note on Tables: Tables are provided to give a broad indication of fleet sizes and other key information but should be regarded only as a general guide. For example, many published sources differ significantly on the principal particulars of ships, whilst even governmental information can be subject to contradiction. In general terms, the data contained in these tables is based on official information updated as of June 2017, supplemented by reference to a wide range of secondary and corporate sources, such as shipbuilder websites.

1 OVERVIEW

INTRODUCTION

'In these matters the only certainty is that nothing is certain' wrote Pliny the Elder (AD 23–AD 79), the Roman officer, fleet commander and author who was to meet an untimely death during the eruption of Vesuvius. The seemingly improbable series of events that have marked the last twelve months bear witness to the truth of this observation. The United Kingdom's vote to leave the European Union (the so-called 'Brexit'), the election of Donald Trump to the American presidency and the emergence of Emmanuel Macron as France's new, progressive leader are just some of the milestones that mark a turbulent and unsettled year.

These political developments will inevitably feed through to defence – and naval – policy. The impact of the new Trump administration in the United States is likely to be the most marked. At one level, the advent of a new administration is apparently being accompanied by a more robust approach to defence spending that has been significantly constrained in recent years by the financial curbs imposed by the 2011 Budget Control Act (BCA). The FY2018 budget request from the US Department of Defense amounts to a total of US$639bn. This is over five percent higher than the previous year's authorisation and includes a base budget set some US$52bn above the BCA cap. However, uncertainties remain. The Trump presidency has yet to translate its campaign promises on defence – including an enlarged, 350-ship navy – into a clear defence strategy. This will only emerge over the next year. It is also unclear to what extent President Trump's rhetoric will be matched by reality. Notably, if a military build-up is to be achieved, an accommodation will need to be reached with the fiscally conservative Republicans who could block spending proposals in Congress.[1]

The new Trump presidency is also significant in terms of its implications for the United States' relations with its global network of alliances. In addition to the inevitable tensions resulting from the President's penchant for 'Twitter diplomacy', a more enduring theme has been his emphasis on a nationalistic 'America First Foreign Policy'. This seemingly views trade agreements and many other forms of international collaboration as a burden. A particular *bête noir* for the new administration is their view that many of America's allies are not paying their fair share of the costs associated with collective security arrangements, such as NATO.[2]

Table 1.0.1: COUNTRIES WITH HIGH NATIONAL DEFENCE EXPENDITURES – 2016

RANK	COUNTRY	TOTAL: US$	SHARE OF GDP: %	CHANGE 2007–16
1 (1)	United States	611bn	3.3%	-4.8%
2 (2)	China	215bn	1.9%	118%
3 (4)	Russian Federation	69.2bn	5.3%	87%
4 (3)	Saudi Arabia	63.7bn	10.0%	20%
5 (7)	India	55.9bn	2.5%	54%
6 (5)	France	55.7bm	2.3%	2.8%
7 (6)	United Kingdom	48.3bn	1.9%	-12%
8 (8)	Japan	46.1bn	1.0%	2.5%
9 (9)	Germany	41.1bn	1.2%	6.8%
10 (10)	South Korea	36.8bn	2.7%	35%
11 (11)	Italy	27.9bn	1.5%	-16%
12 (13)	Australia	24.6bn	2.0%	29%
13 (12)	Brazil	23.7bn	1.3%	18%
14 (14)	United Arab Emirates	22.8bn	5.7%	123%
15 (15)	Israel	18.0bn	5.8%	19%
-	**World**	**1,686bn**	**2.2%**	**14%**

Information from the Stockholm International Peace Research Institute (SIPRI) – https://www.sipri.org/databases/milex/
The SIPRI Military Expenditure Database contains military expenditure data on countries over the period 1949–2016.

Notes:
1 Spending figures are at current prices and market exchange rates.
2 Figures for China Saudi Arabia and the UAE are estimates, with the UAE figures relating to 2014.
3 Data on military expenditure as a share of GDP (Gross Domestic Product) relates to 2016 GDP estimates from the IMF World Economic Outlook and International Financial Statistics database, October 2016.
4 Change is real terms change, ie adjusted for local inflation.
5 Figures in brackets reflect rank in 2015, revised for latest information.

Although previous threats to abandon the alliance made during the presidential election campaign appear to have dissipated, significant pressure is being put on allies to spend more on defence. This appears to be having some effect, as evidenced – for example – by Canada's pledge in early June 2017 to boost its defence budget by nearly three-quarters over the next ten years. An eye-watering CAD$60bn (US$45bn) will be devoted to the wholesale renewal of the Royal Canadian Navy's surface fleet. However, the impact of the burden-sharing approach could eventually prove to be a two-edged sword. A loss of confidence in America as a reliable ally, combined with the growth of more powerfully-armed mid-ranking powers, might well ultimately destabilise the current world order that has the United States' financial and military hegemony at its core.[3]

Some of the difficulties already inherent in maintaining US global influence have been revealed by developments in South East Asia over the past twelve months. Although China's expansionist ambitions in the region received a setback on 12 July 2016 when the Permanent Court of Arbitration in The Hague determined that its claims over much of the South China Sea had no legal basis, an ongoing economic and diplomatic offensive has negated much of the ruling's impact. Notably, a Chinese rapprochement with new Philippine president Rodrigo Duterte have done much to negate previous American efforts to build a network of regional alliances to counter China's claims. The sustained rise in Chinese military spending and capability seen in recent years have already resulted it being regarded as a 'near-peer US competitor'.[4] However, there are some signs that the economic growth that has spurred China's defence expenditure is now slowing as the country starts to reach maturity.

For the time being, the most recent available information for global defence spending is captured by Table 1.0.1. The data, from the respected Stockholm International Peace Research Institute (SIPRI), confirms that overall global expenditure of some US$1,686bn has continued to plateau. The top fifteen highest spenders also saw no change. In spite of the restraint of budgetary control, US defence spending still dominated the global total and increased slightly over 2015 as the previous financial constraints started to ease. Elsewhere, a slowdown in spending by oil-rich countries such as Saudi Arabia as the collapse in world prices started

The US Navy aircraft carrier *Carl Vinson* (CVN-70) pictured operating with the JMSDF destroyer *Samidare* in the East China Sea on 8 March 2017. The advent of the new Trump administration should result in an increase in the size of the overall US fleet but relations with key allies are on a more uneven footing. *(US Navy)*

Asian navies have benefitted from the region's economic success, growing in status and proficiency. The rise of South Korea's Republic of Korea Navy has been comparatively little commented upon but a strong indigenous shipbuilding sector has assisted the production of some capable ships. These images show the newly-completed minelayer *Nampo*, which has some similarities with the FFX series of littoral warfare frigates. Maintaining an appropriate balance between littoral and blue water forces is an ongoing challenge for South Korean Navy admirals. *(Republic of Korea Defence Acquisition Programme Administration)*

The *Arleigh Burke* class destroyers have proved to be a remarkably successful design and remain in production more than twenty-five years after the first was delivered. This view shows one of the earlier members of the class – *John S. McCain* (DDG-56) – operating in the Philippine Sea on 14 June 2017. *(US Navy)*

to bite was counterbalanced by unexpectedly persistent high levels of spending by Russia. This reflected heavy costs associated with Vladimir Putin's apparently decisive intervention in Syria's civil war and a number of one-off adjustments and is a trend that is expected to reverse from now on.[5] A marked reduction in the British total arose from the severe depreciation of sterling after the Brexit vote.

FLEET REVIEWS

Table 1.0.2, estimating major fleet strengths as of mid-2017, shows the broad correlation between high defence spending and naval power. Eleven of the fifteen naval powers listed also feature on the list of the fifteen top military spenders. Notable changes during the past year include the departure of another member of the exclusive 'carrier club' following Brazil's decision to decommission *São Paulo*, a reduction that will soon be balanced by the delivery of Britain's new *Queen Elizabeth* and the restoration of the club's founder to full membership. The UK Government, weakened by a botched general election, is attempting to use the new ship's imminent arrival to herald a rebirth in British naval power following years of managed decline. However, as set out in Richard Beedall's established biennial fleet review of the Royal Navy, the actual position is somewhat more complex. Notably, continued underfunding is exacerbating a shortage of personnel and putting important procurement programmes in jeopardy.

The trajectory followed by many Asian fleets has been far more positive in recent years, with a number achieving a step-change in their status and proficiency. Amongst these, the Republic of Korea Navy's rise has been comparatively little commented-upon, possibly because of the overriding focus on neighbouring North Korea's nuclear ambitions. However, as revealed in the latest of a series of fleet reviews of Indian Ocean and Asian navies by Mrityunjoy Mazumdar, the country has made considerable strides in its ambitions to become a regional 'blue water' naval power. The main challenge is to balance these ambitions with an ongoing

Table 1.0.2: MAJOR FLEET STRENGTHS MID 2017

REGION	**AMERICAS**			**EUROPE**						**ASIA**				**INDIAN OCEAN**	
Country	**Brazil**	**Canada**	**USA**	**France**	**Italy**	**Spain**	**Turkey**	**UK**	**Russia**	**Australia**	**China**	**Japan**	**Korea S**	**India**	**Pakistan**
Aircraft Carrier CVN/CV	–	–	11	1	1	–	–	–	1	–	1	–	–	1	–
Support Carrier CVS/CVH	–	–	–	–	1	–	–	–	–	–	–	4	–	–	–
Strategic Missile Sub SSBN	–	–	14	4	–	–	–	4	13	–	6	–	–	1	–
Attack Submarine SSGN/SSN	–	–	55	6	-	-	-	7	20	-	9	–	–	1	–
Patrol Submarine SSK	5	4	–	–	8	3	12	-	20	6	50	17	15	13	5
Battleships/Battlecruisers BB/BC	–	–	–	–	–	–	–	–	1	–	–	–	–	–	–
Fleet Escort CGN/CG/ DDG/FFG	8	12	87	18	18	11	16	19	25	11	65	36	25	24	10
Patrol Escort DD/FFG/FSG/FS	3	-	9	15	2	-	8	-	40	-	40	6	16	10	–
Missile Attack Craft PGG/PTG	–	–	–	–	–	–	19	-	35	-	75	6	17	10	9
Mine Countermeasures MCMV	4	12	11	14	10	6	11	15	40	6	25	24	11	4	3
Major Amp LHD/LPD/LPH/LSD	1	–	31	3	3	3	–	6	–	3	4	3	3	1	–

Notes:

1 Figures for Russia and China are approximate.

2 Fleet escorts for South Korea include thirteen deployed in littoral warfare roles.

need to counter the largely asymmetric threat to South Korea's littoral posed by the less sophisticated but more numerous Korean People's Army Naval Force. The South Korean Navy also has an important part to play in the evolving three-pronged national defence to potential North Korean missile attacks. Realising the ballistic missile defence potential inherent in the country's Aegis-equipped KDX-III destroyers will form a key part of this strategy.

SIGNIFICANT SHIPS

The KDX-III destroyer design is a development of the US Navy's *Arleigh Burke* (DDG-51) class, a type that has enjoyed considerable longevity since its conception in the later years of the Cold War. Indeed, the failure of the following *Zumwalt* (DDG-1000) design to meet the US Navy's expectations has resulted in the *Burke*s being placed back into construction. These so-called 'Restart' ships are now being delivered and the first two – *John Finn* (DDG-113) and *Rafael Peralta* (DDG-115) – to join the fleet are scheduled for commissioning in July 2017. Transition of construction to an improved Flight III design, featuring the new Raytheon AMDR advanced missile defence radar is also imminent.[6] We have therefore taken the opportunity to ask Norman Friedman, who was a personal advisor to the then Secretary of the Navy, John Lehman, at the time the DDG-51 design was developed, to review the class's origins and review its remarkable success to date. We intend that this will provide a firm foundation from which to carry out an analysis of the evolved, Flight III type in a later edition.

Whilst the *Arleigh Burke* class evidences the US Navy's ongoing focus on high-intensity warfighting, other fleets have adopted a nuanced approach. One case in point is Germany, which has increasingly focused on international stabilisation missions in lower-threat scenarios since the end of the Cold War removed the direct threat to its borders. This emphasis is reflected in its new F125 type *Baden-Württemberg* class frigates, the latest evolution of the highly successful MEKO concept that first originated in the 1970s. These cruiser-sized vessels are the largest surface combatants built in Germany since the end of the Second World War and incorporate a number of design innovations intended to facilitate lengthy overseas deployments. However, their limited armament – possibly driven partly by political considerations – makes them of questionable value in a period of renewed East/West tension.

Lockheed Martin's production line for the F-35 Lightning II Joint Strike Fighter at Fort Worth, Texas is seeing an expansion in activity as the design moves towards maturity. The F-35's powerful suite of sensors is likely to play a key role in the US Navy's development of co-operative engagement capabilities. *(Lockheed Martin)*

An even more modest capability is provided by the Royal New Zealand Navy's *Otago* class OPVs. One of a series of designs developed by what is now Vard Marine, they have greatly enhanced New Zealand's ability to police the vast waters of the South Pacific and Antarctic oceans. Guy Toremans traces the class's acquisition under the somewhat chequered Project Protector and explains the importance of such relatively basic vessels in maintaining a maritime presence at a time when greater awareness of Antarctica's abundant natural resources is driving increased human activity.

TECHNOLOGICAL DEVELOPMENTS

A much greater level of sophistication – and expense – is inherent in much of the equipment considered in our initial reviews of technological developments. David Hobbs' annual overview of world naval aviation notes the delivery of the lead *Ford* class aircraft carrier, *Gerald R. Ford* (CVN-78), to the US Navy and describes some of the progress being made with the Naval Integrated Fire Control-Counter Air (NIFC-CA) system. An iteration of co-operative engagement capability (CEC), NIFC-CA allows suitably-equipped aircraft such as the E-2D Hawkeye and F-35 Lightning II to act as airborne sensors that allow shipboard missiles to engage incoming targets at much greater ranges than hitherto. The greater effectiveness of such networked systems is one possible counter to the anti-access/area denial (A2/AD) strategies being adopted by many of the United States' potential adversaries.

Other technologies that have relevance to countering A2/AD scenarios are considered by Norman Friedman in an assessment of new naval weapons. Focusing on the US Navy's research into laser and rail gun systems, he looks at the progress achieved in bringing these weapons into operational use and some of the challenges that need to be overcome if they are to achieve their theoretical potential.[7] A relatively low-powered laser has already been tested operationally onboard the base ship *Ponce* (AFSB(I)-15) and would be useful against drones and small boats. However, more powerful systems would be needed, for example, in missile defence roles. A number of other navies are looking at how they

Images of a F-35B STOVL variant of the Lightning II undertaking trials on the amphibious assault ship *America* (LHA-6) and the new British aircraft carrier *Queen Elizabeth* at anchor. Although arrivals of new carriers are grabbing the headlines, it is arguable that the advanced technology inherent in systems and planes like the F-35 is of more significance. *(Lockheed Martin / Crown Copyright 2017)*

The Russian light frigate *Boiky* pictured making seemingly heavy weather of a transit of the English Channel in April 2017. Whilst the West has been under pressure, *inter alia*, from a Russian resurgence, there are some indications that its politicians are responding effectively to the challenge. *(Crown Copyright 2017)*

might be able to field weaponised lasers by the early 2020s, not least the Royal Navy under its Dragonfire project.

For the time being, missiles of various types remain the weapon of choice for conducting higher-intensity naval warfare. Returning to the Royal Navy, first-time contributor Richard Scott describes the wide range of new missile systems – largely indigenously developed – that are replacing such familiar Cold War names as Sea Dart and Seawolf. The steady introduction of cutting-edge technology demonstrates the value of the unfortunately comparatively rare partnership between British industry and government that marks UK missile procurement.

We conclude with another first-time contribution. Bruno Huriet, a French merchant navy officer now working in the naval outfitting industry, examines the radical changes in warship habitability that have taken place in recent years. He examines the influence of changing social expectations on crew accommodation, as well as the various technical standards and other considerations that impact how accommodation is laid out and the specification of materials used.

SUMMARY

The frequently unexpected political developments of the past twelve months suggest that we are going through a transitional phase in the world order that has not yet fully run its course. The United States and its global network of allies have appeared to be on something of a back foot in recent years in the face of Chinese expansionism, a Russian resurgence and the persistent threat of global terrorism. Although it would be unwise to suggest that these challenges are now abating, there does seem to be a greater willingness amongst much the West's political leadership to make the investments required to meet these threats. Conversely, however, the loosening of the network of alliances and partnerships marked by developments such as Brexit and the Trump administration's 'America First' policies indicate considerable potential for new instabilities. These have considerable implications for both naval procurement and deployment.

Meanwhile, the technological environment facing navies is also changing. The launch of new warships such as China's first, as yet un-named Type 001A aircraft carrier, is being accompanied by less visible advances in areas such as CEC and new generation weapons that have already been touched upon. Whilst the arrival of new behemoths such as *Gerald R. Ford* and *Queen Elizabeth* are inevitably attracting the headlines, the network-centric capabilities of the F-35 strike fighters that will fly from their decks are, possibly, more indicative of the technological leadership America and its allies are striving to retain.

ACKNOWLEDGEMENTS

As always, this ninth edition of the *World Naval Review* series has only been possible as a result of extensive collaboration. Publishing editor Robert Gardiner has remained unstinting in his support of the title, whilst Steve Dent and John Jordan invariably produce, respectively, design work and drawings of considerable quality. It is pleasing to continue to attract – and retain – writers of the highest distinction. Their contributions are supplemented by a network of photographers, amongst whom Moshi Anahori, Derek Fox, Marc Piché and Devrim Yaylali warrant particular mention. From industry and government, Andrew Bonallack of the RNZN, Marion Bonnet of DCNS, Nathalie van Eeden of Damen Schelde Naval Shipbuilding, Georg Lamerz of TKMS, Esther Benito Lope of Navantia and Craig Taylor of Rolls-Royce have all provided notable assistance at a time when an apparent renewed culture of secrecy means such help is not always assured. Finally, as ever, I thank my wife, Susan, for going far beyond the requirement of spousal duty in her initial proof reading of the text.

Comments and criticisms from readers are always appreciated; please direct them for my attention to info@seaforthpublishing.com

Conrad Waters, Editor
30 June 2017

Notes

1. The Trump administration's 350-ship goal is close to the US Navy's own 2016 Force Structure Assessment, which – doubtless influenced by the incoming President's ambitions – set a requirement of 355 ships. Some of the issues facing President Trump in meeting this – and his wider aspirations for a US military build-up – were assessed by Paul Scharre in 'Why Trump's Military Buildup Is Not As Big As He Promises', carried on the *Fortune.com* website on 2 March 2017 and currently available at: http://fortune.com/2017/03/02/donald-trump-military-defense-spending-budget-address-congress/. Commenting on the initial, outline budget proposal revealed in February 2017, Mr Scharre highlighted both the level of hyperbole in President Trump's presentation of his plans and some of the practical difficulties involved in moving his agenda forward with Congress.

2. A good analysis of some of the concerns relating to President Trump's approach were set out in an article entitled 'Donald Trump seems to see allies as a burden' carried by the Lexington column of web edition of *The Economist* on 4 February 2017 at: http://www.economist.com/news/ united-states/21716034-nato-leaders-make-pitch-president-donald-trump-seems-see-allies-burden. The nationalistic America First Foreign Policy is set out on the White House's website at: https://www.whitehouse.gov/america-first-foreign-policy.

3. An interesting and counter-intuitive assessment of the weaknesses inherent in the Trump administration's approach to burden-sharing was provided by academics Abraham Newman and Daniel Nexon in 'Trump says American allies should spend more on defense. Here's why he's wrong' published on the *Vox* website on 16 February 2017 at: https://www.vox.com/the-big-idea/2017/2/16/14635204/burden-sharing-allies-nato-trump.

4. China's recent diplomatic offence was described by Gabriel Dominguez in 'Turning the tide', *Jane's Defence Weekly* – 22 February 2017 (Coulsdon: IHS Jane's 2017), pp.26–31. The concerns over China as a near-peer competitor were reported in a US Department of Defense news article by Cheryl Pellerin entitled 'DIA Director Testifies on Top Five Global Military Threats' on 23 May 2017. Marine Corps Lieutenant General Vincent R Stewart, Director of the Defense Intelligence Agency, confirmed previous assessments that China, Iran, North Korea, Russia and trans-regional terrorism were currently the most significant threats facing the United States.

5. The view that Russian defence spending has now peaked has been reinforced by Russian treasury figures showing a marked reduction in future military budgets. The headline reduction was highlighted by Craig Caffrey's 'Russia announces deepest defence budget cuts since 1990s', *Jane's Defence Weekly* – 22 March 2017 (Coulsdon: IHS Jane's 2017), p.25. There are, however, good reasons to suggest the actual impact on the armed forces is much less than the headline figure suggests: see Chapter 2.4 for further explanation.

6. AMDR – now officially designated AN/SPY-6 – is a scaleable active phased array assembled from individual building-blocks called radar modular assemblies. Representing a much more recent generation of technology than the AN/SPY-1D series of arrays found on existing Aegis-equipped destroyers, the Raytheon designed and built system will significantly enhance the US Navy's ability to defeat large and complex air attacks.

7. It is noteworthy that systems such as NIFC-CA – coupled with the new Standard SM-6 missile – offer the prospect of longer-range interception of incoming air attacks whereas lasers are effective only at much shorter ranges. This is relevant to an ongoing debate over the optimum mix of missiles carried by US Navy surface combatants, as some argue that filling missile silos with long-range weapons risks negating some of the fleet's offensive potential. An interesting perspective was provided in a 2014 article by Sydney J Freedberg Jr entitled '47 Seconds from Hell: A Challenge to Navy Doctrine' carried by the Breaking Defense website at: http://breakingdefense.com/2014/11/47-seconds-from-hell-a-challenge-to-navy-doctrine/. The article rather overlooks the role of the aircraft carrier as the US Navy's strike weapon of choice by virtue of their much greater munitions-carrying potential.

2.1 REGIONAL REVIEW

Author:
Conrad Waters

NORTH AND SOUTH AMERICA

INTRODUCTION

The key development in the Americas over the past year has inevitably been the steadily-improving outlook for the already globally dominant US Navy. Fleet numbers have already started to slowly rebound from the post-Second World War low-point of c. 270 reached towards the end of FY2015 but there now appears to be a growing consensus for a much bigger fleet. Whilst the general direction of travel is clear, the likely size and structure of the enlarged navy has still to emerge. On the face of it, the target for 355 warships set out in the US Navy's new Force Structure Assessment (FSA) released in December 2016 is closely aligned with the goal of 350 ships set by Donald Trump during his successful presidential campaign. However, the US Navy's figure is based on the national military strategy put in place by the outgoing Obama administration and may need to be revised to reflect a new US defence strategy that is currently being prepared. Various alternatives to the FSA have already been put forward under three Fleet Architecture Studies mandated by Congress in 2015. Perhaps most importantly of all, the financial constraints imposed by the 2011 Budget Control Act (BCA) are still in place and it remains to be seen whether an already damaged Trump administration will have the political authority to achieve any significant relaxation.[1]

An indication of the already disjointed state of US politics was provided by another year of prolonged discussions over the FY2017 budget. Agreement on funding was finally reached at the start of May 2017, some seven months late. The previous practice of using Overseas Contingency Operations (OCO) allocations to avoid BCA restrictions was adopted to fund a total Department of Defense appropriation of US$606bn, some US$23bn higher than initially requested.[2] Of this, US Navy allocations totalled US$174bn compared with a request of US$165bn and the US$169bn approved in FY2016. The extra money helped pay for nine new warships, two more than first planned. The additions comprised a further Littoral Combat Ship (making three rather than two for the year) and another LPD-17 type amphibious transport dock, the thirteenth member of the class. There was also enough money to buy thirty extra aircraft.

The FY2018 budget request released on 23 May 2017 included US$639bn for the Department of Defense. US$574bn (US$523bn) of this related to the base budget and US$65bn (US$83bn) to OCO. Whilst trumpeted as a substantial increase, critics have observed that the proposal represents only a fairly modest uplift over the previous administration's plans. If approved, naval spending will increase by around 3.7 percent to US$180bn, of which US$171.5bn relates to the base budget and US$8.5bn to OCO. Sensibly, the emphasis is on restoring fleet readiness, badly impacted by previous cutbacks. Conversely, shipbuilding shows little change from previous assumptions.

Meanwhile, the Royal Canadian Navy's future prospects have been clarified by the publication of Canada's new defence policy – *Strong, Secure, Engaged* – which is discussed in more detail below.[3] Elsewhere in the Americas, spending in Latin America has been badly impacted by the effect of lower commodity prices on the revenues of major producers such as Mexico, Peru and Venezuela and the major economic and political crisis in Brazil. Brazil, which accounts for almost half of total defence expenditure in South America, was forced to cut defence spending significantly in 2015–16. Although the latest budget shows a material rebound, much of this is swallowed up by personnel and other running costs, leaving comparatively little to spend on procurement.

The position is exacerbated by the cash requirements of a number of existing major programmes, notably the purchase of Saab Gripen jets and Embraer KC-390 transport aircraft for the air force and the giant PROSUB submarine construction and infrastructure project being carried out in conjunction with France's DCNS. Completion of this programme remains the Brazilian Navy's top priority but the funding situation means it is now running badly behind schedule. It is also draining funds from other requirements and will undoubtedly have been a factor in the decision to abandon the planned modernisation of the carrier *São Paulo*. Her planned departure leaves Latin America without an aircraft carrier for the first time since the acquisition of the former HMS *Vengeance* to serve as Brazil's *Minas Gerais* in December 1956.

Table 2.1.1 provides a summary of the region's significant fleets in mid-2017.

The US Navy's latest nuclear-powered attack submarine *Washington* (SSN-787) was delivered in May 2017 after several months of delay. Latest US Navy Force Structure Assessment goals call for a significant increase in submarine numbers. However, even if funding is available, a somewhat depleted industrial base may well struggle to ramp up production. *(Huntington Ingalls Industries)*

Table 2.1.1: FLEET STRENGTHS IN THE AMERICAS – LARGER NAVIES (MID 2017)

COUNTRY	ARGENTINA	BRAZIL	CANADA	CHILE	COLOMBIA	ECUADOR	PERU	USA
Aircraft Carrier (CVN/CV)	–	–	–	–	–	–	–	11
Strategic Missile Submarine (SSBN)	–	–	–	–	–	–	–	14
Attack Submarine (SSN/SSGN)	–	–	–	–	–	–	–	55
Patrol Submarine (SSK)	3	5	4	4	4	2	6	–
Fleet Escort (CG/DDG/FFG)	4	8	12	8	4	2	8	87
Patrol Escort/Corvette (FFG/FSG/FS)	9	3	–	–	1	6	1	9
Missile Armed Attack Craft (PGG/PTG)	2	–	–	3	–	3	6	–
Mine Countermeasures Vessel (MCMV)	–	4	12	–	–	–	–	11
Major Amphibious Units (LHD/LPD/LPH/LSD)	–	1	–	1	–	–	–	31

MAJOR NORTH AMERICAN NAVIES – CANADA

The major development impacting the Royal Canadian Navy over the past year was the publication of the new *Strong, Secure, Engaged* defence policy already referenced above. Widely praised for its realism in recognising that the current defence budget was inadequate to meet existing commitments, the new plan aims to increase military spending from the present c. 1.2 percent of GDP to 1.4 percent of national wealth by 2024/25.[4] Annual cash spending will increase from CAD$18.9bn to CAD$32.7bn over this period. The extra money preserves existing procurement plans and allows some incremental improvements, including a modest rise in overall armed forces personnel numbers and an increase in the air force's planned buy of replacement jet fighters.

From a naval perspective, the defence policy essentially secures the existing force structure rather than heralding any major new developments. There is a commitment to maintaining a blue water naval capability built around an ability to deploy and sustain two naval task groups. These will each comprise up to four surface combatants and a joint support ship, supplemented 'where warranted' by a submarine. The ability to operate in the Arctic remains another important priority. A fleet of fifteen 'Canadian Surface Combatants', two Joint Support Ships, five or six Arctic Offshore Patrol Ships and four modernised *Victoria* class submarines is expected to provide the necessary fleet mix and capacity to sustain these requirements. It is also planned to replace the existing CP-140 Aurora maritime patrol aircraft operated by the Royal Canadian Air Force with a next generation multi-mission capability.

For the Royal Canadian Navy, the most positive news is undoubtedly confirmation of previous plans to build fifteen new surface escorts. Under the terms of Canada's 2010 National Shipbuilding Procurement Strategy, it has already been determined that these will be constructed by Irving Shipbuilding in Halifax, Nova Scotia. An existing warship design and combat system will be used to reduce risk and speed construction. After two extensions in the timetable, proposals from a number of overseas companies are now expected by mid-August 2017 before a selection decision early in 2018. Actual construction is not expected to get underway until the early 2020s and could spread over two decades. The original budget for the Canadian Surface Combatant was set at c. CAD$26.2bn but, following a year-long re-costing exercise, is now set at between CAD$56bn and CAD$60bn for the fifteen ships.[5] In the meantime, as reflected in Table 2.1.2, the current force of surface escorts has been further reduced to twelve ships following the decommissioning of the final *Iroquois* class destroyer, *Athabaskan*, on 10 March 2017. More positively, the CAD$4.3bn *Halifax* Class Modernisation/Frigate Life Extension (HCM/FELEX) programme was brought towards a successful conclusion with the return of *Toronto*, the twelfth and final ship to be modernised, on 29 November 2016. Irving had carried out the work on her and six other east coast-based vessels; work on five west coast-based vessels had been completed earlier in the year by Seaspan at Vancouver.

Irving is currently busy with constructing the new *Harry DeWolf* class Arctic Offshore Patrol Ships. Deliveries are scheduled to commence in 2018 under a landmark programme for Canadian ship-

The Royal Canadian Navy frigate *Montreal* operating in company with the destroyer *Athabaskan* in the course of exercise Spartan Warrior in the Atlantic Ocean in November 2016. The recent Canadian defence review means that fifteen new Canadian Surface Combatants will replace its current surface vessels. *(Canadian Armed Forces)*

Table 2.1.2: CANADIAN NAVY: PRINCIPAL UNITS AS AT MID 2017

TYPE	CLASS	NUMBER	TONNAGE	DIMENSIONS	PROPULSION	CREW	DATE
Principal Surface Escorts							
Frigate – FFG	**HALIFAX**	12	4,800 tons	134m x 16m x 5m	CODOG, 29 knots	225	1992
Submarines							
Submarine – SSK	**VICTORIA** (UPHOLDER)	4	2,500 tons	70m x 8m x 6m	Diesel-electric, 20+ knots	50	1990

building that has effectively involved setting up a new shipyard from scratch. Progress on the lead vessel is now well-advanced and a formal keel-laying ceremony for the second class member, *Margaret Brooke*, was held on 29 May 2017. The Royal Canadian Navy hopes for six members of the class. However, construction of the final vessel depends on the earlier ships being built within budget.

The next twelve months should also see the start of work on the first of two new *Queenston* class Joint Support Ships. Based on the German Type 702 *Berlin* class, these have been allocated to Seaspan on the west coast and are now expected to be delivered from 2021 onwards.[6] Meanwhile, a programme to provide an interim fleet supply capability known as Project Resolve is making good progress in converting the container ship *Asterix* into an auxiliary replenishment vessel. The work is being carried out by Chantier Davie in Quebec, who plan a public unveiling of the converted ship on 20 July 2017 before the commencement of sea trials in the autumn. The ship is being acquired under a CAD$700m contract that includes both the supply and operation of the ship over a five-year period. There is also an option to purchase the ship at the end of the period, a possibility that might ultimately prove attractive given that the new defence policy document makes no reference to the option of a third Joint Support Ship being taken up. The sole-source Resolve project has not, however, been without its controversies and its approval was delayed for a few weeks by the incoming Liberal government in November 2015 after complaints from rival builders. The project has been linked with mysterious suspension of the Vice Chief of the Defence Staff and former head of the Royal Canadian Navy, Vice Admiral Mark Norman, from his post in January 2017. The Canadian press has reported that he has been under police investigation to establish whether he leaked information to the Project Resolve consortium to avoid delays to a programme regarded as vital for naval capabilities.[7]

One new naval investment funded under the 2017 defence policy review is a major life-extension programme for the *Victoria* (former British Royal Navy *Upholder*) class submarines. Confirming plans first mooted in mid-2015, the submarine life extension will modernise the boats' combat systems to allow them to serve into the 2040s. Along with an upgrade of the existing lightweight torpedoes used by the navy and air force, the programme will be part of a CAD$2.9bn investment to meet 'evolving underwater threats'. The decision seemingly confirms growing confidence in platforms that have received considerable criticism – and more than their fair share of problems – since first being acquired in 1998.

Pending orders for new ships, the Royal Canadian Navy's *Halifax* class frigates have been extensively modernised. The work on *Ville de Quebec* was completed at the end of 2015 but she subsequently suffered minor damage from a generator fire whilst undertaking harbour trials. Repairs were quickly effected and she is pictured here in September 2016 whilst undertaking a tour of the Great Lakes and Saint Lawrence Seaway. *(Marc Piché)*

MAJOR NORTH AMERICAN NAVIES – UNITED STATES

The US Navy had an inventory of just over 275 warships as of mid-2017. This represented a slight improvement in the position a year previously. The delivery of the *Gerald R. Ford* (CVN-78) at the end of May 2017 brought carrier numbers back to the Congressionally-mandated eleven. Additionally, completion of the initial 'Restart' DDG-51 class destroyers helped to bolster numbers of major surface combatants whilst ongoing Littoral Combat Ship construction is having a similarly beneficial impact in rebuilding the fleet of frigate-sized ships. The main area of weakness was in numbers of underwater vessels, with construction of *Virginia* (SSN-774) class submarines not quite sufficient to keep pace with the retirement of Cold War-era *Los Angeles* (SSN-688) type boats.

Fleet numbers will continue to grow in the short term – to over 290 vessels during FY2018 – as deliveries of destroyers and Littoral Combat Ships, as well as of amphibious and support shipping, exceed planned withdrawals. Thereafter, much will depend on the extent to which the new December 2016 Force Structure Assessment – or one of the three alternative Fleet Architecture Studies (FASs) published in February 2017 – become a reality. The last-mentioned documents – produced by the Department of the Navy, the Federally-funded MITRE Corporation and the Center for Strategic and Budgetary Analysis (CSBA) – look at potential fleet structures as of 2030.[8] These are summarised in the table and text box below.

The merits of the three studies have been widely debated and elements will undoubtedly find their way into future fleet planning. However, the more immediate question is likely to be the extent to which the requirements of the 355-ship 2016 FSA – effectively the current official view – will be funded. Past history is not promising, as evidenced by current fleet size-compared with previous FSA requirements well in excess of 300 ships. The Congressional Budget Office (CBO) has calculated that the enlarged fleet would cost around US$26.5bn p.a. just to acquire; over US$5bn more each year than its c. US$21bn estimate of the current thirty-year shipbuilding plan and around sixty percent more than the US$16bn or so Congress has been willing to authorise historically.[11] Overall operating costs would also rise significantly. Another question relates to the time required to reach the new 355-ship goal – the CBO looked at different scenarios and concluded that it would not be met before 2035 under even an aggressive shipbuilding programme. Some consideration is therefore being given to reactivating decommissioned ships to accelerate the build-up. Some of the more recently retired FFG-7 class frigates and, more improbably, even the conventionally-powered carrier *Kitty Hawk* (CV-63) are being mentioned in this regard.

For the time being, current major units in the fleet's inventory are set out in Table 2.1.4 and the proposed FY2018 shipbuilding programme is listed in Table 2.1.5. Although the usual full five-year Future Years Defense Program (FYDP) numbers are available from scouring the budget submission, presentations focused only on FY2018 proposals given the high likelihood that plans will change once a new national military strategy has been determined. In a somewhat bizarre last-minute change, the original submission on 23 May 2017 requested eight ships but an additional Littoral Combat Ship was added in an adjustment the following day. This is the only material revision to the FY2018 programme envisaged by the previous administration.

More detailed comments by warship category are set out below.

Table 2.1.3: US NAVY: POTENTIAL FLEET STRUCTURE OPTIONS

SHIP TYPE	ACTUAL MID 2017	TARGET FSA 2014	TARGET FSA 2016	OPTION 2030 NAVY STUDY	OPTION 2030 MITRE STUDY[1]	OPTION 2030 CSBA STUDY
Carrier – CVN/CV	11	11	12	11	14[2]	12
Carrier – CVL	–	–	–	3	–	10[3]
Submarine – SSBN	14	12	12	12	12	12
Submarine – SSN/SSGN/SS	55	48	66	53	74[4]	66
Large Surface Combatant	87	88	104	91	160	74
Small Surface Combatant	9	52	52	48	46	71[5]
Mine Warfare – MCMV[6]	11	–	–	–	–	–
Amphibious Warfare Ship	31	34	38	35	38	29[3]
Expeditionary Fast Transport	9	10	10	10	–[7]	0
Expeditionary Support Base	4	3	6	7	–[7]	2
Combat Logistics Force	29	29	32	30	–[7]	31
Command & Support	16	21	23	21	70	33
TOTAL	**276**	**308**	**355**	**321**	**414**	**340**
Unmanned Vehicles[8]	N/A	N/A	N/A	136	N/A	80

Notes:

1 The structure identified in the MITRE study was based on updating the methodology used in the 2014 FSA to the current world situation and the Defence Strategic Guidance requirement to defeat one near-peer adversary whilst simultaneously deterring another; it was not regarded as realistically achievable and therefore not a recommended structure.

2 MITRE recommended exploring the possibility of building non nuclear-powered carriers.

3 These ships would initially be the existing LHA/LHDs with air groups focused on F-35Bs; they would be replaced by CATOBAR-equipped ships in due course.

4 Some of these boats would use non-nuclear propulsion.

5 These would be supplemented by forty-two additional missile-armed patrol vessels.

6 MCMV capabilities are included within small surface combatants in future plans.

7 Not separately split-out.

8 Not specifically identified in the current fleet, FSAs and the MITRE study.

FUTURE VISIONS FOR THE US NAVY

Alternative visions for the US Navy are summarised in Table 2.1.3. The starting point is the current US Navy fleet structure, as of mid-2017. This can be regarded as a scaled-down version of the balanced fleet of carrier task groups, amphibious forces and submarines that fought and won the Pacific Campaign in the Second World War. The subsequent columns look at various future goals and options, which can be summarised as follows:[9]

FSA 2014 and FSA 2016: Force Structure Assessments are carried out by the Assessment Division (OPNAV 81) of the Office of the Chief of Naval Operations. They are intended to establish the near-term force structure required to meet current US national military strategy. As such, prior to publication of a new strategy by President Trump, they are heavily influenced by the Obama administration's defence strategic guidance, particularly the most challenging requirement that forces should be able to defeat a capable adversary in one region whilst deterring an opportunistic attack by an opportunistic aggressor in another. The FSAs are largely based on existing or planned ship types as well as the current overall fleet architecture; they are therefore largely focused on the number of ships of various categories required to meet the desired objectives. Whereas the 2014 FSA was similar in both total and balance to a series of previous FSAs

Aircraft Carriers: Trials of the new aircraft carrier *Gerald R. Ford* (CVN-78) commenced on 8 April 2017 and she was formally accepted on 31 May 2017. A commissioning ceremony is scheduled for 22 July 2017 but an extensive period of post-acceptance trials and subsequent work-up means that it will not be until early in the next decade that she is ready for operational deployment. Structural work on the second member of the class, *John F. Kennedy* (CVN-79), was reported to be nearly half complete by June 2017 and she is on schedule for launch during 2020. Initial fabrication has also started on the third ship, *Enterprise* (CVN-80), under a contract awarded on 1 February 2017. She will be formally authorised in the FY2018 budget.[12] Latest budget submissions suggest that programme costs are now being reasonably well controlled after earlier increases, although the equipment fit has been simplified on the follow-on-vessels. Attention is turning to whether production needs to be accelerated from the current drumbeat of one ship every five years, as this is insufficient to maintain the fleet at the Congressionally-mandated level of eleven ships beyond FY2040, let alone the new twelve ship target. This may lead to an acceleration of the construction timetable for the as yet unnamed CVN-81.

Surface Combatants: The 'Restart' Flight IIA *Arleigh Burke* (DDG-51) destroyers ordered from FY2010 onwards are now starting to be delivered. *John Finn* (DDG-113) was the first scheduled to commission in mid-July 2017 and *Rafael Peralta* (DDG-115) was expected to follow before the end of the month. *Ralph Johnson* (DDG-114) should also be in the US Navy's hands before 2017 draws to a close. A total of fifteen of the class have been authorised since the decision was taken to recommence production; ten under a multi-year contract. A further two ships of the class have been requested for FY2018 and it is likely that another multi-year purchase will be sought for reasons of efficiency.

In a change of plan, Huntington Ingalls Industries' (HII's) Ingalls facility will now be allocated the first, improved Flight III ship to enter production. This will be *Jack H. Lucas* (DDG-125), the FY2017 ship awarded to the company. The decision will be a disappointment to General Dynamics'

from 2006 onwards that set force structures of between 306 and 328 ships, the 2016 FSA was notable in its higher objective totalling 355 vessels. Key changes were a requirement for a twelfth aircraft carrier (that also impacted numbers of large surface combatants and combat logistics vessels), further additional large surface combatants to provide enhanced air defence and expeditionary ballistic missile defence capabilities, and a very large uptick in the force of attack submarines. These revisions largely reflect the growing capabilities of China's armed forces, particularly its sophisticated anti-surface missiles, as well a desire to counter increasing Russian and Chinese submarine activity.

Navy FAS: The US Navy's internal 25-page Fleet Architecture Study was produced by a project team headed by the OPNAV 81 Assessment Division that produces the FSAs but included representatives from a number of other naval constituencies. Its report was produced before the 2016 FSA. Perhaps unsurprisingly, its proposed 321 manned ship fleet was heavily focused on existing types. The main change of emphasis from the current fleet was the use of technology – including networked systems and large numbers of unmanned vehicles – to create a 'distributed fleet'. This would achieve the twin aims of further strengthening the influence and deterrent effects of peacetime forward presence whilst improving the effectiveness and resilience of warfighting capabilities. Carrier strike groups – augmented by LHA/LHD type amphibious assault ships and new light carriers evolved from the LHA-6 design – would remain central to power projection. However, their air wings would have a greater emphasis on intelligence-gathering and electronic warfare. They would be supplemented by new surface action groups which would deploy increased numbers of strike weapons and complicate enemy targeting. The report acknowledged the industrial challenges involved in developing and maturing the unmanned vehicle capacity required to support the improved situational awareness in the distributed fleet vision. There would also have to be a significant advance in networked weapons control technology over the already cutting-edge Navy Integrated Fire Control-Counter Air (NIFC-CA) capability.[10]

MITRE FAS: MITRE's seventy-page study essentially took OPNAV 81 FSA methodology as its starting point. It uses this to create an updated force structure analysis on the basis of its own analysis of the current global environment, which suggested a requirement for a fleet of around 414 ships. Appreciating such a force would be unaffordable in most likely scenarios, it therefore suggested twelve specific changes to the current shipbuilding plan to improve its overall effectiveness. Prominent amongst these were large changes to the surface fleet that would see Littoral Combat Ship production curtailed to pay for more destroyers, the purchase of a new frigate design and development of an arsenal ship concept to improve overall firepower. New patrol submarines would be purchased to supplement the nuclear-powered attack boats and cheaper design options (e.g. conventionally-powered aircraft carriers) explored to reduce costs. The overall effect would be to increase fleet size by around twenty vessels over previous plans

CSBA FAS: The CSBA's study was the last of the trio to be finished and the only one completed after publication of the FSA 2016 (although it makes comparisons with the earlier 2014 report). It was also the longest, at some 138 pages. Like the navy's internal study, its recommendations were based on a distributed fleet architecture and an enhanced emphasis on forward presence (largely to act as a deterrent). This would be achieved by 'Deterrence Forces' organised into discrete regions. They would be supplemented by a powerful based 'Maneuver Force' where the navy's strike carrier potential would be largely focused. The emphasis on forward presence leads to a much greater focus on smaller surface combatants than other plans, whilst existing amphibious assault ships are largely used as surrogate aircraft carriers in the deterrence fleets. As for the navy's study, there is significant focus on unmanned assets.

Production of 'Restart' DDG-51 destroyers is now well-advanced; this image shows *Paul Ignatius* (DDG-117) being floated out from Huntington Ingalls Industries' (HHI's) Pascagoula facility in the early morning of 12 November 2016. In a change of plan, HHI has now been allocated production of the lead Flight III variant of the class. *(Huntington Ingalls Industries)*

A view of the Littoral Combat Ship *Detroit* (LCS-7) cutting through the waves whilst on acceptance trials in July 2016. A total of nine Littoral Combat Ships have now been delivered to the US Navy and many more are under construction. *(Lockheed Martin)*

Bath Iron Works (BIW) subsidiary, the other facility involved in producing the destroyers and lead yard for the class to date. Press reports suggest that BIW was unwilling to agree a fixed-price contract for building a significantly revamped design.[13] In addition to incorporating the new AN/SPY-6 Advanced Missile Defence Radar (AMDR), the revised design includes greater electrical generating capacity, a new high voltage electrical0distribution network, improved air-conditioning and a number of structural alterations. BIW has also been falling behind schedule with the construction of existing ships, partly due to the complexities inherent in the three *Zumwalt* (DDG-1000) class ships it has been building. *Zumwalt* herself was commissioned at a ceremony in Baltimore on 15 October 2016, subsequently arriving at her allocated homeport of San Diego in December after a voyage that was delayed by a number of propulsion problems relating to contaminated driveshaft bearings. The second ship of the class, *Michael Monsoor* (DDG-1001), was floated out in June 2016 and the keel of the third and final unit, *Lyndon B. Johnson* (DDG-1002), laid on 30 January 2017.

The lower end of the surface combatant mix is seeing continued arrival of Littoral Combat Ships of both the *Freedom* (LCS-1) and *Independence* (LCS-2) variants. Four of the former and five of the latter type were in commission as of June 2017. More deliveries are expected before the year-end. Orders for a total of twenty-six (thirteen of each variant) have been placed to date, with three further ships authorised in FY2017 and two requested for FY2018. The programme has been subject to continuous change in recent years and remains in a state of flux. In December 2015, then Secretary of Defense Ashton Carter directed a reduction in planned Littoral Combat Ship numbers from fifty-two to forty, with previous plans to construct a frigate-derivative of the class accelerated from the thirty-third to the twenty-ninth ship. Subsequent affirmation of the fifty-two small surface combatant requirement in the 2016 FSA, as well as continued requests to approve more of the current variants,

Table 2.1.4: UNITED STATES NAVY: PRINCIPAL UNITS AS AT MID 2017

TYPE	CLASS	NUMBER	TONNAGE	DIMENSIONS	PROPULSION	CREW	DATE
Aircraft Carriers							
Aircraft Carrier – CVN	**FORD** (CVN-78)	1	100,000 tons+	333m x 41/78m x 12m	Nuclear, 30+ knots	4,600	2017
Aircraft Carrier – CVN	**NIMITZ** (CVN-68)	10	101,000 tons	333m x 41/78m x 12m	Nuclear, 30+ knots	5,200	1975
Principal Surface Escorts							
Cruiser – CG	**TICONDEROGA** (CG-47)	22	9,900 tons	173m x 17m x 7m	COGAG, 30+ knots	365	1983
Destroyer – DDG	**ZUMWALT** (DDG-1000)	1	15,800 tons	186m x 25m x 8m	IEP, 30+ knots	175	2016
Destroyer – DDG	**ARLEIGH BURKE** (DDG-51) – Flight II-A	36	9,400 tons	155m x 20m x 7m	COGAG, 30 knots[1]	320	2000
Destroyer – DDG	**ARLEIGH BURKE** (DDG-51) – Flights I/II	28	8,900 tons	154m x 20m x 7m	COGAG, 30+ knots	305	1991
Littoral Combat Ship – FS	**FREEDOM** (LCS-1)	4	3,500 tons	115m x 17m x 4m	CODAG, 45+ knots	<50[2]	2008
Littoral Combat Ship – FS	**INDEPENDENCE** (LCS-2)	5	3,000 tons	127m x 32m x 5m	CODAG, 45+ knots	<50[2]	2010
Submarines							
Submarine – SSBN	**OHIO** (SSBN-726)	14	18,800 tons	171m x 13m x 12m	Nuclear, 20+ knots	155	1981
Submarine – SSGN	**OHIO** (SSGN-726)	4	18,800 tons	171m x 13m x 12m	Nuclear, 20+ knots	160	1981
Submarine – SSN	**VIRGINIA** (SSN-774)	14	8,000 tons	115m x 10m x 9m	Nuclear, 25+ knots	135	2004
Submarine – SSN	**SEAWOLF** (SSN-21)	3[3]	9,000 tons	108m x 12m x 11m	Nuclear, 25+ knots	140	1997
Submarine – SSN	**LOS ANGELES** (SSN-688)	34	7,000 tons	110m x 10m x 9m	Nuclear, 25+ knots	145	1976
Major Amphibious Units							
Amph. Assault Ship – LHD	**AMERICA** (LHA-6)	1	45,000 tons	257m x 32/42m x 9m	COGAG, 20+ knots	1,050	2014
Amph Assault Ship – LHD	**WASP** (LHD-1)	8[4]	41,000 tons	253m x 32/42m x 9m	Steam, 20+ knots	1,100	1989
Landing Platform Dock – LPD	**SAN ANTONIO** (LPD-17)	10	25,000 tons	209m x 32m x 7m	Diesel, 22+ knots	360	2005
Landing Ship Dock – LSD	**WHIDBEY ISLAND** (LSD-41)	12[5]	16,000 tons	186m x 26m x 6m	Diesel, 20 knots	420	1985

Notes:

1 Some being fitted additionally with hybrid electric drive.
2 Plus mission-related crew.
3 Third of class, SSN-23 is longer and heavier.
4 LHD-8 has many differences.
5 Includes four LSD-49 HARPERS FERRY variants.

have effectively overtaken this direction. It now seems that the navy intends to transition to the new frigate in FY2020. Whilst production of derivatives of the existing types remains possible, other designs will be considered.

Meanwhile, significant changes to the way the navy plans to operate the Littoral Combat Ship type were announced in September 2016. The original concept relied heavily on the type's modularity to allow rapid switches between missions by embarkation of specialised equipment and supporting

The Littoral Combat Ship *Coronado* (LCS-4) pictured whilst undertaking a South East Asian deployment in June 2017. She is seen carrying a trial installation of Harpoon surface-to-surface missiles as part of efforts to increase the type's firepower. Boeing has now withdrawn Harpoon from the competition to provide smaller surface combatants with an over-the-horizon anti-surface capability. *(US Navy)*

Table 2.1.5: US NAVY PLANNED WARSHIP PROCUREMENT: FY2016–FY2018

WARSHIP TYPE	FY2016[1] AUTHORISED	FY2017[2] REQUESTED	F2017[2] AUTHORISED	F2018[3] REQUESTED
Aircraft Carrier (CVN-78)	Nil	Nil	Nil	1(1)
Attack Submarine (SSN-774)	2	2	2	2 (2)
Destroyer (DDG-51)	3[4]	2	2	2 (2)
Littoral Combat Ship (LCS-1/2)	3	2	3	2 (1)
Amphibious Assault Ship (LHA-6)	Nil	1	1	Nil
Amphibious Transport Dock (LPD-17)	1	Nil	1	Nil
Expeditionary Fast Transport (T-EPF-1)	1	Nil	Nil	Nil
Expeditionary Support Base (T-ESB-3)	1	Nil	Nil	Nil
Fleet Oiler (TAO-205)	1	Nil	Nil	1(1)
Towage, Salvage and Rescue Ship (T-ATS(X))	1	Nil	Nil	1(1)
TOTAL	**13**	**7**	**9**	**9 (8)**

Notes:

1 FY2016 numbers relate to the authorised procurement programme. This varied significantly from the initial Presidential budget request, with Congress approving funds for one additional expeditionary fast transport, one additional expeditionary support base and one towage, salvage and rescue ship.

2 Numbers for 2017 relate to the initial FY2017 budget plans requested by the Obama administration and the numbers actually authorised in May 2017.

3 Numbers in brackets reflect purchases for FY2018 previously envisaged in the Future Years Defence Programme numbers provided in the FY 2017 budget.

4 One of the three DDG-51 destroyers listed in FY2016 was approved as an additional ship to meet the terms of a historical agreement between the US Navy, General Dynamics' Bath Iron Works (BIW), and what is now Huntington Ingalls Industries (HHI) relating to the allocation of destroyer and amphibious transport dock production. The approval of a twelfth member of the LPD-17 class in FY2016, to be built by HHI, meant that BIW was entitled to a contract for a further destroyer under the terms of the deal. The additional destroyer did not appear in the FY2016 budget request, nor in many other budget documents.

A computer-generated image of the new amphibious assault ship *Bougainville* (LHA-8), which was ordered on 16 June 2017. She is a modified variant of the *America* class, reverting to the well-deck arrangement seen in earlier US Navy amphibious assault ship classes and incorporating a number of other changes to increase the size of her flight deck. *(Huntington Ingalls Industries)*

personnel. Under new arrangements, ships will be assigned to a division focused on just one single mission and crewed accordingly. The first four ships, regarded as prototypes, will be dedicated to training and evaluation. The next twenty-four vessels will be split equally between six mission-focused divisions assigned to mine countermeasures, anti-submarine and anti-surface warfare. To assist with logistical support, all the *Freedom* variants will be based at Mayport in Florida, whilst the *Independence* class type will have their home-base in San Diego, California. The majority of ships will be assigned dual 'Blue' and 'Gold' crews in a similar arrangement to that adopted for strategic missile submarines.

Amphibious and Support Shipping: No major amphibious warships have been delivered to the US Navy in the last twelve months. However, a strong order book has been maintained by the award of contracts to HII for the third *America* (LHA-6) class amphibious assault ship, *Bougainville* (LHA-8), on 16 June 2017 and the twelfth *San Antonio* (LPD-17) class amphibious transport dock, *Fort Lauderdale* (LPD-28), on 19 December 2016. *Bougainville* incorporates a number of design revisions when compared with her two sister-ships, including a reversion to the well-deck arrangement seen in previous amphibious assault ship classes. Her flight deck has also been enlarged through use of a smaller island structure and an additional sponson to facilitate operations by F-35B Lightning II strike fighters and Osprey V-22 tilt-rotors. Construction is expected to begin late in 2018 for delivery in 2024. Meanwhile her sister, *Tripoli* (LHA-7), was launched on 1 May 2017, some thirteen weeks ahead of schedule.

Fort Lauderdale (LPD-28) is a transitional design incorporating some cost-saving measures that will be used in follow-on classes. It was anticipated that she would be the final member of the LPD-17 type. However, funding for an additional, thirteenth ship, has been approved by Congress. This decision is partly a reflection of the increase from thirty-four to thirty-eight ships in the 2016 FSA, thereby moving closer to meeting a long-standing US Marine Corps' requirement for sufficient shipping to lift the assault echelons of two full Marine Expeditionary Battalions (MBEs) simultaneously.[14] Another influence is a desire to maintain continuity of production of the LPD-17 type pending transition to the new LX(R) type, which will replace the twelve existing LSD-41/LSD-49 series of dock landing ships. This will essentially be a less expensive (and less capable) variant of the *San Antonio* class, built on the same hull. HHI has been awarded the bulk of design work for the LX(R) but General Dynamics' National Steel & Shipbuilding Company (NASSCO) is also involved in the programme and could play a role in construction.

The navy is keen to maintain production at both companies and NASSCO has been awarded construction of the first six vessels of a new class of fleet oilers as part of this strategy. The *John Lewis* (TAO-205) class will replace the existing fifteen *Henry K. Kaiser* (TAO-187) fleet oilers as they reach retirement .However, as many as twenty TAO-205s may be acquired to achieve the 2016 FSA goals.

Another major programme being led by NASSCO is construction of the series of ESD/ESB expeditionary transfer docks/expeditionary support bases. These are modified from a commercial tanker design to act as auxiliary amphibious ships. The keel of the fourth, *Hershel "Woody" Williams* (T-ESB-4) was laid on 2 August 2016 and fabrication of a fifth, the as-yet unnamed T-ESB-5, commenced on 26 January 2017. A commercial design has also formed the basis of the *Spearhead* (T-EPF-1) expeditionary fast transports – the former Joint High Speed Vessels – that are used to support rapid intra-theatre transfers of personnel and equipment. Eight are now in service. The ninth ship, *City of Bismarck* (T-EPF-9), was launched on 7 June 2017 and builder Austal USA was awarded contracts for T-EPF-11 and T-EPF-12 in September 2016. Whilst the type has found favour as a relatively cost-effective means of boosting force manoeuvrability, their lightweight aluminium construction and lack of defensive systems leave them vulnerable to damage in all but the lowest intensity scenarios. This was demonstrated by the effective destruction of the former HSV-2 *Swift*, a very similar vessel, by a missile attack off the coast of Yemen in October 2016 whilst in UAE service.[15]

Submarines: A decline in submarine numbers,

The US Navy is expected to switch from building the current LPD-17 type amphibious transport docks to a new class of LX(R) amphibious ships over the next few years. Huntington Ingalls Industries is heavily involved in both programmes and is leading a design effort to adapt the LPD-17 design to form the basis for the new ship. This artist's impression has, rather confusingly, been used to illustrate both the LX(R) concept and *Fort Lauderdale* (LPD-28), the twelfth member of the LPD-17 class and a transitional ship between the two programmes. *(Huntington Ingalls Industries)*

Construction of the new *John Lewes* (T-AO-205) class of replenishment oilers is due to start at General Dynamics' NASSCO division in San Diego during the course of 2018. As many as twenty ships may eventually be built. *(General Dynamics NASSCO)*

The *Virginia* class submarine *Indiana* (SSN-789) in the course of being floated out from the Newport News shipbuilding facility on 9 June 2017. The US Navy is currently ordering two *Virginia* class boats each year and needs to build more to meet force structure targets. However, production of new *Columbia* class strategic submarines is also set to begin in the next few years and this will put a major strain on capacity. *(Huntington Ingalls Industries)*

On 17 June 2017 the US Navy destroyer *Fitzgerald* (DDG-62) was in collision with the Philippine-flagged container ship *ACX Crystal* and suffered significant damage. Flooded compartments included two berthing areas and seven crew members lost their lives in the incident, the reasons for which have still to be determined. The accident will put further pressure on an already overly-stretched US Navy fleet. *(US Navy)*

reflecting lower levels of procurement following the end of the Cold War, is one of the major challenges currently facing the US Navy. Although the 2016 FSA increased the targeted attack submarine force from forty-eight to sixty-six boats, actual fleet numbers could fall to just over forty in the next decade on the basis of current shipbuilding activity. Orders for the existing *Virginia* (SSN-774) class have been running at a steady two p.a. since FY2011 under what has been considered a very successful programme. The two yards – General Dynamics Electric Boat and HII Newport News – involved in construction have managed to deliver many of the latest boats under target cost. In addition, construction times have been steadily falling from the eighty-four months required for the earliest members of the class to around sixty-six months currently. However, there have recently been signs that the 2011 ramp-up in production has been straining both the building yards and the supporting industry base, including the supply of defective parts and – most recently – delays to delivery of the latest, fourteenth boat *Washington* (SSN-774).[16] A further complication is the planned start of construction of the new *Columbia* (SSBN-826) class strategic missile submarines with effect from the FY2021 programme. These will be built by the same yards constructing the *Virginia* class and *Virginia* production is currently scheduled to fall to one p.a. in years when a *Columbia* is ordered. The navy is investigating the potential of boosting capacity to a total of three boats p.a. but, as demonstrated by the problems already referenced, this may not be easy.

Operationally, the US Navy continues to be challenged by the demands of a forward presence strategy that is seeing around 100 vessels of the c. 275-ship fleet deployed at any one time and average deployment length remaining stubbornly over 200 days. This high operational tempo – as well as past cutbacks to maintenance budgets – appear to be having some impact on availability. For example,

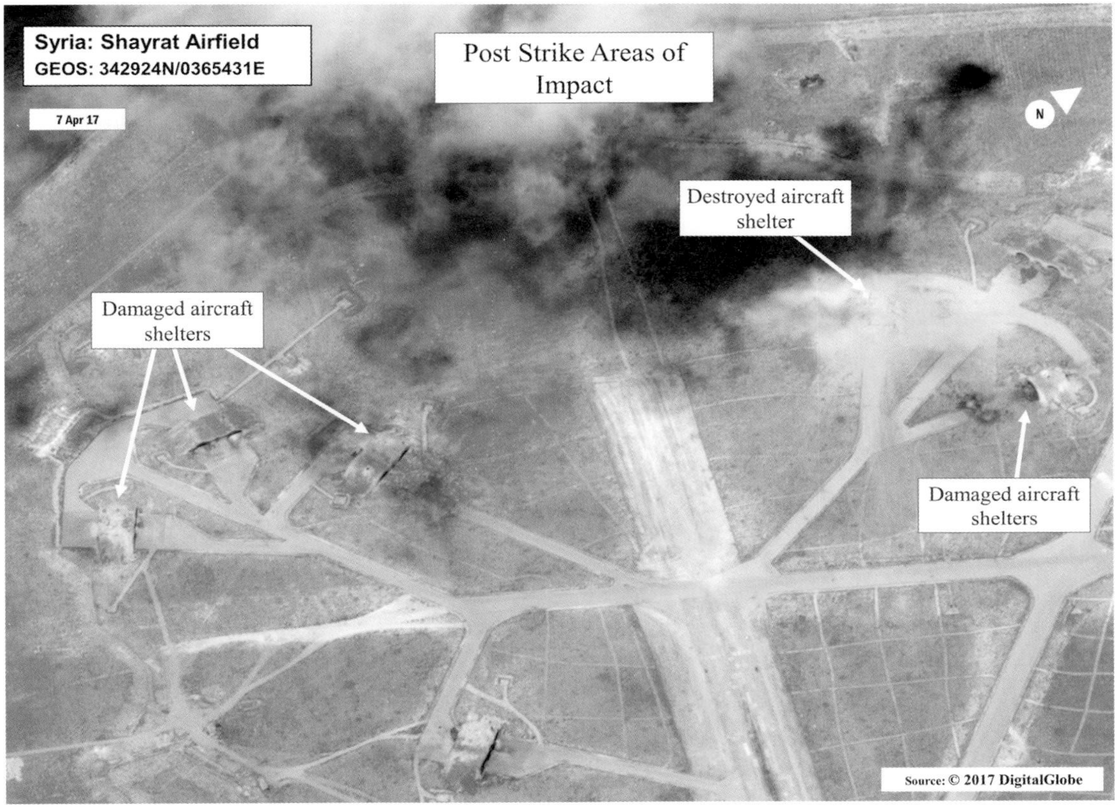

The US Navy has seen considerable active service in the Middle East over the last twelve months. This 7 April 2017 image is one of a series released by the US Navy after cruise missile strikes on the Syrian Shayrat Airfield by the destroyers *Ross* (DDG-71) and *Porter* (DDG-78) in response to the Syrian government's alleged use of chemical weapons. *(US Navy)*

it has been reported that the key US Central Command was left without a US Navy carrier presence for around two months at the start of 2017. In the first week of 2017, no US Navy carrier was deployed anywhere for the first time since the Second World War.[17] A major problem has been the build-up of maintenance work required as a result of extended deployments and a lack of resources at the naval yards – Norfolk, Virginia and Puget Sound, Washington – that perform carrier overhauls to meet the backlog. Given this backdrop, the emphasis placed on bolstering maintenance – including a continued increase in numbers of shipyard personnel – in the FY2018 budget request is a positive development.

Pending any revisions resulting from the new administration's review of national defence strategy, the Pacific remains the navy's main area of focus. Two – China and North Korea – of the five main threats the United States believes it faces are located in the region and around half the forward deployed fleet is located there at any one time. China's ongoing maritime expansion and North Korea's nuclear weapons programme have inevitably continued to exert a major influence on fleet activity. This has included an increased presence in the South China Sea to supplement the freedom of navigation operations (FONOPs) initiated by President Obama.[18] To date, physical confrontation has been avoided. The navy did, however, suffer seven fatalities and severe damage to the destroyer *Fitzgerald* (DDG-62) after a collision with the Philippine-flagged container ship *ACX Crystal* on 17 June 2017.

Conversely, the navy is seeing considerable active service in the Middle East. Here enduring operations against Islamic State forces across the region are being complicated by skirmishes with Houthi insurgents targeting shipping off the coast of Yemen

The sixth 'Legend' class national security cutter *Munro* (WMSL-755) – seen here on builders' trials in August 2016 – was commissioned into US Coast Guard service on 1 April 2017. The design of a new, intermediate-sized offshore patrol cutter was decided during the course of the last year. *(Huntington Ingalls Industries)*

and some more significant clashes with Russian-backed loyalist forces engaged in Syria's civil war. A major cruise missile strike involving fifty-nine Tomahawk missiles was launched against the Al-Shayrat airbase by the destroyers *Ross* (DDG-71) and *Porter* (DDG-78) on 7 April 2017 in response to the alleged use of chemical weapons by Syrian government forces. Subsequently, on 18 June 2017, a F/A-18E Super Hornet operating from *George H. W. Bush* (CVN-77) shot down a Syrian government Su-22 fighter-bomber attacking US-backed Syrian Democratic Forces. Both actions have put renewed strain on American relations with Russia, which had previously look set to improve under the new Trump presidency. The US Navy also continues to remain active in operations intended to reassure NATO allies facing renewed Russian assertiveness in spite of the pressure the Trump administration is exerting on Europe to spend more on its own defence.

The last year has seen the US Coast Guard continue to take delivery of 'Legend' class national security cutters and 'Sentinel' fast response cutters whilst finalising the design of the new offshore patrol cutters that will slot between these two existing types in size and capability. A contract has been awarded to Eastern Shipbuilding of Panama City, Florida for up to nine of what should ultimately be a total of twenty-five ships that are needed to replace twenty-seven elderly medium endurance cutters. The winning design is based on Vard Marine's 'Vard 7-110' patrol vessel concept. This is part of a series that also encompasses the Irish Naval Service's *Samuel Beckett* and *Róisín* classes, as well as New Zealand's *Otago* and *Wellington*. The new ships will displace around 4,000 tons, have an overall length of 110m and be capable of speeds up to 22 knots. There is a hangar and flight deck capable of supporting a MH-60R medium helicopter, three RHIBs and a 57mm Mk 110 gun. Construction of the lead ship should start in 2018 for delivery during 2021.

The Royal Bahamas Defence Force is completing a major modernisation of its fleet with a range of vessels provided by Dutch-based Damen under Project Sandy Bottom. The project also involves considerable investment in shore-side infrastructure which is not yet fully complete. This October 2016 image shows much of new fleet at US Naval Air Station Key West in October 2016 whilst sheltering from Hurricane Matthew. *(US Navy)*

OTHER NORTH AND CENTRAL AMERICAN NAVIES

The countries immediately to the south of the United States continue their focus on constabulary operations in the absence of any major external threat. Although **Mexico** has aspirations to commence domestic construction of a new class of light frigates to replace its elderly collection of former US Navy ships, its current focus is inevitably on offshore and inshore patrol vessels. The last year has seen good progress achieved with a batch of improved *Oaxaca* class OPVs, which are being built by the local ASTIMAR naval shipyards under the 2013–18 modernisation programme. The first of these, *Chiapas*, was commissioned on 23 November 2016 and a second, named, *Hidalgo* was launched on 9 August. Two further units are currently under construction and authorisation for a fifth ship was announced on 10 April 2017. The new ships have a revised hull form that makes them slightly broader than the initial four *Oaxaca* class vessels that were completed in pairs during 2003 and 2010, as well as a new bulbous bow to improve speed and stability. Other changes include specification of a 57mm Bofors main gun – directed by an indigenous fire-control system – in lieu of the 76mm Oto Melara Compact found in the earlier ships. All the class have facilities to operate and stow helicopters, including the new AS565MBe Panthers that are now in the course of delivery.

The inshore counterparts of the *Oaxaca* class are

Two recent images of the advanced state of construction of the new Brazilian submarine base at Itgauaí near Rio da Janeiro, from where the Brazilian Navy's first 'Scorpène' type submarine will be launched in the course of 2018. Please refer to *Seaforth World Naval Review* editions 2014 and 2016 for earlier views of the site. The giant US$9.5bn PROSUB project is draining money from other Brazilian Navy programmes. *(DCNS)*

the *Tenochtitlan* class inshore patrol vessels that are also being built by ASTIMAR to the Damen Stan Patrol 4207 design. Twenty of these are planned under the fleet renewal plan, of which ten have been ordered to date. The ninth, *Bonampak*, was launched on 12 January 2017 and the tenth, *Chichén Itzá*, followed on 27 June. The design has found favour with several other regional fleets, including **Jamaica**. It concluded an interesting deal with Damen towards the end of 2016 that saw the part-exchange of three old and poorly-maintained Stan Patrol 4207s for two new vessels. The three ships acquired by Damen were collected by heavy lift vessel and returned to the Netherlands for refurbishment and resale. The 4207 type also formed a significant part of the 'Sandy Bottom' project Damen has carried out for the **Bahamas.** The project was due to be largely concluded during 2017 following the delivery of a final pair of four smaller Stan Patrol 3007 patrol ships ordered under the programme and the return in May of the existing *Bahamas* class patrol vessel *Nassau* from a refit in the Netherlands. However, her sister-ship *Bahamas* suffered damage in a fire in October 2016 whilst undergoing a similar refurbishment in a local yard and this will delay completion. The project also involves significant upgrades to supporting base infrastructure and some of this has yet to be concluded.

MAJOR SOUTH AMERICAN NAVIES – BRAZIL

The *Marinha do Brasil* has been hard hit by the country's financial and political crisis. Although the most recent defence budget reversed previous cuts, there remains a major imbalance between an ambitious modernisation programme and available funding given most defence spending is allocated to personnel costs. The result has been the ongoing postponement of all but the highest-priority projects. Moreover, fleet readiness has been heavily impacted by lack of money for maintenance; it has been reported that up to half the fleet has been out of service awaiting refit or repair.[19]

Table 2.1.6: BRAZILIAN NAVY: PRINCIPAL UNITS AS AT MID 2017

TYPE	CLASS	NUMBER	TONNAGE	DIMENSIONS	PROPULSION	CREW	DATE
Principal Surface Escorts							
Frigate – FFG	**GREENHALGH** (Batch I Type 22)	2	4,700 tons	131m x 15m x 4m	COGOG, 30 knots	270	1979
Frigate – FFG	**NITERÓI**	6	3,700 tons	129m x 14m x 4m	CODOG, 30 knots	220	1976
Corvette – FSG	**BARROSO**	1	2,400 tons	103m x 11m x 4m	CODOG, 30 knots	145	2008
Corvette – FSG	**INHAÚMA**	2	2,100 tons	96m x 11m x 4m	CODOG, 27 knots	120	1989
Submarines							
Submarine – SSK	**TIKUNA** (Type 209 – modified)	1	1,600 tons	62m x 6m x 6m	Diesel-electric, 22 knots	40	2005
Submarine – SSK	**TUPI** (Type 209)	4	1,500 tons	61m x 6m x 6m	Diesel-electric, 22+ knots	30	1989
Major Amphibious Units							
Landing Ship Dock – LSD	**BAHIA** (FOUDRE)	1	12,000 tons	168m x 24m x 5m	Diesel, 20 knots	160	1998

The Brazilian Navy's lead *Inháuma* class corvette was decommissioned in November 2016, reducing the fleet of surface escorts to eleven ships. Financial difficulties have significantly delayed renewal of the surface fleet and its size is shrinking as a result. *(Brazilian Navy)*

A 2005 image of Brazil's sole aircraft carrier *São Paulo* (the former French *Foch*) operating in company with the replenishment tanker *Marajó*. The carrier has suffered from poor availability since her acquisition and, in early 2017, it was decided to abandon plans for the expensive and time-consuming modernisation that would have been required to return her to operational service. Meanwhile, *Marajó* was decommissioned on 21 November 2016. *(Brazilian Navy)*

The top priority remains the giant PROSUB project that is being overseen by France's DCNS under a contract awarded in 2009. The programme involves the construction of a new submarine base and assembly plant at Itgauaí near Rio de Janeiro under a joint venture with local conglomerate Odebrecht; the local construction of four submarines to DCNS' 'Scorpène' design; and technical support for the non-nuclear elements of Brazil's first nuclear-powered attack submarine. Total costs are estimated to be in the region of US$9.5bn and the financial crisis has meant that activity has had to be slowed for lack of funds. The fact that Odebrecht is also one of the focal points for the corruption investigations that have rocked the Brazilian political establishment is also an unhelpful backdrop. As of mid-2017, the outfitted sections of the lead boat, *Riachuelo*, were almost ready to be transferred to the final assembly hall for integration prior to a launch scheduled in the second half of 2018. The second member of the class, *Humaitá*, is also in the course of outfitting prior to launch in the autumn of 2020, whilst construction of sections for the final two boats is underway. Deliveries are currently scheduled for 2020 through to 2025, several years later than first planned. Construction of the new nuclear-powered submarine is even further into the future and there have been reports that the project – first launched as long ago as 1978 – may be further delayed.[20] However, the Brazilian Navy was quick to counter the rumours, stating in a press release that the nuclear boat would be launched, on schedule, in 2027.

Next on the navy's priority list is renewal of the rapidly-shrinking escort fleet. This has been further reduced to a total of eight frigates and three corvettes following the decommissioning of the lead *Inháuma* class corvette on 25 November 2016. The original plan was to undertake a major domestic construction programme similar to PROSUB under the PROSUPER banner. This was to encompass five 6,000-ton frigates, five oceanic patrol vessels and a replenishment ship. However, such a project is no longer feasible in the current financial climate. It has therefore slowly been merged with another plan to build four *Tamandaré* class corvettes, which may now become the first of two batches of a programme that will eventually comprise eight ships. Designed with the assistance of Vard Marine, the new 103m ships are broadly similar in size to the existing *Barroso* but will be somewhat broader (12.8m compared with 11.4m) and heavier (2,800 tons compared with 2,400 tons). The specification of an all-diesel propulsion system also suggests an emphasis on endurance rather than speed. The ships are likely to be assembled at the naval arsenal in Rio de Janeiro with the assistance of a foreign yard under a contract that should be signed during 2018.

The *Marinha do Brasil's* other priority remains naval aviation in spite of the decision to abandon plans to modernise the existing *São Paulo* when the true cost of refurbishment became apparent. Although she officially remains on the navy's books, the carrier will never go to sea again in an operational capacity and the existing plan to refurbish navy AF-1 Skyhawks will be cut back to maintain a minimal cadre of fixed-wing pilots with the assistance of friendly navies. It is hoped that this will ease the transition to two new large aircraft carriers that still form part of long-term plans. This ambition appears to be something of a pipe dream given the major funding challenges involved in achieving renewal of the submarine and escort fleets.

OTHER SOUTH AMERICAN NAVIES

Of the other two South American 'ABC' navies, the prognosis for **Chile** remains by far the most positive.[21] Possibly the most efficient Latin American navy, it is built around a relatively small but effective front-line fleet of eight surface escorts, four submarines and an amphibious transport dock. These are supplemented by a range of second-tier vessels of increasingly indigenous construction. The surface vessels were all acquired second-hand from European fleets but have benefitted from periodic modernisation. The three Type 23 frigates are the main focus of current attention. Previous purchases of Thales Type 2087 towed arrays are being followed by a major combat systems upgrade awarded to Lockheed Martin Canada. The new system will be similar to that installed as part of the Royal Canadian Navy's *Halifax* class upgrades and will integrate Hensoldt's TRS-4D multi-function radar and the MBDA Sea Ceptor missile into the frigates' combat suite. The next major programme on the horizon will focus on replacement of the two German-built Type 209/1400 submarines commissioned in the early 1980s. France's DCNS, which built the newer pair of 'Scorpène' type boats that form the balance of the submarine flotilla, must be regarded as a strong contender for the new order.

Meanwhile, domestic warship construction is

Chile's Type 23 frigate *Almirante Cochrane* (the former British *Norfolk*) pictured during the 'Rim of the Pacific' exercise in July 2016. The three Chilean Navy vessels are being modernised with a Lockheed Martin Canada combat management system, Hensoldt TRS-4D radar and MBDA surface-to-air missiles. *(US Navy)*

The Colombian National Navy's *Nariño* – formerly South Korea's *Donghae* class corvette *Anyang* – pictured on a port call to Pearl Harbor during her delivery voyage to her new home in September 2014. Korea is keen to pursue shipbuilding partnerships with Latin American countries and is transferring surplus equipment to regional navies as a means of building goodwill *(US Navy)*

focused on licence-built patrol vessels of Fassmer OPV-80 design, also known as the PZM, by ASMAR. *Cabo Odger*, second of a second pair being built to a more heavily armed and ice-strengthened specification, was launched on 3 August 2016 and will be commissioned in the summer of 2017. A further two members of the class are planned. It is also possible the design could be developed into a more capable, corvette-type vessel to replace the three aging Saar 4 *Casma* class fast attack craft that are approaching the end of their service lives. Meanwhile, ASMAR has also started construction of a new, 13,000-ton icebreaker to replace the existing 1960s-era *Almirante Óscar Viel*, which is scheduled to decommission in 2020. The US$210m vessel is yet another being built to a Vard design. With a length of 111m and beam of 21m, the ship will be capable of operating on 1m thick, year-old ice at a speed of up to 3 knots and will be capable of operating at 15 knots in normal conditions. She will be able to accommodate a crew of 120 personnel and support an embarked helicopter.

By contrast, **Argentina's** navy continues to suffer from the decades of neglect that have been the Argentine armed forces' lot since the end of the Falklands War. Although various renewal programmes have been discussed, there is little sign of tangible progress. Indeed, there has been some talk of further cuts, including suspension of the mid-life refit currently being carried out on the submarine *Santa Cruz*. A rare piece of good news has been completion of the refit of the icebreaker *Almirante Irízar* at the Tandanor shipyard, some ten years after she was disabled by a serious fire in 2007. Initial sea trials are reported to have been successful and she should be ready in time to resupply Argentine outposts in Antarctica during the Antarctic summer.

Elsewhere in Latin America, **Colombia** benefits from the second-largest defence budget in South America, although much of this has been devoted to countering the lengthy insurgency by the FARC revolutionary movement, a campaign that has only recently ended. The navy's need to operate in both the Pacific and Caribbean, as well as along the country's extensive internal river system, is another complicating factor. In general terms, a broadly similar approach to that of Chile has been adopted, with a small force of modernised front-line ships acquired from overseas being supplemented in secondary roles by vessels built by a growing domestic shipbuilding industry. This similarity extends to the selection of the Fassmer OPV-80 design for domestic construction. The third of six ships – renamed *Victoria* in commemoration of the end of the insurgency – was delivered by local shipbuilder COTECMAR in mid-2017. In the same fashion as the later Chilean ships, she features an upgraded armament centred on a 76mm Leonardo gun. The intention appears to base the OPV-80 class vessels on the Caribbean coastline and build a larger, long-range variant of around 93m in length for service in the Pacific. COTECMAR will also be involved in meeting a requirement for a new class of frigates to replace the existing *Almirante Padilla* class, probably through licensed assembly of a foreign type.

Peru's overall fleet renewal programme has many similarities with that of its neighbours. The existing focus is on upgrading the capabilities of the best of existing surface warships and submarines at local facilities whilst rebuilding domestic shipbuilding capabilities with the construction of simpler ships. The former encompasses four of the *Lupo*/*Carvajal* class frigates and the four most modern Type 209 submarines, the latter upgrade being conducted with assistance from Germany's TKMS. However, it is South Korea that is supporting much of the new construction programme, which includes two *Makassar* type amphibious transport docks and a number of patrol vessels for the coast guard. Construction of the former vessels is proving somewhat slower than envisaged but the lead ship, *Pisco*, was floated out on 25 April 2017, allowing a start to

Notes

1. An in-depth analysis of the US Navy's historic, current and possible future force structure is provided by the Congressional Research Service's veteran specialist in naval affairs, Ronald O'Rourke, in the recently-updated *Navy Force Structure and Shipbuilding Plans: Issues for Congress – RL32665* (Washington: Congressional Research Service, 7 June 2017). He is author of a number of regularly-updated reports that are of immense help in clarifying current US Navy procurement issues. The reports are not made directly available to the public but many are hosted by the Federation of American Scientists (FAS) as a public service at https://fas.org/sgp/crs/

2. Overseas Contingency Operations (OCO) funding is supposedly used to support the costs of war-related activity (e.g. in Afghanistan and Iraq) that is additional to 'normal' defence requirements. Since the 11 September 2001 terrorist attacks, over US$1.6 trillion has been spent on OCO or similar 'emergency' type funding. Such emergency expenditure is not subject to the sequestration process imposed by the 2011 Budget Control Act and has been increasingly seen as a 'safety-valve' or 'slush fund' to avoid the worst effects of the budget cap. A good overview of a complex issue is contained in the latest CRS report on the issue by Lynn M Williams and Susan B Epstein entitled *Overseas Contingency Operations Funding: Background & Status – R44519* (Washington: Congressional Research Service, 7 February 2017).

3. See *Strong, Secure, Engaged: Canada's Defence Policy* (Ottawa; Department of National Defence, 2017) at: http://dgpaapp.forces.gc.ca/en/canada-defence-policy/docs/canada-defence-policy-report.pdf

4. Perhaps taking a leaf from the British Ministry of Defence's book, the GDP-related percentages have been recalculated to add some previously-excluded expenditure to the figures. The changes appear to add around 0.2 percent to the calculations.

5. There is some uncertainty over the basis of the revised estimated programme cost, although it would appear to be the cash outlay associated with acquiring (but not operating) the ships over a lengthy procurement timescale. If so, it correlates quite closely with figures contained in a report by Jean-Denis Fréchette, Canada's Parliamentary Budget Officer, published on 1 June 2017 entitled *The Cost of Canada's Surface Combatants* (Ottawa: Office of the Parliamentary Budget Officer, 2017). It suggests cash costs of CAD$61.8bn for a fifteen-ship programme, equating to CAD$39.9bn in terms of FY2017 Canadian dollars. It also indicates that a quarter of this budget could be saved if the ships were built in a foreign yard. A copy can currently be found at: http://www.pbo-dpb.gc.ca/web/default/files/Documents/Reports/2017/CSC percent20Costing/CSC_EN.pdf

6. Under the terms of the 2010 National Shipbuilding Procurement Strategy, west coast-based Seaspan is responsible for construction of non-combatant vessels for the navy and coastguard, with east coast-based Irving being allocated major warship work.

7. See for example David Pugliese's 'Why a trial for Vice-Admiral Mark Norman could prove embarrassing to the Liberals', *National Post* – 27 April 2017 (Toronto: Postmedia Network Inc., 2017). To date, no charges have been laid against the Vice-Admiral.

8. The three studies became public at the same time but were completed at various dates. The MITRE report – *Navy Future Fleet Platform Architecture Study* (McLean VA: The

be made on the second vessel, *Paita*. South Korea has transferred the *Po Hang* class corvette *Gyeongju* as part of the deal; she is now in service under the name *Ferré*. South Korea's ultimate aim is to secure pole position for the next stage of Peru's renewal programme, which will involve the replacement of the remaining unmodernised *Carvajal* class frigates and *Velarde* class fast attack class with a new surface combatant.[22]

Of the other Latin American navies, both **Ecuador** and **Uruguay** are focused on acquiring constabulary assets. The former is another regional fleet attracted to the Damen Stan Patrol concept and is in the course of taking delivery of two Stan Patrol 5009 type vessels of the *Isla San Cristobal* class for coast guard service. Uruguay has hopes of acquiring new offshore patrol vessels – possibly of the Fassmer OPV80 type – to replace its current motley collection of elderly second-hand warships but funding appears to be a problem. Meanwhile, **Venezuela's** *Armada Bolivariana de Venezuela* is unlikely to see any new major procurement in the light of the country's near economic collapse. However, the fourth Navantia-designed BVL type patrol vessel – contracted to local builder DIANCA – appears to be progressing towards sea trials some six years after her sisters were completed in Spain. If all goes well, the renamed *Comandante Eterno Hugo Chávez* (formerly *Tanamaco*) should be delivered before the end of 2017. The status of the larger Navantia-built corvette *Warao*, damaged by grounding in 2012, remains uncertain.

The Peruvian Navy *Lupo/Carvajal* class frigate *Villavisiencio* pictured off Vancouver in the summer of 2015. Although some of this elderly class are being modernised, Peru hopes to start work on a new frigate class in the near future. *(Marc Piché)*

MITRE Corporation, 2016) was first out of the blocks on 1 July 2016. This was followed by the internal navy analysis – *Report to Congress: Alternative Future Fleet Platform Architecture Study* (Washington DC: Navy Project Team, 2016) followed on 27 October 2016. The CSBA study – Bryan Clark, Peter Haynes, Bryan McGrath, Craig Hooper, Jesse Sloman and Timothy A. Walton's *Restoring American Seapower: A New Fleet Architecture for the US Navy* (Washington DC: CSBA, 23 January 2017) completed the trio. These – and the US Navy's 2016 FSA – can all currently be found by searching the web.

9. A detailed analysis of these various studies and the 2016 FSA is beyond this annual's scope. In addition to Ronald O'Rourke's CRS report on naval force structure (op cit), detailed analysis has been provided by Sam LaGrone and Megan Eckstein in several reports on the *USNI News* website: https://news.usni.org/

10. NIFC-CA is more fully described in Chapter 4.1.

11. See *Costs of Building a 355-Ship Navy* (Washington D.C.; Congressional Budget Office, April 2017) at: https://www.cbo.gov/publication/52632

12. The US Navy funds carrier construction over a number of years. Advanced procurement in the years before the award of a full-ship construction contract to pay for design costs and long-lead items is followed by incremental funding that allows payment to be authorised over a number of years (currently up to a maximum of six) after contract award. Reflecting the huge costs of carrier construction, incremental funding is an exception to the normal Department of Defense full funding policy that requires capital investments to be fully funded in the year they are contracted. For more information, see another of Ronald O'Rourke's reports, *Navy Ford (CVN-78) Class Aircraft Carrier Program: Background & Issues for Congress – RS20643,* (Washington: Congressional Research Service, 16 June 2017), pp.4–7.

13. See Christopher P Cavas' 'Navy, GD Hit Crossroads in Destroyer Negotiations', in *Defense News* on 8 January 2017 at http://www.defensenews.com/articles/tough-destroyer-negotiations-between-navy-gd for further background. At that stage, a decision on Flight III allocation had yet to be taken.

14. The goal of providing enough amphibious shipping to land the assault echelons of two MEBs was set in 2006, reflecting a reduced requirement from a 2.5 MEB lift requirement set in 1991. This translates into a requirement for thirty-eight operational ships, with a further three or four ships in maintenance. In practice, financial constraints meant that a force level of thirty-three ships was set (eleven LHA/LHDS + eleven LPDs + eleven LSDs) to provide thirty operational ships, with some cargo and vehicles being allocated to less survivable auxiliaries in the follow-on echelon. This requirement was increased to thirty-four ships (eleven LHA/LHDS + twelve LPDs + eleven LSDs) when *Fort Lauderdale* (LPD-28) was authorised and further increased to thirty-eight ships (twelve LHA/LHDS + thirteen LPDs + thirteen LSDs) under the 2016 FSA. This structure still does not fully meet the two MBE requirement as it does not take account of ships in maintenance.

15. Please refer to Chapter 2.3 for a more detailed description of this incident.

16. The emergence of glitches in the *Virginia* class construction programme was covered by Christopher P Cavas in 'US Navy submarine program loses some of its shine' on the *Defense News* website on 13 March 2017, currently available at: http://www.defensenews.com/articles/us-navy-submarine-program-loses-some-of-its-shine

17. See Jamie McIntyre's 'No US carriers were at sea for the past week. That hasn't happened since World War II' in the *Washington Examiner* – 5 January 2017 (Washington DC: Media DC, 2017).

18. There is a school of thought that the high profile FONOPs have had mixed effect and that maintain a quieter but more enduring presence in the South China Sea may bring better results. See, for example, Ankit Panda's 'Making the US Navy's South China Sea Presence Count' in *The Diplomat* – 19 June 2017 (Tokyo: Trans-Asia Inc, 2017).

19. This figure was quoted by José Higuera in 'Briefing Brazil: Feeling the Pinch', *Jane's Defence Weekly* – 8 March 2017 (Coulsdon: IHS Jane's, 2017), pp.22–9. This section has also benefitted from the ongoing coverage of the Brazilian Navy in the Portuguese language *Poder Naval* website at: http://www.naval.com.br/blog/

20. See José Higuera 'Brazil's nuclear submarine project faces postponement', *Jane's Defence Weekly* – 21 June 2017 (Coulsdon: IHS Jane's, 2017), p.5.

21. Good sources of information on naval developments in the Spanish-speaking South American countries include the Spanish language *infodefensa.com* and *defensa.com* websites.

22. A good assessment of Peru's naval objectives were provided by Guy Toremans' 'Interview: Admiral Edmundo Deville' in *Jane's Defence Weekly* – 25 January 2017 (Cousldon: IHS Jane's, 2017), p.26.

2.2 REGIONAL REVIEW

ASIA AND THE PACIFIC

Author:
Conrad Waters

INTRODUCTION

Naval developments in Asia and the Pacific continue to be driven by the twin challenges to regional security posed by Chinese 'expansionism' in the South China Sea and other neighbouring waters and by the ongoing advancement of North Korea's nuclear ambitions. Neither situation has developed particularly favourably over the past year.

Efforts to counter Chinese actions in the South China Sea started well enough. On 12 July 2016, the Permanent Court of Arbitration in The Hague backed a case brought against China by the Philippines under the United Nations Convention on the Law of the Sea (UNCLOS). The court ruled that China's 'nine-dash line' claim to much of the South China Sea had no legal basis and that China's actions – including its island-building activities – had breached Philippine sovereign rights in a number of respects. The binding – but not enforceable – ruling was swiftly denounced by China as being null and void. It had boycotted the proceedings in spite of being a UNCLOS signatory.

Although the tribunal's decision undoubtedly represented a major loss of face for China, it did little to alter the reality of a considerable and growing Chinese military presence amongst the artificial islands it has created on top of some of the sea's many shoals. Just as significantly, China has started to achieve results from a diplomatic and economic offensive to disrupt the United States' strategy to counter China's growing regional dominance. This American strategy has essentially been based on strengthening military and political ties with a network of countries fearful of China's expansion whilst bolstering its own regional military presence through the so-called 'Pivot to the Pacific'.

Perhaps ironically, it has been the Philippines themselves that have proved to be the weak link in this approach. Possibly influenced by American criticism of human rights violations that saw over 3,000 violent deaths – many of them extra-judicial killings – in his first 100 days in office, the new Philippine president, Rodrigo Duterte, has made a decisive political shift in China's favour.[1] Threatening to break his country's long-running defence ties with the United States, he has actively sought increased trading and economic ties with China in a courtship that China has been only too happy to reciprocate. China has quickly become the Philippines' largest external trading partner in a relationship described as a 'golden period of fast development' by China's foreign minister.[2] This period of détente has been accompanied by a reversal of recent moves to bolster the Philippines' external defences – including an order for light frigates from South Korea – in favour of a renewed emphasis on internal security. China has already donated light weaponry and ammunition to 'extend a helping hand to its close neighbour' in support of this revised focus.

The blossoming of the relationship between China and the Philippines is reflected in other steps China has been taking to bolster military collaboration with countries across the region. In April 2017, Malaysia signed a contract with the China Shipbuilding Industry to acquire four littoral mission ships, two of which would be assembled locally at the Boustead Naval Shipyard. During the same month, Thailand's military government approved the c. US$400m acquisition of the first of three planned Chinese submarines under a highly controversial deal. Each of these developments further undermines prospects of creating a united front against Chinese expansionism.

The action that the United States and its allies might take to counter China is further complicated by North Korea's ongoing ballistic missile and nuclear programme. The pace of testing has been picking up speed and there are increasing concerns that Kim Jong-un's unpredictable regime may be capable of launching a nuclear strike against the American mainland within the next few years. The threat facing key regional allies such as South Korea and Japan is even greater. In the absence of direct military action, China is the only country with sufficient diplomatic and economic influence to bring North Korea's leadership to heal. The United States is therefore in the unwelcome and conflicted position of attempting to secure China's help to reduce the risk from North Korea whilst simultaneously trying to thwart China's own ambitions in support of a revised focus that is very much in its own interests.[3]

Table 2.2.1 provides a summary of the more significant regional fleets as of mid-2017.

The PLAN Type 054A frigate *Hengshui* pictured operating with the US Navy destroyer *Stockdale* (DDG-106) during the RIMPAC 2016 exercise on 28 July 2016. In spite of occasional collaboration, countering Chinese 'expansionism' in the Pacific is one of the main strategic challenges facing the United States and its allies. *(US Navy)*

Table 2.2.1: FLEET STRENGTHS IN ASIA AND THE PACIFIC – LARGER NAVIES (MID 2017)

COUNTRY	AUSTRALIA	CHINA	INDONESIA	JAPAN	S KOREA	SINGAPORE	TAIWAN	THAILAND
Aircraft Carrier (CV)	–	1	–	–	–	–	–	–
Support/Helicopter Carrier (CVS/CVH)	–	–	–	4	–	–	–	1
Strategic Missile Submarine (SSBN)	–	6	–	–	–	–	–	–
Attack Submarine (SSN)	–	9	–	–	–	–	–	–
Patrol Submarine (SSK/SS)	6	50	2	17	15	4	4	–
Fleet Escort (DDG/FFG)	11	65	7	36	25	6	26	7
Patrol Escort/Corvette (FFG/FSG/FS)	–	40	24	6	16	7	1	11
Missile Armed Attack Craft (PGG/PTG)	–	75	20	6	17	–	c.30	6
Mine Countermeasures Vessel (MCMV)	6	25	11	24	11	4	10	6
Major Amphibious Units (LHD/LPD/LSD)	3	4	5	3	3	4	1	1

Notes: Chinese numbers approximate; Some additional Indonesian patrol gunboats are able to ship missiles; South Korean fleet escorts include thirteen deployed in littoral warfare roles; Taiwan's submarines are reported to have limited operational availability.

The new air-warfare destroyer *Hobart* was delivered to the Royal Australian Navy on 16 June 2017 after an extensive series of sea trials that commenced in September 2016. Two additional ships will join the fleet before 2020 in a major boost to Australian naval capabilities. *(AWD Alliance/Royal Australian Navy)*

MAJOR REGIONAL POWERS – AUSTRALIA

Table 2.2.2 summarises Royal Australian Navy fleet composition as at mid-2017. There have been no material changes year-on-year. However, fleet structure and composition is set for significant revision in the years ahead.

The Royal Australian Navy's future direction is being driven by the latest in a series of defence white papers, published in February 2016. Displaying strong threads of continuity with its immediate predecessors, it places significant emphasis on a maritime strategy focused towards the security of South East Asia and the Pacific.[4] A robust funding plan – that will take defence spending to just over two percent of GDP by 2020/21 – will support a major programme of naval investment estimated to total around AU$90bn (US$70bn) over the coming decades. Under these plans, the fleet will eventually comprise twelve new conventional submarines and a total of twelve major surface combatants, supported by twelve new offshore patrol ships. These warships will be supplemented by the recently modernised, three ship amphibious flotilla and two new fleet replenishment vessels.

Fleet expansion is being supported by a *Naval Shipbuilding Plan* published on 16 May 2017.[5] This aims to achieve a continuous programme of domestic construction for submarines, major surface combatants and minor naval vessels stretching out to the 2050s. Assembly of submarines and major combatants will be focused on the north (submarines) and south (surface vessels) yards at Osborne Naval Shipyard in Adelaide, Southern Australia. Construction of minor warships will take place at the Henderson facility near Perth in Western Australia. Investment of around AU$1.3bn (US$1bn) will be required before the commencement of the more important components of the construction programme to make these yards fit for their allocated roles.

Selection of the French DCNS 'Shortfin Barracuda' design for the Future Submarine Program (Project SEA 1000) was announced in April 2016. The detailed design phase of the programme commenced at the end of September in the same year. Lockheed Martin Australia was selected as combat systems integrator for the new boats at the same time. It is envisaged that design work will run in parallel with the required upgrades to the yard at Adelaide, with actual construction

Table 2.2.2: ROYAL AUSTRALIAN NAVY: PRINCIPAL UNITS AS AT MID 2017

TYPE	CLASS	NUMBER	TONNAGE	DIMENSIONS	PROPULSION	CREW	DATE
Principal Surface Escorts							
Frigate – FFG	**ADELAIDE** (FFG-7)	3	4,200 tons	138m x 14m x 5m	COGAG, 30 knots	210	1980
Frigate – FFG	**ANZAC**	8	3,600 tons	118m x 15m x 4m	CODOG, 28 knots	175	1996
Submarines							
Submarine – SSK	**COLLINS**	6	3,400 tons	78m x 8m x 7m	Diesel-electric, 20 knots	45	1996
Major Amphibious Units							
Amph Assault Ship – LHD	**CANBERRA** (JUAN CARLOS I)	2	27,100 tons	231m x 32m x 7m	IEP, 21 knots	290	2014
Landing Ship Dock – LSD	**CHOULES** (LARGS BAY)	1	16,200 tons	176m x 26m x 6m	Diesel-electric, 18 knots	160	2006

commencing in 2022–3. This will allow the first submarine to be commissioned in the early 2030s. Deliveries will run through to the early 2050s, by which time AU$50bn (U$38bn) will have been spent on the programme. The protracted timescale associated with the vast project means that sustainment of the existing *Collins* class will need to be carefully managed if a capability gap is to be avoided. Over AU$9bn has been allocated to ongoing upgrade and sustainment efforts over the remaining lives of the six *Collins* class boats to mitigate this risk.

Surface warship construction is currently dominated by work on the three *Hobart* class air-warfare destroyers based on Navantia's F-100 type frigate (Project SEA 4000). The programme, now expected to cost over AU$9bn (US$7bn), has previously been beset by a number of problems. However, the lead ship was accepted from the Air Warfare Destroyer (AWD) Alliance on 16 June 2017 after conclusion of a successful programme of trials that commenced in September 2016. Formal commissioning is expected to follow during September 2017. The second member of the class, *Brisbane*, was launched on 15 December 2016, with the third and final ship, *Sydney*, expected to enter the water during 2018. By that time a decision should have been taken on the design of a new class of nine anti-submarine Future Frigates that will ultimately replace the existing *Anzac* class under Project SEA 5000. Proposals from BAE Systems (Type 26), Fincantieri (FREMM) and Navantia (an F-100 derivative) are currently being evaluated. Construction is expected to commence at Adelaide during 2020 for delivery of the first ship within a 2027–30 timescale. Subsequent ships are likely to follow at two-yearly intervals under a programme valued at AU$35bn (US$27bn). The ultimate aim of the twelve-ship surface force is to have one surface task group of a destroyer and three frigates at immediate readiness and a further, similar task group available within ninety days.

The third element of proposed new construction relates to minor naval vessels under Project SEA 1180. Local shipbuilder Austal is currently contracted to design and construct nineteen 'Guardian' class patrol vessels for smaller Pacific nations at its Henderson facility under an AU$300m deal awarded in March 2017.[6] This programme will continue until 2025. During this period, Austal will transition to constructing the new offshore patrol vessels intended to replace the existing *Armidale* class in the future fleet. Given the demands of the 'Guardian' programme and the need to improve infrastructure at Henderson, the initial pair of the twelve offshore patrol vessels ultimately envisaged will be built at Osborne. Damen, Fassmer and Lürssen designs have been shortlisted for the contract. A decision is expected before the end of 2017. It is envisaged that all twelve vessels will be completed by 2030 at a cost of AU$3–4bn. Henderson will then be allocated other construction work, including new hydrographic and mine warfare vessels, as well as constabulary assets for other government agencies.

Meanwhile, another major project (SEA 1654 Phase 3) got underway on 17 June 2017 when Navantia started fabrication of the first of two new replenishment ships based on the Spanish Navy's *Cantabria* ordered under a contract signed in May 2016. Unlike the other warships, capacity issues and lack of suitable infrastructure led the Australian government to the conclusion that it was more appropriate for the ships to be built abroad. The two ships will be delivered in 2019 and 2020 under a deal valued at AS$646m. Opponents of the deal will not have been reassured by the emergence of propulsion problems with the *Juan Carlos I* type *Canberra* class amphibious assault ships, also partly built by Navantia. The issue apparently emerged during first of class flight trials with a range of helicopters on Canberra in March 2017 and relates to the Siemens podded propulsors fitted to both ships. The navy is also suffering from the need to carry out a mid-life refit programme on the current *Armidale* class patrol

A view of the Royal Australian Navy's amphibious assault ship *Adelaide* in dry dock in Sydney in June 2017. Both members of the *Canberra* class were laid up for inspections and rectification work after the discovery of problems with their podded propulsion units in March 2017. *Canberra* returned to sea at the end of June 2017 and *Adelaide* was due to follow shortly afterwards. *(Royal Australian Navy)*

Twenty-five Type 054A frigates have now entered service with the PLAN and a further two have been launched. Representing an intermediate level of capability between the fleet's larger Type 052 series destroyers and the littoral warfare-orientated Type 056 corvettes, these frigates are the workhorses of the blue water fleet, These pictures show *Huangshan*, one of the earlier members of the class, on exercises off the Chinese coast in April 2017. *(Royal Australian Navy)*

vessels. Their aluminium hulls have been suffering badly from fatigue. Two border force 'Cape' class patrol boats were operated on loan during 2015–16 to supplement patrol force availability. Subsequently, two similar boats have been acquired on lease as a longer-term solution pending delivery of the new offshore patrol vessel capability.[7]

MAJOR REGIONAL POWERS – CHINA

China's maritime ambitions continue to be supported by an ongoing construction programme that is unequalled globally in terms of quantity and surpassed in scope only by that of the United States. One measure of the speed with which new vessels are entering service is the estimate that a new Type 956 corvette was commissioned, on average, once every six weeks during 2016. An interesting status check on China's maritime ambitions was provided by the American CNA research centre in a report entitled *Becoming A Great 'Maritime Power': A Chinese Dream* published in mid-2016.[8] The report confirmed previous analysis that suggested the People's Liberal Army Navy (PLAN) is steadily transitioning from a single-minded focus on offshore waters defence (otherwise known as an anti-access/area denial or A2/AD strategy). Instead, the continued importance placed on defending the near seas will be balanced by a greater emphasis on blue water power projection. As part of this trend, the PLAN will complete a process of overhauling other major 'blue water' navies such as those fielded by France, Russia and the United Kingdom to become the second largest ocean-going fleet by 2020.

This trend is supported by Table 2.2.3, which highlights the PLAN's most important warship classes. Additional detail on recent developments with respect to key types is provided below.[9]

Aircraft Carriers and Amphibious Ships: A highlight of the last year was China's launch of its first indigenously-constructed Type 001A aircraft carrier from the Dalian Shipbuilding Industry Company's yard on 26 April 2017. The new ship's design is based closely on that of the existing Soviet-origin, Project 1143.5/6 *Kuznetsov* class carrier *Liaoning* (ex *Varyag*). However, reports suggest that she will incorporate a number of detailed improvements. Completion of fitting out is likely to take two–three years and it therefore seems it will not be until the 2020s that she will join *Liaoning* in operational service. In the meantime, China is said to be

Table 2.2.3: PEOPLE'S LIBERATION ARMY NAVY: PRINCIPAL UNITS AS AT MID 2017

TYPE	CLASS	NUMBER	TONNAGE	DIMENSIONS	PROPULSION	CREW	DATE
Aircraft Carriers							
Aircraft Carrier – CV	Project 1143.5/6 **LIAONING** (Kuznetsov)	1	60,000 tons	306m x 35/73m x 10m	Steam, 32 knots	Unknown	2012
Principal Surface Escorts							
Destroyer – DDG	Type 052D **KUNMING** ('Luyang III')	6	7,500 tons	156m x 17m x 6m	CODOG, 28 knots	280	2014
Destroyer – DDG	Type 051C **SHENYANG** ('Luzhou')	2	7,100 tons	155m x 17m x 6m	Steam, 29 knots	Unknown	2006
Destroyer – DDG	Type 052C **LANZHOU** ('Luyang II')	6	7,000 tons	154m x 17m x 6m	CODOG, 28 knots	280	2004
Destroyer – DDG	Type 052B **GUANGZHOU** ('Luyang I')	2	6,500 tons	154m x 17m x 6m	CODOG, 29 knots	280	2004
Destroyer – DDG	Project 956E/EM **HANGZHOU** (Sovremenny)	4	8,000 tons	156m x 17m x 6m	Steam, 32 knots	300	1999
Destroyer – DDG	Type 051B **SHENZHEN** ('Luhai')	1	6,000 tons	154m x 16m x 6m	Steam, 31 knots	250	1998
Destroyer – DDG	Type 052 **HARBIN** ('Luhu')	2	4,800 tons	143m x 15m x 5m	CODOG, 31 knots	260	1994
Plus c. 5 additional obsolescent destroyers of Type 051 **JINAN** ('Luda') class							
Frigate – FFG	Type 054A **XUZHOU** ('Jiangkai II')	25	4,100 tons	132m x 15m x 5m	CODAD, 28 knots	190	2008
Frigate – FFG	Type 054 **MA'ANSHAN** ('Jiangkai I')	2	4,000 tons	132m x 15m x 5m	CODAD, 28 knots	190	2005
Frigate – FFG	Type 053 H3 **LIANYUNGANG** ('Jiangwei II')	10	2,500 tons	112m x 12m x 5m	CODAD, 27 knots	170	1992
Frigate – FSG	Type 056/056A **BENGBU** ('Jiangdao')	32+	1,500 tons	89m x 12m x 4m	CODAD, 28 knots	60	2013
Plus c.10 additional obsolescent frigates of Type 053 H1/H1G/H2 **TAIZHOU** ('Jianghu II–V') classes							
Submarines							
Submarine – SSBN	Type 094 ('Jin')	c.5	9,000 tons	133m x 11m x 8m	Nuclear, 20+ knots	Unknown	2008
Submarine – SSBN	Type 092 ('Xia')	1	6,500 tons	120m x 10m x 8m	Nuclear, 22 knots	140	1987
Submarine – SSN	Type 093/093G ('Shang')	c.6	6,000 tons	107m x 11m x 8m	Nuclear, 30 knots	100	2006
Submarine – SSN	Type 091 ('Han')	3	5,500 tons	106m x 10m x 7m	Nuclear, 25 knots	75	1974
Submarine – SSK	Type 039A/039B (Type 041 'Yuan')	12+	2,500 tons	75m x 8m x 5m	AIP, 20+ knots	Unknown	2006
Submarine – SSK	Type 039/039G ('Song')	13	2,300 tons	75m x 8m x 5m	Diesel-electric, 22 knots	60	1999
Submarine – SSK	Project 877 EKM/636 ('Kilo')	12	3,000 tons	73m x 10m x 7m	Diesel-electric, 20 knots	55	1995
Plus c.15 obsolescent patrol submarines of the Type 035 ('Ming' class). A Type 032 'Qing' trials submarine has also been commissioned.							
Major Amphibious Units							
Landing Platform Dock	Type 071 **KULUN SHAN** ('Yuzhao')	4	18,000 tons	210m x 27m x 7m	CODAD, 20 knots	Unknown	2007

working on a follow-on Type 002 carrier design that will incorporate a conventional catapult-assisted take-off but arrested recovery arrangement.

Renewed construction of amphibious ships also reflects an expeditionary focus. A fifth Type 071 amphibious transport dock was launched from Shanghai's Hudong-Zhongua yard in June 2017and a sixth vessel is said to be under construction. There have also been reports that work has begun on a long-awaited Type 075 of broadly similar size to the US Navy's LHA/LHD types.[10]

Surface Combatants: PLAN surface combatant construction continues to encompass three major elements. The most potent of these is a series of general-purpose destroyers, of which the Type 052D variant is the latest to enter service. Six of the class have been commissioned since 2014, with another seven launched to date by the Jiangnan-Changxing yard in Shanghai and by Dalian Shipbuilding. The 7,500-ton ships are now being followed into production by a new class of 10,000-ton Type 055 'Renhai' destroyers. The first of these was launched from Jiangnan-Changxing on 28 June 2017. Although official details are sparse, it appears that the new ships will be equipped with similar sensors and weapons to those found on previous Chinese destroyers. However, their missile capacity may be expanded from sixty-four to as many as 128 vertical launch cells. To date, two pairs of Type 055s have been observed under construction at Jiangnan-Changxing and Dalian Shipbuilding but it would seem likely that many more will be ordered in due course.[11]

The second strand of construction relates to the intermediate frigate-sized vessels of the Type 054A 'Jiangkai II' class. Twenty-five of these have entered service since 2008, with the most recent, *Wuhu*, commissioning on 29 June 2017. As for the destroyers, construction has been split between two yards: the Hudong-Zhongua facility in Shanghai and the Huangpu Shipyard in Guangzhou (Canton). Both of these have each launched a further vessel of the class. Like the destroyers, the Type 054A frigates have a blue water, general-purpose orientation, albeit their diesel propulsion may limit their effectiveness in an anti-submarine role. Development of a more silent diesel-electric or integrated electrical propulsion variant has been rumoured for some time but, if correct, this has yet to produce a tangible result. Some commentators have suggested that the PLAN is waiting for domestically-produced technology to reach sufficient maturity before switching to construction of a new class.

The final class of surface combatant under assembly is the Type 056/Type 056A 'Jiangdao' corvette, the latter being distinguished by a modification of the ship's stern to accommodate a towed array. Over thirty of these diminutive littoral warfare vessels have been delivered from four shipyards since 2013 and some analysts believe that as many as sixty may ultimately be completed.

Submarines: Tangible information on the development of the PLAN's submarine force remains difficult to obtain. This is particularly the case with respect to the nuclear-powered force of strategic and attack submarines, about which only vague details exist. Most commentators believe that four or five Type 094 'Jin' class strategic submarines were completed in the first decade of the millennium and all these are now regarded as being operational. They are reported to be noisier even than the oldest Russian strategic submarines remaining in service, negating much of their second-strike potential. A follow-on Type 096 'Tang' class, armed with an improved JL-3 ballistic missile and presumably equipped with better noise-reduction technology, has been reported as being under development for some time. However, American intelligence assessments suggest that construction will only begin in the early 2020s.[12] Meanwhile, the two initial Type 093 'Shang' class attack submarines delivered in 2006–7 have recently been joined by a quartet of improved Type 093A/Type 093G 'Shang II' variants. These are believed to have been lengthened to incorporate vertical launch tubes for cruise missiles. As for the strategic submarines, it appears experience gained from operating these boats will be used to inform the design of a further class – variously reported as the Type 093B or Type 095 – that will probably also not be built until the 2020s. A number of older nuclear-powered submarines of the 091 and 092 types remain in commission. Their operational status and utility must be regarded as being marginal.

There is slightly greater clarity around the flotilla of conventional submarines, which form one of the most potent weapons in the PLAN's A2/AD arsenal. Three main classes of submarine – the domestic Type 039 'Song' and Type 039A/B 'Yuan' series and the Russian-built 'Kilos' – form the operational force. These are backed by a steadily diminishing force of Type 035/035G 'Ming' submarines based on the 1950s-era Russian 'Romeo', which are now used largely for training. Recent reports suggest that construction of 'Yuan' class boats has now resumed after a considerable gap. At least three new submarines are expected to join the existing thirteen members of the class.[13] The reason for the hiatus in production since the end of 2013 is not known for certain but may relate to the development of improved AIP technology for this latest batch.

Other Vessels: The construction of warfighting vessels is being supplemented by a wide range of minor warships and auxiliaries. These range from a sail training ship – the PLAN's first – to a new 50,000-ton semi-submersible vessel broadly similar to the US Navy's *Montford Point* (T-ESD-1). Making good the previous deficiency in replenishment vessels remains an area of focus. With the eight-strong Type 903/903A class now completed, attention has turned to the new Type 901 'comprehensive supply ship'. The 45,000-ton design is reportedly powered by gas turbines to keep pace with a carrier task force. It has multiple fuel hoses to allow simultaneous provision of both marine and aviation fuel as well as a significant dry stores transfer capability. The first member of the class commenced an extensive period of sea trials in December 2016. A second member of the class will also soon be ready for launch.

The JMSDF's second *Izumo* class helicopter-carrying destroyer *Kaga* seen entering the naval base at Yokosuka before commissioning into the navy on 22 March 2017. *(JMSDF)*

MAJOR REGIONAL POWERS – JAPAN

Although Japan is seeing modest growth in its defence budget after a lengthy period of decline in the post-Cold War era, the FY2017 budget of Yen 4.9 trillion (c. US$45bn) is still a little lower than at the turn of the Millennium.[14] One result is that the navy is apparently struggling to meet the force level goals of twenty-two operational submarines and fifty-four major surface combatants in spite of some significant qualitative improvements.

A summary of current fleet strength is set out in Table 2.2.4. The main development in the past year has been delivery of the second *Izumo* class helicopter carrier *Kaga* (DDH-184). She was commissioned on 22 March 2017, replacing the elderly *Shirane* class helicopter destroyer *Kurama* (DDH-144), which was withdrawn on the same day. No orders for major surface vessels were included in the FY2017 defence budget. This possibly reflected the

Table 2.2.4: JAPAN MARITIME SELF-DEFENCE FORCE: PRINCIPAL UNITS AS AT MID 2017

TYPE	CLASS	NUMBER	TONNAGE	DIMENSIONS	PROPULSION	CREW	DATE
Support and Helicopter Carriers							
Helicopter Carrier – DDH	**IZUMO** (DDH-183)	2	27,000 tons	248m x 38m x 7m	COGAG, 30 knots	470	2015
Helicopter Carrier – DDH	**HYUGA** (DDH-181)	2	19,000 tons	197m x 33m x 7m	COGAG, 30 knots	340	2009
Principal Surface Escorts							
Destroyer – DDG	**ATAGO** (DDG-177)	2	10,000 tons	165m x 21m x 6m	COGAG, 30 knots	300	2007
Destroyer – DDG	**KONGOU** (DDG-173)	4	9,500 tons	161m x 21m x 6m	COGAG, 30 knots	300	1993
Destroyer – DDG	**HATAKAZE** (DDG-171)	2	6,300 tons	150m x 16m x 5m	COGAG, 30 knots	260	1986
Destroyer – DD	**AKIZUKI** (DD-115)	4	6,800 tons	151m x 18m x 5m	COGAG, 30 Knots	200	2012
Destroyer – DDG	**TAKANAMI** (DD-110)	5	6,300 tons	151m x 17m x 5m	COGAG, 30 knots	175	2003
Destroyer – DDG	**MURASAME** (DD-101)	9	6,200 tons	151m x 17m x 5m	COGAG, 30 knots	165	1996
Destroyer – DDG	**ASAGIRI** (DD-151)	8	4,900 tons	137m x 15m x 5m	COGAG, 30 knots	220	1988
Destroyer – DDG	**HATSUYUKI** (DD-122)	2 (3)	3,800 tons	130m x 14m x 4m	COGOG, 30 knots	200	1982
Frigate – FFG	**ABUKUMA** (DE-229)	6	2,500 tons	109m x 13m x 4m	CODOG, 27 knots	120	1989
Submarines							
Submarine – SSK	**SORYU** (SS-501)	8	4,200 tons	84m x 9m x 8m	AIP, 20 knots+	65	2009
Submarine – SSK	**OYASHIO** (SS-590)	9 (2)	4,000 tons	82m x 9m x 8m	Diesel-electric, 20 knots+	70	1998
Major Amphibious Units							
Landing Platform Dock – LPD	**OSUMI** (LST-4001)	3	14,000 tons	178m x 26m x 6m	Diesel, 22 knots	135	1998

Note: Figures in brackets refer to trials or training ships.

heavy cost of the additional pair of Aegis-equipped destroyers ordered in previous years. Relatively small amounts continue to be allocated to life-extension activity relating to existing ships but the pace of new construction will have to pick up if overall surface ship numbers are to increase. There have been reports that Japan is considering ordering a new class of smaller frigate-type vessels to close the gap. Up to eight ships might be ordered at the rate of two p.a.[15]

Expansion of the submarine flotilla has also been taking place at a relatively pedestrian rate. A one-boat increase to seventeen submarines (plus two training units) was achieved in 2016 but this improvement was not sustained in 2017. The arrival of the eighth *Soryu* class boat, *Sekiryu* (SS-508), in March 2017 was counterbalanced by the assignment of a second *Oyashio* class submarine to training duties. This allowed the last *Harushio* class submarine, *Asashio* (TSS-3601), to be retired. Four further members of the *Soryu* type have been approved, the last pair of which will be equipped with lithium-ion batteries for increased underwater endurance. The FY2017 budget included the first of a new class of slightly larger 3,000-ton (surfaced displacement) submarines with greater surveillance capacity. Yen 80bn (c. US$725m) – including non-recurring development costs – was allocated to acquire the new boat. The increasing importance attached to countering Chinese submarine activity was also reflected in FY2017 budget provision for a third ocean surveillance ship of the *Hibiki* class. Based on the US Navy's *Victorious* (T-AGOS-19) surveillance vessels, these ships use a catamaran-hull design as a stable platform from which to deploy towed-array sonar.

One factor impacting the seeming difficulty in increasing fleet size may be lack of personnel, with the number of sailors remaining static at around 45,000. There are some signs that this may be being managed by reducing secondary units in commission. For example, the fleet of mine countermeasures vessels (MCMVs) has taken a noticeable dip in the last year due to retirement of all three 1990s-era *Yaeyama* class oceanic minesweepers without direct replacement. March 2017 did, however, see the commissioning of *Awaji*, first of a new class of fibre-reinforced plastic, MCMV. A second member of the class, *Hirado*, was launched in February 2017. The FY2017 budget made provision for a third ship.

MAJOR REGIONAL POWERS – SOUTH KOREA

A detailed review of the Republic of Korea Navy is contained in Chapter 2.2A.

OTHER REGIONAL FLEETS

Indonesia: The last year has seen Indonesia make progress with two projects aimed at recapitalising its relatively small fleet of front-line units. The contract for South Korean Type 209/1400 'improved *Chang Bogo*' class submarines saw the first boat, *Nagapasa* (formerly *Nagabanda*), commence sea trials from Daewoo Shipbuilding & Marine Engineering's (DSME's) Okpo facility in September 2016. The second of class, *Ardadedali*, was launched from the same facility on 24 October; both boats are expected to be operational by the end of 2017. A third unit is being assembled under licence at a new facility at PT PAL's yard in Surabaya. The navy ultimately wants to expand its submarine force to between ten and twelve submarines. Whilst it would make sense to standardise on a single supplier, it appears a number of options are being examined before a planned second batch of three boats is ordered. Further construction is urgently needed given that the country's two existing Type 209/1300 submarines were delivered in the early 1980s and cannot be expected to remain in service indefinitely.

Above the waves, the first 'Sigma' 10514 frigate *Raden Eddy Martadinata* commissioned on 7 April 2017 after a lengthy series of sea trials. She is the first of a class being assembled by PT PAL from blocks

The Indonesian Navy's *Raden Eddy Martadinata* is the first of a new class of frigates being assembled in Indonesia to the modular 'Sigma' 10514 design. The 10514 designation indicates that the ship is 105m long and has a beam of 14m. The Indonesian Navy already operates four smaller 'Sigma' corvettes, bringing a welcome degree of homogeneity to a disparate fleet. *(Damen Schelde Naval Shipbuilding)*

built both in-country and in the Netherlands with support from the Dutch Damen group. Some weapons-integration work is still outstanding and the ship will not be fully operational until 2019. A second member of the class, *I Gusti Ngurah Rai*, was launched on 29 September 2016 and Damen hopes for further orders. The new frigates provide welcome homogeneity with the four smaller 'Sigma' 9113 class corvettes delivered from the Netherlands between 2007 and 2009.

Much of the rest of the fleet of frigates and corvettes is quite elderly, although three *Bung Tomo* class corvettes originally ordered by Brunei from BAE Systems were delivered in 2014. The bulk of the force comprises former East German Project 1331 'Parchim' class anti-submarine corvettes acquired after German unification and now essentially used as large patrol vessels. Sixteen were originally purchased but two have been decommissioned after accidents. The latest to go was *Pati Unus*, formally retired in April 2017 after previously suffering significant damage in a grounding.

Domestic industry continues to work on a wide range of secondary missile-armed fast attack craft and patrol vessels, which sometimes share the same basic hull. Given Indonesian topography, amphibious vessels also have a relatively high priority. The *Teluk Bintuni* LST design that first entered service in 2015 has been selected for extended production, with seven vessels authorised as of mid-2017. An improved variant of the *Makassar* class amphibious transport dock class also entered production at PT PAL early in the year.

Malaysia: The Royal Malaysian Navy is currently focused on implementing a '15 to 5' Transformation Plan.[16] This aims to improve both operational and cost effectiveness by rationalising the fleet around five major warship types. If achieved, this would ultimately provide a fleet structure targeted on:

- Four 'Scorpène' type submarines, two of which are already in service.
- Six to twelve littoral combat ships of the DCNS 'Gowind' type. Six of these were authorised in 2011, of which two are currently under construction following the start of work on the second on 28 February 2017.
- Eighteen 'New Generation Patrol Vessels' of the existing MEKO 100 *Kedah* design, six of which are currently in service.

- Eighteen littoral mission ships designed for inshore patrol.
- Two or three multi-role support ships.

Whilst this represents a logical approach that appears to have political support, budgetary headwinds suggest the plan is unlikely to be achieved in its entirety. Malaysia's 2017 defence budget experienced a sharp, thirteen percent drop as falling commodity prices impacted government revenues and is clearly unlikely to achieve the government's objectives.

Against this backdrop, the selection of a Chinese design to meet the littoral mission ship requirement appears to be a sound move, getting another aspect of the transformation plan underway at what is likely to be an affordable price. The deal was first announced by Malaysian Prime Minister Najib Razak on 1 November 2016 and subsequently confirmed in April 2017. It initially involves four vessels. The first pair will be built in China for delivery during 2019–20. Construction will then shift to the Boustead Naval Shipyard, which are also building the 'Gowind' type. The new c. 70m vessels are expected to displace around 700 tons and carry a light, gun-based armament.

New Zealand: The New Zealand *Defence White Paper 2016* published in June 2016 was followed by a *Defence Capability Plan 2016* in the autumn.[17] The new white paper placed increased emphasis on policing Antarctic waters, an emphasis reflected in plans for a third, ice-strengthened offshore patrol vessel and an ice-strengthened replacement for the existing fleet tanker *Endeavour*. The new replenishment tanker was subsequently ordered from Hyundai Heavy Industries on 18 July 2016 at a cost of NZ$493m (US$360m). It was subsequently announced she will be named *Aotearoa* after the Maori name for New Zealand. The 24,000-ton tanker is being built to the Rolls-Royce 'Environship Leadge bow' design. She will utilise combined diesel and diesel-electric propulsion, the latter allowing more cost-effective operation at low speed. Images of the ship suggest that *Aotearoa* will be lightly armed with self-defence equipment but will be fitted with a hangar and flight deck for helicopter operation. When commissioned, she will be the largest ship ever operated by the Royal New Zealand Navy.

The new ice-strengthened offshore patrol vessel

The missile-armed fast attack craft *Halasan* is the third member of the KCR-60 or *Sampari* class. A fourth ship is currently under construction by leading domestic shipbuilder PT PAL, which is also responsible for the 'Sigma' type frigates. *(Singapore Ministry of Defence)*

The Royal Malaysian Navy frigate *Jebat* (foreground) operating in company with the Royal New Zealand Navy's replenishment tanker *Endeavour* and the corvette *Katsuri* in May 2017. The Malaysian surface vessels will ultimately be replaced under the '15 to 5' Transformation Plan whilst *Endeavour* will retire when the new replenishment vessel *Aotearoa* is delivered in 2020. *(Royal Australian Navy)*

A computer-generated image of the new replenishment vessel *Aotearoa*, which is being built by Hyundai Heavy Industries in South Korea to a Rolls-Royce design. The ship will be ice-strengthened as part of a higher priority being given to support naval constabulary and supply missions in Antarctic waters. *(Rolls-Royce)*

The Philippine Navy acquired a third *Hamilton* class high endurance cutter from the US Coast Guard in July 2016. This picture shows the new *Andrés Bonifacio* – the former *Boutwell* (WHEC-719) on the day of transfer. All eight units decommissioned from American service have found new homes so far (two in Bangladesh, two in Nigeria, three in the Philippines and one in Vietnam) in spite of being around fifty years old. *(US Coast Guard)*

will be ordered in 2019 for delivery by 2023. She will effectively replace the four existing relatively modern inshore patrol vessels delivered as part of Project Protector. The other Project Protector vessels – the multi-role vessel *Canterbury* and the two existing *Otago* class offshore patrol vessels – will receive mid-life upgrades in due course. Acquisition of a new littoral operations support vessel that can serve in low- to medium-threat environments will also go ahead. The two frigates – *Te Mana* and *Te Kaha* – will be replaced by the early 2030s. A final decision will be taken in 2026.

North Korea: North Korea's determination to develop an effective submarine-launched nuclear missile capability is being backed by a rising tempo of test missile launches. 2016 also saw nuclear tests in January and September, the fourth and fifth to date and the first since February 2013. An underwater launch of a KN-11 ballistic missile from what is believed to be North Korea's only 'Sinpo' class strategic submarine in August 2016 achieved what is widely regarded as the country's first completely successful test-firing to date. The weapon travelled over 300 miles before falling into the Sea of Japan. Whilst the 'Sinpo' SSB is effectively a test vehicle with probably only one missile tube, there are reports of a larger strategic submarine under development. New, larger submarine pens – heavily protected against bomb and missile attack – are also been constructed, presumably to house the new vessels.[18]

The Korean People's Army Naval Force's (KPAN's) conventional force structure is based on an asymmetric structure. This encompasses large numbers of small submarines and fast attack craft, as well as a significant amphibious force. Recent reports of the development of new c. 80m corvettes are therefore noteworthy, not least because of uncertainty as to how they fit with this established strategy. It appears that at least four of the corvettes – built in two different variants – have been launched since 2014. Both incorporate significant amounts of stealth technology. They are regarded by some analysts as sharing many features with Myanmar's latest surface combatants, which are said to have benefitted from North Korean-supplied equipment.[19]

The Philippines: The future direction of the Philippine Navy is somewhat unclear as a result of the marked change in strategic alignment seen under the new Duterte administration. More specifically, the new president's desire to focus the country's armed forces towards internal security may terminate the development of an embryonic warfighting capability that was being achieved by the previous regime.

The most significant development in recent years has been the acquisition of second-hand *Hamilton* class high endurance cutters from the United States Coast Guard. Although already fifty years old, the ships are still considerably more modern than much of the Philippines' Second World War-era corvette force. A third ship, *Andrés Bonifacio* (FF-17) – the former *Boutwell* (WHEC-719) – was commissioned on 21 July 2016. She joins the previously acquired *Gregorio de Pilar* (FF-15) and *Ramon Alcaraz* (FF-16) to form the core of a revamped surface fleet.

A further stage in the fleet's renewal will be achieved around the turn of the decade with the delivery of two new frigates, likely to be the first missile-armed ships in the Philippine fleet. Although

India's Garden Reach Shipbuilders & Engineers (GRSE) submitted the lowest bid with a variant of their Project 28 design, they subsequently failed the post-bid qualification process. As a result, South Korea's Hyundai Heavy Industries (HHI) were awarded a c. US$337m contract for the two ships, which are derived from the *Incheon* class frigate, on 24 October 2016. A digital image of the successful design suggests that the ships will be armed with a medium-calibre gun, surface-to-air and surface-to-surface missiles and be capable of supporting an embarked helicopter. South Korea will also transfer a decommissioned *Pohang* class corvette to help familiarise the Philippine Navy with the significant enhancement in expertise required to operate these sophisticated ships.

Meanwhile, another significant enhancement in the navy's capacity has been provided by the arrival of the two new *Tarlac* class amphibious transport docks. The lightly-armed ships have been built by Indonesia's PT PAL and are based on the *Makassar* design. The lead ship was formally commissioned on 1 June 2016, being followed into service by *Davao del Sur* on 31 May 2017. As for Indonesia, the Philippines archipelagic nature places a premium on strategic sealift and the vessels should prove valuable whatever future direction the fleet takes.

Singapore: The Republic of Singapore Navy (RSN) celebrated its fiftieth anniversary in 2017. It marked the occasion with an international fleet review on 15 May 2017. Twenty-eight warships from twenty foreign fleets jointed sixteen ships from the RSN and two police guard boats in Singapore's first such event. The new littoral mission vessel *Independence* had previously been commissioned on 5 May, Singapore's Navy Day and the actual anniversary of the day the RSN ensign was hoisted for the first time. The second member of the class, *Sovereignty*, was delivered at the end of 2016. Two further units of the eight ship class have been launched to date. The new littoral mission vessels will replace the *Fearless* class large patrol vessels. Ten of these now remain following the decommissioning of the previous *Independence* on 6 March 2017.

May 2017 also saw Singapore commit to acquiring a further pair of ThyssenKrupp Marine Systems (TKMS) Type 218SG submarines. The AIP-equipped boats will join the original pair that were ordered in 2013. The new class will start to

The Republic of Singapore Navy held an inaugural fleet review in May 2017 to celebrate its fiftieth anniversary. This picture shows the *Formidable* frigate *Stalwart* leading one of a number of columns of participants in the run-up to the review. *(Singapore Ministry of Defence)*

The Republic of Singapore Navy's first littoral mission ship *Fearless* was commissioned in May 2017. A second member of the eight-ship class has also been delivered, whilst the third and fourth units have been launched. *(Republic of Singapore Navy)*

replace Singapore's existing second-hand Swedish submarines from 2021 onwards. The c. 70m, 2,000-ton Type 218 incorporates the polymer electrolyte membrane (PEM) fuel cell system of the Type 214 submarine already in service with several navies. It also has a similar armament of eight forward-firing torpedo tubes. However, other design features have been lifted from the German Navy's Type 212A design, notably its X-shaped rudder configuration. This arrangement is particularly well-suited for the confined littoral waters in which the submarines are likely to operate.

Taiwan: The Taiwanese Republic of China Navy announced ambitious plans for a twenty-year programme of force structure modernisation in 2014. This essentially aims to modernise and upgrade the existing fleet structure, under which submarines and light surface forces such as missile-armed fast attack craft would be reinforced by surface-action groups of larger destroyers and frigates in the decisive stage of any conflict with mainland China. This historic strategy has been criticised in light of China's steadily growing military superiority. Some commentators suggest that investment would be best channelled into a 'small ship navy' with integrated co-operative engagement capability similar to that being developed by the navy as part of a more survivable 'distributed lethality concept'.[20]

Irrespective of the direction of this debate, it is questionable whether Taiwan has either the money or technology to achieve all elements of the existing modernisation plan. 2017 is expected to see a start on the first production batch of three *Tuo Chiang* (*Two River*) catamaran missile-corvettes and a new amphibious transport dock. A new type of fast minelayer may also start construction. However, more significant units – a fleet of eight patrol submarines and new destroyers equipped with Aegis-like combat systems – face significant developmental hurdles. For example, there are a number of key elements of submarine technology manufactured by only a handful of nations that Taiwan is currently struggling to obtain.[21] The fact that the only material enhancement to force structure achieved in the past year was the May 2017 arrival of two 1980s-era FFG-7 type frigates transferred from the US Navy shows something of the challenge faced.

Thailand: One country that is moving ahead with long-standing submarine acquisition plans is Thailand. A c. US$390m contract for the first of three planned Chinese S26T boats was signed on 5 May 2017 after approval by the country's ruling military junta. Described as being '… not for battle, but so that others may be in awe of us …' by Thai Prime Minister Prayuth Chan-ocha, the purchase is controversial given that Thailand's shallow coastal waters are not ideally suited for

The Vietnamese frigate *Dinh Tien Hoang* is the first of up to six Russian-built Project 1166.1 'Gepard' class frigates being delivered from the inland Zelenodolsk yard in Tatarstan as a counter to Chinese 'expansionism'. Two further ships of the class are in service with the Russian Navy in the Caspian Sea and at least one of these has been used to conduct cruise missile strikes against Syrian rebels. The two ships delivered to Vietnam to date are focused on anti-surface warfare, with eight SS-N-25 'Switchblade' missiles (broadly similar to the US Navy's Harpoon) located amidships. *(Singapore Ministry of Defence)*

Notes

1. Rodrigo Duterte became President of the Philippines on 30 June 2016, succeeding the pro-American Benigno Aquino III. Although he has adopted a more favourable approach to China, his position with respect to the South China Sea dispute has been more nuanced, increasing the Philippine presence on some islands. Some of his antagonism towards the United States also results from a poor relationship with former American President Barack Obama. Outside the presidency, institutional and military ties with the United States remain largely strong.

2. Foreign Minister Wang Yi was quoted by Christian Shepherd in a report entitled 'China hails "golden period" in relations with Philippines' posted to the *Reuters* website on 29 June 2017.

3. For further analysis of the United States' strategic bind, see Megan Eckstein's 26 April 2017 article 'PACOM to Conduct South China Sea FONOPs 'Soon' But Also Needs China To Help With North Korea' posted to the *USNI News* website at: https://news.usni.org/2017/04/26/pacom-to-conduct-south-china-sea-fonops-soon-but-also-needs-china-to-help-with-north-korea

4. The *2016 Defence White Paper* was accompanied by a *2016 Integrated Investment Program* and a *2016 Defence Industry Policy Statement* (all Canberra: Commonwealth of Australia, 2016). All three documents can be accessed at: http://www.defence.gov.au/WhitePaper/

5. See *Naval Shipbuilding Plan* (Canberra: Commonwealth of Australia, 2017) currently available at: http://www.defence.gov.au/navalshipbuildingplan/Docs/NavalShipbuildingPlan.pdf

submarine operations.[22] It is anticipated that the lead boat – based on China's 'Yuan' design – will be in service by 2023. Further orders should ensure the desired three-strong force structure is in place by 2026.

Investment is also being maintained in surface forces. *Tachin*, the first of two DWH300H frigates being built to a variant of the KDX-I destroyer design by South Korea's DSME, was launched on 23 January 2017. Delivery is scheduled for 2018. She will be followed by an additional vessel constructed domestically under licence. Licensed production is also being adopted for the construction of offshore patrol vessels based on the British 'Improved River' class design. A keel-laying ceremony for a second member of the class was held on 23 June 2017; she should follow the existing *Krabi* into service by 2019. At least two further vessels of the type are planned in due course.

Vietnam: The Vietnam People's Navy commissioned the final pair of six Project 636.1 'Kilo' class submarines constructed by Russia at the end of February 2017, bringing to a close a procurement exercise that commenced in 2009. The development of a first-time underwater capability is a major step forward in the navy's structure and should provide a significant deterrent to China's expansionary tendencies. Russian equipment is also being used to modernise and expand surface capacity. At least two more Project 1166.1E 'Gepard' class light frigates are in an advanced stage of construction at Russia's Zelenodolsk yard in Tatarstan and should join the initial pair within the next year. There have also been reports of orders for a further pair, taking the total front-line frigate force to six vessels. Licensed construction of smaller missile-armed fast attack of Russian origin also continues.

Whether or not Russia's domination of Vietnamese procurement will continue into the future remains a moot point. June 2017 saw the navy take delivery of the former South Korean *Pohang* type corvette *Gimcheon*. Her arrival follows the delivery of former Japanese patrol vessels and, perhaps most significantly of all, the former American *Hamilton* class cutter, *Morgenthau* (WHEC-722), to the Vietnamese coast guard. Such strategic cooperation is likely to continue in the face of ongoing concerns over Chinese foreign policy and will probably lead to more significant transactions in the years ahead.

---

6. The new patrol vessels are a replacement for the twenty-two 'Pacific' class patrol boats built by Australia from 1985 to 1997 and donated to twelve small South Pacific countries to enhance maritime security in their exclusive economic zones.

7. The two new patrol boats are named *Cape Fourcroy* and *Cape Inscription*. They have been funded by the National Bank of Australia under an AUS$63m contract and chartered to the navy for a minimum period of three years from mid-2017. Unlike the *Armidale* class, they are not commissioned and take an ADV (Australian Defence Vessel) prefix.

8. See Rear Admiral Michael McDevitt, USN (retired): *Becoming A Great 'Maritime Power': A Chinese Dream* (Arlington VA: CAN, 2016) available at: https://www.cna.org/CNA_files/PDF/IRM-2016-U-013646.pdf

9. Web-based sources the *China Air and Naval Power* blog at http://china-pla.blogspot.co.uk/ and the *China Defense Blog* at http://china-defense.blogspot.co.uk/ remain useful sources of information on the PLAN, although the former has not been active during 2017.

10. Please refer to Chapter 4.1 for further discussion on the PLAN's aircraft and helicopter carriers.

11. News of the launch of the new ship was carried by many journals. Of these, Mike Yeo's 'China launches its most advanced homegrown class of guided-missile destroyers' carried on the *Defense News* website on 28 June 2017 at http://www.defensenews.com/articles/china-launches-first-type-055-class-guided-missile-destroyer provides much additional analysis on the new design.

12. This assessment was contained in the *Annual Report to Congress: Military and Security Developments Involving the People's Republic of China 2017* (Washington DC: Office of the Secretary of Defense, 2017), p.24. The report provides a wide ranging overview of current Chinese strategic objectives, modernisation goals and resources from a US viewpoint. A copy can currently be found at: https://www.defense.gov/Portals/1/Documents/pubs/2017_China_Military_Power_Report.PDF?ver=2017-06-06-141328-770.

13. Further analysis is provided in Andrew Tate's 'China resumes production of Yuan-class subs', *Jane's Defence Weekly* – 11 January 2017 (Coulsdon: IHS Jane's, 2017), p.8.

14. More detailed information on the budget is provided in *Defense Programs and Budget of Japan: Overview of FY2017 Budget* (Tokyo: Ministry of Defense, 2017) currently available at: http://www.mod.go.jp/e/d_budget/pdf/290328.pdf

15. This possibility was reported by Nobuhiro Kubo in 'Exclusive: Japan to speed-up frigate build to reinforce East-China Sea – sources' carried on the *Reuters* website on 17 February 2017.

16. A detailed review of the Malaysian Navy was provided by Mrityunjoy Mazumdar in 'The Royal Malaysian Navy: A difficult balancing act ahead, *Seaforth World Naval Review 2017* (Barnsley: Seaforth Publishing, 2016), pp.42–53.

17. See *Defence White Paper 2016* and *New Zealand Government Defence Capability Plan 2016* (both Wellington: Ministry of Defence, 2016). Current web references are provided in notes 11 and 12 of Chapter 3.3.

18. See Nick Hansen's 'North Korea building new, larger submarine pens', *Jane's Defence Weekly* – 27 July 2016 (Coulsdon: IHS Jane's, 2016), p.6. H I Sutton's *Covert Shores* website at: http://www.hisutton.com/ is a good source of reference in broader KPAN developments.

19. Details of the corvettes were provided by Chad O'Carroll on 8 November 2016 in an article entitled 'Exclusive: New low-visibility corvette spotted in North Korea' on the *NK News.org* website. See: https://www.nknews.org/

20. An interesting review of current Taiwanese Navy strategy and future options is provided by Michal Thim and Liao (Kitsch) Yen-Fan in 'Taiwanese Navy Plans to Enhance Fleet Air Defence', *China Brief Volume XVI, Issue 7* of 21 April 2016 (Washington DC: The Jamestown Foundation, 2016), pp.11–14.

21. See Mike Yeo's 'Taiwan struggles to acquire 5 types of submarine tech for local programme' carried on the *Defense News* website on 7 April 2017 at http://www.defensenews.com/articles/taiwan-struggles-to-acquire-5-types-of-submarine-tech-for-local-program. The know-how to produce crucial submarine systems such as torpedo-handling equipment and AIP is only held by a very few nations. These may well be reluctant to risk offending China by helping Taiwan.

22. Prime Minister Chan-ocha's remarks were made in 2016 and quoted by Marwaan Macan-Markar in an article entitled 'Thailand and China: Brothers in Arms', *Nikkei Asian Review* – 2 February 2017 (Tokyo: Nikkei Inc., 2017).

2.2A Fleet Review

Author:
Mrityunjoy Mazumdar

REPUBLIC OF KOREA NAVY

Balancing blue water ambitions with regional threats

The sinking of the South Korean corvette *Cheonan* on the night of 26 March 2010 by what South Korea claimed was a surprise torpedo attack from a North Korean midget submarine provided a stark reminder of the fragile state of peace that exists between the two Koreas. The heavily militarised Korean peninsula – with the dystopian, unpredictable and isolated communist-ruled Democratic People's Republic of Korea (DPRK) in the north and the economically prosperous Republic of Korea (ROK) in the south – remains an active 'hotspot'. The ramping-up of the DPRK's nuclear weapons programme – and simultaneous development of a submarine-launched ballistic missile capability – has only served to exacerbate tensions on the peninsula. It also impacts the complex and often delicate web of geopolitical and economic relations that exist between the other main regional players, notably North Korea's benefactor, China; South Korea's main ally, the United States; and Japan. Whilst South Korea's new president, Moon Jae-in, campaigned for a moderate approach towards the DPRK before his May 2017 election win, his acknowledgement that '... there is a high possibility of conflict ...' with the North illustrates the risk of war.[1]

The loss of the corvette *Cheonan* to an alleged North Korean torpedo attack in 2010 was something of a wake-up call to the ROKN, demonstrating the need to balance blue water expansionist abilities with the littoral warfare capabilities needed to counter potential DPRK aggression. The recovered hull of *Cheonan* is now a memorial to the forty-six crew members killed in the incident. *(Republic of Korea Armed Forces)*

Indeed, since the end of the Korean War in July 1953, skirmishes and covert infiltration attempts between the two countries – mostly along the de facto maritime border known as the Northern Limit Line (NLL) – have not been uncommon.[2] However, a full-scale renewal of hostilities would be disastrous for both Koreas – a risk that the more developed and wealthy ROK has attempted to deter through its continued mutual defence treaty with the United States and considerable military investment of its own. With a population of fifty-one million and an economy that ranks amongst the world's twelve largest – and fourth in Asia – South Korea is able to afford a large military budget. Defence spending in 2017 will amount to 40.34 trillion won (c.

US$35bn), a c. 4 percent increase on 2016. Amounting to over 2.5 percent of GDP and more than 10 percent of all government spending, this is the largest (nominal) amount the ROK has ever spent on defence and confirms South Korea's position as one of the top ten military spenders globally.

MARITIME SECURITY THREATS TO SOUTH KOREA

Given the tensions on the peninsula, it is inevitable that North Korea is the primary military threat facing South Korea, both in the maritime domain and more broadly. Whilst the considerable US military presence in support of its ROK ally remains the ultimate counterweight to this menace, the South is steadily developing its capabilities to respond to – and even pre-empt – any hostilities. Given the DPRK's nuclear ambitions, increasing emphasis is being placed on establishing a three-pronged national defence to missile attacks based on (i) a 'Kill Chain' pre-emptive strike system aimed to attack relevant North Korean installations before missiles are launched, (ii) a Korea Air & Missile Defence (KAMD) to destroy incoming missiles, and (iii) a Korea Massive Punishment & Retaliation targeting the North Korean leadership in Pyongyang. The Republic of Korea Navy (ROKN) has a major part to play in this strategy, notably in terms of land-attack weapons deployed from surface vessels or submarines and the missile-defence potential of the country's Aegis destroyers. Recent pronouncements suggest work on the triad of systems is being accelerated in the light of recent DPRK missile and nuclear tests.[3]

In a more conventional conflict, the ROKN would be severely tested in its ability to counter successfully the likely asymmetric and unpredictable warfighting tactics adopted by the relatively low-tech but numerically larger Korean People's Army Naval Force (KPAN). The 60,000-strong KPAN is thought to operate seventy submarines of various types and, while there are very few large surface combatants of any modernity, there are approaching 500 smaller craft of which many are missile or rocket armed. Other challenges are poised by numerous land-based anti-ship missiles, mines and large numbers of amphibious and infiltration craft. The KPAN is widely dispersed across numerous naval bases and operating facilities and is well-versed in the arts of deception and concealment.

It is important to note that North Korea is not the exclusive maritime security challenge faced by the ROK. For example, China's rising military power and regional aspirations – including claims over the entire South China Sea and the 2013 creation of an East China Sea air defence identification zone (ADIZ) encroaching on South Korean territorial interests – have clearly been a cause for concern. Another issue is the territorial dispute with Japan over the Dokdo/Takeshima islands and a legacy of bitterness dating from Imperial Japan's occupation of the country. Although the greater threat to both countries from the DPRK has tended to keep resulting tensions in the background, rivalry with Japan has probably been one influence on the desire to develop a 'blue water' navy that has been a key influence on ROKN strategy in recent years. Another has been Korea's development as an international trading nation and the growing importance of safeguarding the country's maritime trade routes.

North Korea's KPAN remains the ROKN's most likely adversary. There have been periodic clashes between the two navies, particularly around the Northern Limit Line (NLL) that forms the de facto maritime boundary between the two countries. This picture shows a squadron of new PKX-A missile-armed fast attack craft on exercises near the NLL in 2016. *(Republic of Korea Armed Forces)*

HISTORICAL BACKGROUND

Korea has a long and rich history of naval warfare. Notably, their famous Joseon era *kobukson* armoured 'turtle ships' and smaller cannon-armed *panokson* defeated numerically superior Japanese forces on several occasions under the leadership of Admiral Yi Sun Shin in the sixteenth century, most notably at the Battle of Hansan Island in 1592. Korean power subsequently waned and the country was ultimately annexed by Japan in 1910. Locals who served in the Imperial Japanese Navy or who had merchant marine experience would form the core of the ROKN following the end of occupation after the Second World War and the country's partition into the two Koreas.

The current ROKN traces its roots to the Maritime Affairs Association, which was established shortly after the liberation from Japan. Led by former merchant mariner Son Won-il, subsequently the ROKN's first head, it evolved into the Marine Defense Group on 11 November 1945 (marked as the navy's founding day) and then into the Korean Coast Guard. The organisation was renamed the Republic of Korea Navy with the foundation of the Republic of Korea in 1948.

The new navy initially equipped itself largely from US Navy sources, receiving its first 'proper' warship, the PC-461 type submarine chaser *Baekdusan* (*Pak Tu San*) – the former *Ensign Whitehead* (PC-823) – in October 1949.[4] Other vessels quickly followed. By the time the Korean War broke out on 25 June 1950, the ROKN comprised c. 7,000 personnel and over thirty ships. There were also c. 1,000 marines in two battalions. Although outnumbered by its northern counterpart, this small fleet played its part in the conflict, losing several vessels in the process. A notable contribution was made by *Baekdusan*, which thwarted a surprise attack on Busan by sinking a North Korean transport ship. More – and larger – warships were transferred by the US Navy during the war, a process that continued after the signing of an armistice on 27 July 1953. In 1963 the first destroyer, the former US Navy *Erben* (DD-631), was taken into service as *Chung Mu*. In all, a dozen units of the *Fletcher*, *Gearing* and *Allen M Summer* classes were acquired from 1963 to 1981, forming the fleet's core until the end of the Cold War.

By the 1970s, the ROKN was starting to evolve into a multi-dimensional force, albeit one still tethered firmly to the littoral. A naval air wing was formed with the acquisition of Grumman S-2 Tracker aircraft and the separate ROK Marine Corps (ROKMC) was formally consolidated into the naval service in 1973. Efforts were also made to establish an underwater capability. By the early 1980s, the first of three German-designed but locally-built *Dolgorae* class midget submarines had entered service. This last-mentioned development also reflected an increasing focus on creating a domestic naval shipbuilding capability, initially through the licensed construction of foreign designs with appropriate technical assistance. However, the ultimate aim was to develop a comprehensive indigenous design and build capability, an intent evidenced by the delivery of the first locally-developed high speed patrol craft in the early 1970s. By the 1980s the indigenous *Ulsan* class frigates and smaller *Donghae/Pohang* class corvettes were in series production and embryonic plans were being laid for what were to become the KDX series of destroyers.

The post-Cold War period brought further qualitative gains and a move towards developing a blue water naval capacity. By 1992, the navy had acquired its first full-sized Type 209 submarine from Germany. It was followed by eight locally-built examples. Lockheed P-3C Orion aircraft replaced the elderly S-3 Trackers, while Lynx helicopters expanded the capabilities of progressively larger surface combatants, all designed and built 'in-country'. The ROKN's high seas ambitions were reflected in the 1999 document *Navy Vision 2020*, which set out the strategic mobile concept of a flotilla of sophisticated warships able to conduct independent regional operations.[5] Ships such as *Dokdo* – the first 'flat top' built in Asia since the Second World War – and the large KDX-III *Sejong Daewang* Aegis destroyers were a key part of this vision. Also significant was the expansion of infrastructure – notably the controversial Jeju Civilian-Military Complex Port (finally officially opened in February 2016) – to support the blue water fleet. The loss of *Cheonan* appears to have produced some soul-searching as to whether littoral warfare capabilities had been downgraded as a result of the increased high-seas focus. However, current construction plans largely maintain the previous trend.

The ROKN relied heavily on second-hand US Navy equipment during its formative years, including twelve former Second World War destroyers that formed the core of its Cold War fleet. This picture shows the last ship transferred, the *Gearing* class *Jeon Ju* – previously *Rogers* (DD-876) – in Puget Sound in August 1981 shortly before the delivery voyage to her new home. *(Courtesy Marc Piché)*

THE ROKN TODAY

The ROKN – also known as *Daehanminguk Haegun* or more commonly the *Haegun* – numbers around 70,000 personnel. Some 41,000 – around 6,700 officers, 17,300 non-commissioned officers, 17,000 enlisted personnel and 4,000 civilians – are in the navy proper; the balance being in the marines. The majority of enlisted sailors are conscripts serving for two years and there is a large pool of reserves – possibly around 100,000 – to draw on. The ROKN operates over 150 commissioned ships, including c. fifteen submarines and approaching sixty major surface combatants and missile-armed fast-attack craft. There are also large numbers of patrol craft and amphibious ships, all supported by a flotilla of yard craft estimated at between 150–200 vessels.

Organisational Structure: The ROKN's organisational structure has been evolving in recent years. As of 2016, main elements encompassed Navy Headquarters, its subordinate commands and several direct reporting units (DRUs). The former include (i) the ROK Fleet, also referred to as the Operations Command, (ii) the ROK Marine Corps/Northwest Islands Defence Command, (iii) the Naval Education and Training Command, (iv) the Naval Logistics Command, and (v) the Naval Academy. DRUs include the Jinhae Naval Base Command, naval intelligence and support functions such as the naval military police, naval court service, information management systems group and maritime medical centre. The ROKN is headed by the four star Chief of Naval Operations – currently Admiral Um Hyun-Seong – from Navy Headquarters, located within the tri-service headquarters of South Korea's overall armed forces at Gyeryong.

Fleet operations are directed by the ROK fleet commander (COMROKFLT) – a three-star admiral – based at ROK Fleet Headquarters at Busan. He is responsible for three numbered fleets, each led by a rear admiral and responsible for distinct geographical areas in the ROK's littoral. The First Fleet at Donhgae covers the eastern naval sector; the Second Fleet at Pyongtaek covers the western naval sector; and the Third Fleet at Mokpo covers the southern naval sector. Each fleet typically comprises a three-squadron Maritime Combat Group under the command of a deputy fleet commander, two or more patrol squadrons and a number of other specialist squadrons and support units.

The three fleets are supplemented by additional numbered flotillas not tied to a given geographical area of operation. The Fifth Flotilla encompasses mine warfare (Squadron 52), amphibious (Squadron 53) and salvage (Squadron 55) disciplines; Air Wing Six is the naval air arm; Readiness Flotilla Eight is the operational training and evaluation group; and the newly-formed Submarine Command (formerly Submarine Flotilla 9) takes responsibility for all underwater operations.[6] A particularly noteworthy development was the creation of Maritime Task Flotilla Seven in 2010 to act as a strategic manoeuvre unit in line with the mobile fleet concept. Moved from Busan to the new base at Jeju Island at the end of 2015, it comprises some of the fleet's most powerful units, such as KDX-II and KDX-III series destroyers.

Naval Bases and Infrastructure: Major naval bases include Pusan (home of ROK Fleet Headquarters), the three regional fleet bases at Donhgae, Pyongtaek and Mokpo, the new facility at Seogwipo in the southern part of Jeju Island and the location of the former fleet headquarters at Jinhae. Jinhae – an important facility built during the Japanese occupation – remains home to a number of support and naval special warfare units, not least the Korea Naval Academy. Smaller facilities, including Pohang (also home to naval aviation) and Incheon, are supplemented by forward operating bases and floating outposts, particularly around the NLL.

Given the ongoing risk of infiltration and sabotage, base security is an important consideration. One important defence has been the installation of harbour underwater surveillance system (HUSS) technology, with at least five WSD-120K HUSS sets supplied by local contractor LIG Nex1 in conjunction with South Korea's Agency for Defense Development (ADD). Its HUSS is a three-layered surveillance and detection system that fuses data

The ROKN's organisational structure includes three numbered fleets, each focused on littoral operations in a distinct geographical area. This picture shows the new FFX type frigate *Gyeong Ji* leading older-generation combatants during a Second Fleet exercise off Korea's west coast in January 2016. *(Republic of Korea Armed Forces)*

A South Korean Type 214 AIP-equipped submarine pictured running on the surface. A new Submarine Command established in 2015 now has responsibility for all underwater forces. *(Republic of Korea Armed Forces)*

from radar, electro-optical tracking equipment and sonars to detect ships, submersibles and combat swimmers at various stages of entry. The initial stage of a wider integrated coastal surveillance system is now operational in some vulnerable areas where the threat of infiltration is greatest. This links existing surveillance equipment over an extensive network of communication lines to provide a better overall picture.

ROKN EQUIPMENT – WARSHIPS

A summary of the main ROKN warships is provided in Table 2.2A.1 and should be referred to in conjunction with the following analysis.

Submarines: The ROKN currently has fifteen full-sized submarines and is ultimately targeting a force structure of at least eighteen boats. The oldest were acquired under a '1,000-ton' submarine project codenamed KSS-I for which the German Type 209/1200 design was selected in 1987. The first of these – *Chang Bogo* – was built in Germany and

Table 2.2A.1: REPUBLIC OF KOREA NAVY – PRINCIPAL UNITS AS AT MID 2017

TYPE	CLASS	NO.	YEAR [1]	TONNAGE	DIMENSIONS	SPEED	PRINCIPAL ARMAMENT
Submarines (15+5)[2]							
Submarine (SSK)	Type 214 (KSS-II)	6+3	2007	1,900 tons	65m x 6m x 6m	20+ knots	8 x 533mm torpedo tubes for torpedoes and anti-surface missiles
Submarine (SSK)	Type 209/1200 (KSS-I)	9	1993	1,300 tons	56m x 6m x 6m	22 knots	6 x 533mm torpedo tubes for torpedoes and anti-surface missiles
The first three of a planned nine KSS-III class boats are under construction for delivery from 2022 onwards, probably replacing the KSS-1 type. There are also an unspecified number of midget submarines.							
Fleet Escorts (12)							
Destroyer (DDG)	KDX-III – Batch 1	3	2008	10,500 tons	166m x 21m x 6m	30 knots	128 Mk 41/K-VLS cells, 8 x SSM, 1 x RAM, 1 x 127mm, 1 x CIWS, 6 x TT, 2 x helicopter
Destroyer (DDG)	KDX-II	6	2003	5,500 tons	150m x 17m x 5m	30 knots	32 Mk 41 VLS cells, 8 x SSM, 1 x RAM, 1 x 127mm, 1 x CIWS, 6 x TT, 1 x helicopter
Destroyer (DDG)	KDX-I	3	1998	3,900 tons	135m x 14m x 4m	30 knots	16 Mk 48 VLS cells, 8 x SSM, 1 x 127mm, 2 x CIWS, 6 x TT, 1 x helicopter
A further three KDX-III destroyers are being designed to an improved Batch 2 specification, with construction expected to start in 2018. They will possibly replace the KDX-I class.							
Patrol Escorts (29+4)							
Frigate (FFG)	FFX – Batch 1	6	2013	3,000 tons	114m x 14m x 4m	30 knots	8 x SSM. 1 x RAM, 1 x 127mm, 1 x CIWS, 6 x TT, 1 x helicopter
Frigate (FFG)	**ULSAN**	7	1981	2,300 tons	102m x 12m x 4m	35 knots	8 x SSM, 2 x 76mm, 6/8 x 40mm/30mm, 6 x TT
Corvette (FSG)	**POHANG**	16	1984	1,200 tons	88m x 10m x 3m	32 knots	2–4 x SSM, 1-2 x 76mm, 4 x 40mm/30mm, 6 x TT
A further four enlarged FFX-Batch 2 frigates are under construction, with a further four Batch II and, possibly, eight Batch III ships expected. They are progressively replacing the older types.							
Missile Armed Fast Attack Craft (17+1)							
FAC (PGG)	PKX-A	17+1	2008	600 tons	63m x 9m x 5m	40+ knots	4 x SSM, 1 x 76mm, 2 x 40mm
The first four units of a smaller, guided-rocket armed PKX-B design are under construction and orders for four more were placed in June 2017. Current patrol forces include c. fifty PKM CHAMSURI fast attack craft and numerous YUB-P patrol boats.							
Mine-Countermeasures Vessels (11+3)							
Minelayer (ML)	MLS-II **NAMPO**	1	2017	4,200 tons	114m x 17m x 5m	23 knots	4 x K-VLS cells, 1 x 76mm gun, 6 x TT, 1 x helicopter
Minelayer (ML)	MLS- I **WONSAN**	1	1997	3,300 tons	104m x 15m x 4m	22 knots	1 x 76mm, 2 x 20mm, 6 x TT, 1 x helicopter (platform)
MCMV	**YANGYANG**	3	1999	900 tons	59m x 11m x 3m	15 knots	1 x 20mm
MCMV	**GANGGYEONG**	6	1986	550 tons	50m x 8m x 3m	15 knots	1 x 20mm
A further three mine-countermeasures vessels are reportedly under construction. At least one more member of the NAMPO class is planned.							
Major Amphibious Vessels (7+3)							
LHD	LPX-I **DOKDO**	1	2007	18,900 tons	200m x 32m x 7m	22 knots	1 x RAM, 2 x CIWS, c. 10 x helicopters, well-deck
LPD	LST-II **CHEON WANG BONG**	2+2	2014	7,000 tons	127m x 19m x 5m	23 knots	1 x 40mm, 1 x helicopter (platform), well-deck
LST	LST-I **GO JUN BONG**	4	1994	4,300 tons	113m x 15m x 3m	16 knots	2 x 40mm/30mm, 2 x 20mm, 1 x helicopter (platform)
A further, improved *Dokdo* type amphibious assault ship is under construction. There are numerous smaller landing craft and hovercraft in service.							
Other ROKN vessels include 3 AOE-I *Chun Jee* replenishment ships (shortly to be joined by the first of the new AOE-II *Soyang* class), 3 submarine rescue and salvage ships, and at least one trials ship. A new training ship and an additional trials vessel are under construction. The fleet is supported by large numbers of tugs and other yard craft							

Notes :

1 Date relates to the date the first ship of the class entered service with the ROKN. In some cases, these ships are now retired.

2 Second number refers to ships of the type/class currently under construction.

delivered in 1992. Subsequent boats, delivered between 1994 and 2001, were built by Daewoo Shipbuilding & Marine Engineering (DSME) at Okpo with progressively greater levels of indigenous input. Press reports suggest the class have been extensively modernized with Korean equipment throughout their service lives and some may have been subject to a hull plug extension. Three modified variants of the type are currently being completed for the Indonesian Navy.

German technology was also selected for the follow-on KSS-II programme, with an order for an initial three Type 214 AIP-equipped submarines placed in 2000. These were assembled in Korea from the outset, with the first – *Son Won-il* – entering service at the end of 2007. A further six were ordered in 2008. Six have been delivered to date and the seventh is expected to enter service in the second half of 2017. Production has been split between Hyundai Heavy Industries (six hulls) and DSME (three hulls), with the final delivery anticipated in 2019.

By late 2014, construction was underway on the first boat in the new indigenous KSS-III series. It is envisaged that nine of these new submarines will be built in three batches of three, each batch incorporating progressive improvements on the previous. The initial batch are large submarines displacing around 3,700 tons in submerged condition and incorporating six vertical launch system (VLS) tubes for indigenously-developed cruise missiles. The lead boat is being built by DSME and should be delivered around 2020. DSME have also been contracted to undertake design development of the second batch, which will reportedly be larger and include ten VLS tubes capable of launching (conventionally-armed) ballistic missiles.

The ROKN has also been a major operator of midget submarines. Although the last c. 200-ton *Dolgorae* class submarines built by what was then Korea Takoma (Now part of Hanjin Heavy Industries) were retired in June 2016, various sources suggest a number of Cosmos SX series submersibles remain in service to support Special Forces operations.[7]

The ROKN is one of the few submarine operators to have avoided any major operational accidents at sea.

The lead ROKN KDX-III type destroyer *Sejong Daewang* arrives at Pearl Harbor on 30 June 2016 for the RIMPAC exercises. Based on the US Navy *Arleigh Burke* (DDG-51) class, the Aegis-equipped KDX-IIIs are larger and more heavily armed. *(US Navy)*

Major Surface Combatants: The centrepiece of the surface warfare force is comprised of the three KDX-III Batch 1 Aegis-equipped destroyers of the *Sejong Daewang* class that entered service between 2008 and 2012. Based on the US Navy's *Arleigh Burke* (DDG-51) design, they have been enlarged to house no fewer than 128 VLS cells as well as sixteen inclined tubes for 'Haesung' surface-to-surface missiles. Forty-eight of the VLS cells are of an indigenous K-VLS design and can house Red Shark (K-ASROC) rocked-launched anti-submarine torpedoes and indigenous cruise missiles. The KDX-III class ships have yet to be fitted with a ballistic missile defence (BMD) capability like their US and Japanese Aegis-equipped counterparts but 2016 reports suggest this is under active consideration. There are also six KDX-III series general purpose destroyers dating from 2003–8 and three comparatively elderly KDX-I series ships delivered between 1998 and 2000.

The ROKN's frigates and corvettes are in the middle of a renewal programme. The legacy *Donghae/Pohang* class corvettes and *Ulsan* class frigates – a total of thirty-seven hulls – that date from the country's first major indigenous construction programme are being progressively replaced by the larger and more potent FFX series frigates. Six c. 3,000-ton FFX Batch 1 *Incheon* class frigates have already been delivered and construction is now focused on the improved Batch II *Daegu* class. The lead ship is currently undergoing trials and a further three of an expected eight-ship batch have been ordered. The 122m ships are 8m longer than the first Batch and displace around 3,500 tons full load. The incorporation of a sixteen-cell K-VLS in the design allows a wider range of missiles to be carried and they also incorporate a towed-array sonar and improved CODLAG propulsion system. Design of the third phase FFX-III ships has already been contracted to Hyundai and up to eight may be delivered towards the latter years of the 2020s. All three frigate batches are focused on littoral warfare, including anti-submarine, anti-surface and land attack missions.

Patrol Forces: The ROKN's patrol forces are also in the middle of major fleet renewal. Around 100 *Chamsuri* (PKM) class hulls in three variants once formed the core of patrol capabilities but only around half remain in service. They are being replaced by two distinct types of fast attack craft – a larger, missile-armed *Gumdoksuri* (PKX-A) type and a smaller, gun and rocket-armed *Chamsuri II* or PKMR (PKX-B) variant. The first PKX-A type commissioned in 2008 but subsequent teething troubles delayed the programme. The eighteenth and final vessel should be delivered before the end of 2017.[8] Meanwhile, the lead PKX-B was launched by Hanjin in 2016 and orders for seven more have already been placed. Production is likely to be split into two separate design batches, with the latter batch receiving an anti-submarine focus.

The fast attack craft are supplemented by numbers of non-commissioned patrol vessels designated YUB or yard utility boats. The current type is the YUB-P, typically armed with a 20mm gun, lighter weapons and depth charges and used largely for harbour defence in conjunction with the PKMs. Trials are also underway of a new fast unmanned surface vessel (USV), the 40-knot, waterjet-powered 'Haegeom', under a programme initiated by Korea's Defence Acquisition Programme Administration (DAPA) in 2015.

Mine Warfare Forces: Mine warfare forces currently include two large minelayers and nine mine-countermeasures vessels (MCMVs). The two minelayers were both constructed by Hyundai Heavy Industries, with the 1990s-era *Wonsan* recently being joined by the slightly larger *Nampo*. The latter has similarities with the FFX frigate design and is reportedly expected to be followed by three or four sisters. The MCMV force comprises six *Ganggyeong* and three larger, more recent *Yangyang* vessels, all delivered by Pusan based Kangnam Corporation and having a distinct resemblance to the Italian *Lerici* type. Three further improved *Yangyang* vessels were reportedly ordered in 2010 but there have been no deliveries to date.

Amphibious Forces: Amphibious assets are headed by the largest LHD type 'flat top' in Asia, the Hanjin Heavy Industries-built *Dokdo* that was delivered in 2007.[9] Hanjin commenced production of a long-awaited second amphibious assault ship (LPX-II) in November 2016 ahead of a formal keel laying ceremony on 28 April 2017. She will be launched in 2018 and delivered in 2020. Improvements over *Dokdo* reportedly include a K-VLS launcher, sensor enhancements and upgraded flying-control arrangements. Often reportedly called *Marado*, her name will be confirmed three months before the launch ceremony.

Other major amphibious units include the four 1990s-era *Go Jun Bong* tank landing ships (LSTs), which were delivered as replacement for former USN LSTs from the Second World War. They are being followed by the larger *Cheon Wang Bong* class, which are essentially small amphibious transport docks with a well deck and helicopter flight facilities aft. Two ships have been delivered to date and at least four are expected.

Smaller amphibious vessels include numbers of utility and personnel landing craft, many based on US Navy designs. The smaller types are regarded as service craft and referred to as harbour utility boats (HUBs). There are also five hovercraft. Three are Russian 'Murena E' types and two are indigenous LSF-II craft based on the USN's LCAC and built to deploy from *Dokdo.* At least two more will be built for the second LHD by Hanjin.

The PKX-A type fast attack craft *Deokchil* pictured in August 2016. The eighteen members of the PKX-A or *Yun Youngha* class form the 'high end' of the ROKN's patrol force mix. *(Republic of Korea Armed Forces)*

Fleet Support Vessels: Support vessels are headed by the three small *Chun Jee* (AOE-I) replenishment ships delivered in the 1990s. These are barely adequate to support current operations – let alone the navy's blue water ambitions – and a new AOE-II programme is underway. The lead ship, the 10,000-ton *Soyang*, should be delivered in the next twelve months. She is able to carry more than twice as much cargo as the older ships and is equipped with multi-purpose logistic support facilities. A hybrid diesel-electric propulsion plant provides a respectable 24-knot top speed and the incorporation of a hangar is an improvement of the helicopter facilities of the earlier class. Local media reports suggest a class of three is planned.

Other important auxiliary assets include two new *Tongyeong* salvage and submarine rescue ships that entered service in 2014 and 2016, joining the older *Cheon Hae Jin* delivered in 1996. The ships are able to support a comprehensive range of deep diving equipment.

The ROKN also operates a large fleet of auxiliaries and support vessels classified as yard craft. In addition to the previously mentioned 26m YUB-P patrol craft, there are numbers of slightly smaller 22.4m YF ferry-cum-patrol craft as well as small YO oilers, YWS water supply barges and various diving tenders and dive boats. Other yard craft include several classes of YTL harbour tugs, a number of catamaran hulled oil skimmer vessels (OS), floating outposts of the YPK-A and YPK-B types, smaller patrol boats and various accommodation, crane and sullage barges.

Other Naval Craft and Coast Guard: In addition to the vessels previously described, the ROKN operates a fleet of specialised research vessels, technology demonstration vessels, and intelligence-gathering ships for various agencies. These encompass a mix of grey and white-hulled trials ships deployed in conjunction with the ADD and a number of intelligence-gathering vessels operated for South Korea's National Intelligence Service (NIS). The latter are believed to include at least two tenders that act as mother ships for midget submarines and other covert craft.

No account of South Korea's maritime capabilities would be complete without reference to the Korea Coast Guard. After a major re-organisation in 2014, it is now a sub-agency of the newly formed Ministry of Public Safety and Security. With some 10,000 personnel, around twenty-four aircraft and over 300 patrol and other vessels, the Coast Guard makes an important contribution to overall maritime surveillance and stability. Its largest patrol vessels exceed 6,000 tons full load displacement and are more heavily-armed and capable than the constabulary assets operated by many front-line navies.

The front-line fast attack craft are supplemented by numbers of non-commissioned Yard Utility Boats armed with light weapons and depth charges. This picture shows YUB-P 853 exercising with a RHIB in March 2016. *(Republic of Korea Armed Forces)*

ROKN EQUIPMENT – NAVAL AVIATION

The ROKN's first foray into naval aviation occurred in 1951 when an intrepid officer with a passion for flying acquired a crashed US Air Force AT-6 Texan, successfully repairing and converting it into a float-plane using parts from former Japanese aircraft. A more formal Fleet Air Corps – still operating a motley collection of aircraft – was formed in 1957 but disbanded early in 1963. The naval air service was reactivated some ten years, later being renamed Navy Air Wing Six under a 1986 re-organisation.

Navy Air Wing Six currently operates around twenty fixed-wing aircraft and sixty helicopters under the following Air Group structure:

- **Air Group 61:** This operates fixed-wing maritime patrol and training aircraft, viz. sixteen P-3C and P-3CK Orions and five Reins F406 Caravan IIs. It currently comprises 611 Squadron and 613 Squadron at Pohang and 615 Squadron at Jeju.
- **Air Group 62:** This operates shipborne Super Lynx and Wildcat helicopters. It encompasses 625, 627 and 629 Squadrons located at Jinhae, Pyongtaek and Pohang. About twenty Super Lynx Mk 99As delivered in the 1990s remain in service, with eight new AW159 Wildcats joining them during 2016.
- **Air Group 63:** This operates UH-60P Black Hawk and U1-1H Iroquois helicopters in the amphibious support role through 631 and 633 Squadrons. Both types are also operated by other branches of the armed forces and estimates of numbers allocated to the navy vary.

The ROKN's small mine-countermeasures force includes two minelayer/support vessels. The oldest is the *Wonsan*, commissioned in 1998 and pictured operating with the Royal Australian Navy MCMV *Gascoyne* in 2014. South Korea's MCMVs are similar to the RAN's *Lerici* type. *(Royal Australian Navy)*

An Orion PC-3K maritime patrol aircraft of the ROKN's Air Group 61. A total of sixteen P-3C and P-3CK aircraft are currently in service. *(Republic of Korea Armed Forces)*

Flight training takes place at the Naval Flight Training Centre with 609 Flight Training Squadron at Mokpo. Naval aviators undergo their initial flight training with the ROK Air Force before proceeding for advanced training on navy aircraft types. There are also two other non-air groups, viz. Logistics Group 65 and Base Group 66.

THE MARINES AND SPECIAL FORCES

The ROKMC or *Daehanminguk Haebyeongdae* is a semi-autonomous body that is responsible for amphibious operations under the nominal control of Naval Headquarters. On a historical note, the ROKMC saw extensive combat in Vietnam, mostly around the major city of Da Nang.

The ROKMC sees itself as a national 'strategic manoeuvre' element and has been improving its infrastructure and operational structure to prepare for a rapidly-evolving security situation. It is headquartered at Hwaseong under the leadership of the ROKMC Commandant, a three-star lieutenant general. He is also responsible for the North Western Island Defence Command (NWIDC), which was created in 2011 after North Korea's 2010 shelling of Yeonpyeong.

Organisational structure is focused on two divisions and two independent brigades. The 1st *Hae Ryong* (Sea Dragon) Division is headquartered at Pohang and the 2nd *Cheongryong* (Blue Dragon) Division is based at Gimpo. Each is structured as a light division with a headquarters battalion, a tank battalion, an amphibious assault vehicle battalion, a reconnaissance battalion, a support battalion, an artillery regiment of three battalions and three infantry regiments of three battalions (i.e. nine battalions in all). The Sixth *Heukryong* (Black Dragon) Brigade is based at Baekryong Island near the NLL and the Ninth *Baekryong* (White Dragon) Brigade at Jeju Island. These independent brigades are typically comprised of one or more island garrisons plus artillery and reconnaissance support. There is also a Yeonpyeong (Y-P) garrison, an Education and Training Group, and a Landing Support Group. A recent development has been the creation of the rapid-response, airmobile 'Spartan 3000' regiment that can be deployed within 24 hours, for example to attack high-value targets in North Korea or respond to an internal emergency.

Equipment includes around 500 armoured fighting vehicles, including indigenous K1 tanks (replacing veteran M-48 Pattons), K9 self-propelled artillery and KAAVP 7A1 amphibious assault vehicles. The ROKMC has also recently achieved greater autonomy from the navy in terms of operating its own air wing, with thirty domestic Surion twin-engine utility helicopters under order for its own organic air corps.

Turning to naval Special Forces, two well-known specialist units are the Naval Special Warfare Flotilla (NSWF) and the Ship Salvage Unit (SSU). Modelled after the US Navy's SEALs, the NSWF is headquartered at Jinhae with units posted to the main fleet commands. Tracing its origins to the 1950s, the NSWF includes separate specialist teams for reconnaissance, underwater demolition, ordnance disposal and maritime counter terrorism. Members of the CT Team have been continuously deployed with the *Cheonghae* anti-piracy group off the Horn of Africa since March 2009. NSWF equipment includes a range of RHIBs and special delivery vehicles, some based on captured North Korean examples.

The SSU is also based at Jinhae and reports to the Fifth Flotilla. Subject to a similar training regime as the NSWF, the SSU is responsible for deep sea salvage and rescue. They came to prominence during the salvage operations to recover the *Cheonan*. The SSU personnel like to showcase their bare-body winter swimming evolutions in icy waters as a mark of their physical fitness, stamina and endurance. The SSU use a number of diving tenders (YDTs), smaller RHIBs and torpedo retrievers in support of their work.

FUTURE PROCUREMENT

ROKN fleet modernisation is ongoing, with upgrades to existing ships being supplemented by a major new construction programme. The downturn in commercial ship construction that has badly hit South Korea's shipbuilding industry has resulted in the acceleration of some naval orders to alleviate the crisis impacting a major economic sector.[10] The most important programmes are summarised below:

Submarines: New submarine construction is dominated by the KSS-III programme, already previously described. The 3,700-ton Batch I design developed by DSME features a PEM fuel cell-based AIP system and locally sourced lithium-ion batteries for extended range. Some sources suggest an underwater endurance of fifty days may be achievable but this is unlikely to be at the top underwater speed of 20 knots. The sonar system is based on LIG Nex1 technology whilst the combat management system is to be supplied by the former Hanwha Thales joint venture now wholly owned by the Korean partner. However, the difficulties in achieving full local sourcing are reflected in the specification of a British Babcock weapons handling technology, French Sagem optronic masts and a Spanish Indra Pegaso ESM system. The first batch of three boats is likely to be delivered by 2023, with the full nine-hull programme scheduled for completion before 2030. If previous precedent is followed, construction is likely to be split between DSME and Hyundai Heavy Industries, with the former probably being allocated four of the initial six orders as design developer for the first two batches. A mooted longer-term plan for a new type SSX submarine may see a transition to nuclear propulsion.

Surface Combatants: The most significant surface warship contract is for three KDX-III Batch II Aegis-equipped destroyers. This project is being led by Hyundai Heavy Industries under a design and engineering development contract signed with the DAPA in June 2016, with an order for the first vessel likely to be placed in 2018. The new design will incorporate significant improvements over the first batch, including hybrid gas turbine/electric propulsion, to reduce manning and other operating costs. Sensor enhancements are likely to encompass a BMD capability and a low-frequency towed array.

Plans for a follow-on KDDX destroyer, which is also likely to incorporate Aegis technology, are likely to be delayed until the current FFX programme is completed in the mid-2020s. Similarly, a relatively cheap anti-submarine warfare focused PCX patrol vessel may follow the current PKX-B fast attack craft programme.

The amphibious assault ship *Dokdo* is the ROKN's largest warship. This picture shows her operating with a range of smaller amphibious assets, including LSF-II hovercraft and Marine Corps KAAVP 7A1 assault vehicles. The ROKMC also has thirty indigenous Surion utility helicopters on order as the basis of its own air corps, which is also likely to operate with the ship. *(Republic of Korea Armed Forces)*

Other Vessels: A broad range of other construction is led by ongoing amphibious ship programmes, including the second amphibious assault ship and further vessels of the LST-II *Cheon Wang Bong* class. There has been a long-held desire to follow the amphibious assault ships with a transition to light aircraft carriers but this ambition does not yet appear to be funded. Further AOE-II type replenishment ships are a more urgent priority.

More bespoke projects include a dedicated ATX training ship finally ordered from Hyundai after being stalled for lack of finds and scheduled for delivery in 2020. Possibly the lead of a class of two vessels, the lightly armed but helicopter equipped ship is reportedly based on the KDX-II hull form and can accommodate up to 300 trainees. Meanwhile, DSME is working on a larger, 6,300-ton submarine rescue ship specifically designed to support the KS-III type and Hanjin on a catamaran-hulled trials ship. Additional programmes will include new insertion craft and may also extend to encompass a new generation of minehunter and modern midget submarines.

Many upgrades of existing ships are being carried out under so-called Performance Improvement Projects (PIP) that modernise or replace combat system technology. For example, a number of the KSS-I Type 209 submarines are being retrofitted by DSME with a new combat system being supplied by

ROKMC marines practice a beach landing. The force sees itself as the strategic manoeuvre element of the Korean Armed Forces. *(Republic of Korea Armed Forces)*

A picture of the KDX-II type destroyer *Choi Young* welcoming in the 2016 New Year. The six members of the class are scheduled to benefit from a Performance Improvement Project that will focus on improving their anti-submarine warfare capabilities. *(Republic of Korea Armed Forces)*

LIG Nex1 and a PIP contract for the KDX-I destroyers was awarded by the DAPA to Hanwha Thales in 2016. This will be followed by a similar programme for the KDX-II class, with both upgrades focused on improving anti-submarine capabilities. A major programme to develop a range of indigenous missiles is also likely to see variants fitted to existing platforms, not least the Haeseong II and Hyeonmu series land-attack missiles that form an important part of the three-pronged national defence strategy.

Aircraft acquisitions are likely to be headed by replacements for the current elderly P-3C Orion maritime patrol aircraft (MPA) and legacy Lynx helicopters, as well as an expanded overall helicopter requirement. Boeing is touting its increasingly popular P-8 as the future MPA, whilst the Wildcat is competing with the Sikorsky MH-60R and NH Industries NH90 for an expected order for a dozen additional helicopters. It would seem likely that further variants of the indigenous KAI Surion will feature in the helicopter mix following the ROKMC's initial order for thirty in December 2016.

Daegu is the first of a second batch of FFX type frigates that could ultimately extend to eight ships. The main picture shows her on launch day and the smaller image during the course of sea trials. The second batch are larger and more heavily-armed than the initial ships and incorporate a CODLAG propulsion system built around the Rolls-Royce MT-30 gas turbine. It is not clear whether they will embark the Wildcat helicopter type ordered for the first batch or competing Sikorsky or NH Industries proposals. *(Republic of Korea Armed Forces / Rolls-Royce)*

CONCLUSION

The ROKN and South Korea's supporting naval construction and equipment industry have seen a massive uptick in technical capabilities over the two decades since the end of the Cold War. Whilst the US Navy remains both an important influence and ally, providing vital capabilities in areas such as ballistic missile defence, the degree of reliance is lessening. The ROKN undoubtedly enjoys a marked technological edge over the KPAN, its most likely adversary, and has made significant progress in achieving its blue water naval aspirations. It also has an important role to play in the developing strategic three-pronged response to the North Korean missile threat. An ambitious future procurement programme points to further advances.

There are, however, concerns. The 2010 *Cheonan* sinking revealed gaps in the navy's ability to counter KPAN submarines and other covert craft, whilst a weakness in real-time intelligence was suggested by an apparent failure to anticipate a massive KPAN submarine deployment in 2015. The accidental sinking of the ferry MV *Sewol* with heavy loss of life in April 2014 has also given rise to the effectiveness of command and control procedures in handling unexpected situations. Long-running ethical issues that have seen allegations of corruption at the highest level are another problem.

Ultimately, it is the sheer unpredictability of the asymmetrical warfare tactics likely to be adopted by the DPRK and the KPAN that pose the greatest challenge.[11] The ROKN therefore needs to strike a careful balance between further developing its littoral warfare capabilities to counter the omnipresent North Korean threat and pursuing its ambitions to become an important player on the global maritime stage.

Notes

1. The President's remarks were widely reported. For example see Christine Kim's 'South Korea's Moon says "high possibility" of conflict with north as missile crisis grows', *Reuters* – 17 May 2017.

2. As a formal peace treaty has never been signed, the two countries remain technically at war. Incidents since 1953 have included the loss of the patrol boat *Danigpo* to Korean artillery in 1967, two major naval clashes off the Yeonpyeong Islands in 1999 and 2002, the shelling of the islands in 2010, another clash off Daecheong Island in 2009, the *Cheonan* sinking and several groundings of midget submarines and infiltration craft.

3. For example, see Hiroshi Minegishi's 'South Korea to bolster Defences against North', *Nikkei Asian Review* – 19 October 2016 (Tokyo: Nikkei Inc., 2016).

4. ROKN ship names are frequently followed by *a –ham* which is an honorific and thus not technically part of the name. English-language names also frequently vary because of translation/transliteration issues, for example B and P as in Busan/Pusan, G and K (Gimpo/Kimpo), and C and J (Chinhae/Jinhae) are used interchangeably. Official translations of names have been used where possible.

5. See ROKN Headquarters' *Navy Vision 2020* (Gyeryong-dae: Republic of Korea Navy, 1999).

6. The establishment of the independent submarine command under a two-star admiral took place in February 2015. Its headquarters are at Jinhae. The move reflects the increased strategic importance of the submarine force, not least in offering the potential for pre-emptive and retaliatory strikes on DPRK missile sites.

7. Some of these boats may be quite elderly and it is not clear to what extent replacements in the form, for example, of Hyundai's HDS-400 and HDS-500 series have been procured. In August 2016 there were reports of an explosion on a thirty-year old, 70-ton boat undergoing repairs that killed three personnel. See 'Aged submersible vehicle explodes during inspection', *The Dong A-Ilbo* – 18 August 2016 (Seoul: Dong A Media Group, 2016).

8. The last vessel of the class was seriously damaged by a typhoon whilst under construction at STX and has been delivered late and out of sequence. Some sources suggest the original vessel was effectively a total loss and the current hull is essentially a new ship.

9. For further detail see Guy Toremans' '*Dokdo* Class Assault Ships', *Seaforth World Naval Review 2017* (Barnsley, Seaforth Publishing, 2016), pp.120–35.

10. All of South Korea's major shipbuilding companies have been posting large losses. The fourth largest group, STX Offshore & Shipbuilding, filed for receivership in 2016 whilst top three group DSME has received over US$2.5bn in state support to avoid collapse. The largest shipbuilding group, Hyundai Heavy Industries, completed a major restructuring in May 2017 that saw non-shipbuilding activities split from the group to improve the financial strength of the core group.

11. North Korea military analysts Joost Oliemans and Stijn Mitzer – both affiliated with NKNews.org – provided the following analysis for *World Naval Review*: 'The KPAN is and has historically been absolutely unique in the tactics and overall doctrine it employs to counter its technologically advanced adversaries. As such, it defies typical classification, and will pose a highly unpredictable opponent in the event of a full scale war. On the offensive side, naval craft are expected to provide fire support for the North's immense hovercraft invasion fleet, which, despite being the largest of its kind in the world, is lightly armed and only aimed at landing large volumes of infantry ashore. Due to the difficulties faced with producing large ships and their vulnerability to their more advanced Southern equivalents, little attempt is made to gain the upper hand in symmetrical naval battles, and instead great quantities of torpedo boats and missile craft are maintained and produced to this day. Although most are armed with older torpedoes and anti-ship missiles that are by now thoroughly obsolete, newer torpedo systems as well as the introduction of a modern anti-surface missile derived from the Russian Kh-35 has certainly evolved the threat coming from the KPAN's fast attack craft and coastal defence systems. The situation is further complicated by the fact that most of the North's newest naval designs have been highly experimental in nature, specifically aiming to exploit the advantages of asymmetrical warfare. Their massive submarine fleet mirrors this philosophy closely, and will be similarly tough to counter, with the addition of the first (but definitely not last) SSBN changing the stakes completely. The fact that new large naval craft are still being designed and produced at the same time is a solid indicator that innovation is ongoing, and as such that the North's naval threat continues to evolve.' Their new book, *North Korea's Armed Forces: On the Path of Songun* is due to be published by Helion & Co. in January 2018.

12. The dearth of English-language material and official bilingual websites means that information of both Korean navies is sparse and not readily reported outside of a narrow audience. In addition to the sources already cited, the English language *Korea Times* http://www.koreatimes.co.kr/www2/index.asp regularly carries defence news stories whilst the Tokyo-based *The Diplomat* at http://thediplomat.com/ provides a wider regional perspective. H I Sutton's *Covert Shores* website at http://www.hisutton.com/ often provides valuable insights into KPAN submarine programmes. Reference can also be made to *The Evolution of the Maritime Security Environment in Northeast Asia and ROKN-USN Cooperation* (Seoul, Korean Institute for Maritime Strategy, 2016) at http://file.kims.or.kr/KIMS%20CNA%20Maritime%20Security.pdf for more detailed perspectives on the relationship between the two navies.

2.3 REGIONAL REVIEW

Author:
Conrad Waters

THE INDIAN OCEAN AND AFRICA

INTRODUCTION

On 12 March 2017 the Bangladesh Navy held a ceremony at Chittagong to commission the submarines *Nabajatra* and *Joyjatra* into its fleet.[1] The second-hand Type 035G 'Ming' class boats provide only a modest level of underwater capability. However, the ceremony was notable from both a national and regional perspective. For Bangladesh, the ceremony marked a further step towards achieving the ambition – first announced in 2009 – of developing a modern, three-dimensional naval force operating on, above and beneath the waves. The attendance of Bangladesh's Prime Minister Sheikh Hasina at the ceremony demonstrated its political significance. In a regional context, the submarines' arrival was seen by many as marking another stage in the ongoing power struggle between India and China that is seeing the latter expand its military and economic presence across the Indian Ocean.

There has been much debate about the significance of Bangladesh's acquisition to the regional balance of power. At one level, the purchase can be seen simply as the Bangladesh Navy selecting the best economic option in pursuing its force objectives. The country also enjoys relatively good relations with its larger Indian neighbour, with whom it has made progress recently in resolving outstanding boundary disputes. Other observers – gripped, perhaps, by a degree of paranoia – point to a growing web of Chinese influence ranging from massive infrastructure investment in Africa to strengthening military ties across the region.[2] For example, Pakistan is also acquiring Chinese submarines, whilst the People's Liberation Army Navy (PLAN) is building links with other regional forces. It recently held its first joint naval exercise with Myanmar. The most material development – certainly of greater moment than Bangladesh's submarines – is China's establishment of a naval support facility in Djibouti, its first overseas. Due to become operational during 2017, it is located four miles from the American base at Camp Lemonnier, the only permanent US facility in Africa. Reports suggest a degree of American concern about the proximity of their new neighbours.[3]

One reason for Indian sensitivity over Bangladesh's submarine purchase is the increased importance of the underwater domain to India's strategic defences at a time when it is struggling to rebuild its own submarine force after a period of neglect. Although there has been no official confirmation, press reports suggest that India's first strategic missile submarine, *Arihant*, was commissioned in 2016. She is the first of what is believed to be a planned force of four boats that will provide India with an enhanced second-strike capability, greatly enhancing its security. The prospect of Bangladesh Navy submarines – doubtless supported by embarked Chinese trainers – operating close to India's strategic submarine bases in the Bay of Bengal is therefore an unappealing prospect. In the absence of a significant force of nuclear-powered attack submarines, the new 'Scorpène' type *Kalvari* class submarines are likely to play an important role protecting the nuclear deterrent. The lead vessel, first of six boats, has been undergoing trials over the past year. She is expected to be delivered shortly.

One of the ostensible reasons for the PLAN's increased Indian Ocean presence is its desire to protect a key line of communication for its maritime trade. The threat of piracy off the Horn of Africa had abated in recent years. However, a spate of attacks in early 2017, including the first successful hijacking since 2012, evidence a recent resurgence.[4] To the north, Yemen's coastline has also become more hazardous as a result of its ongoing civil war. Widely seen as an extension of the wider struggle for regional dominance between Saudi Arabia and Iran, the protracted conflict has drawn in other groups such as ISIS and Al-Qaeda. Despite intervention by a powerful Saudi-led coalition of Sunni Arab states in support of pro-government forces, the war continues. At the start of October 2016, the former US Navy-operated high speed vessel *Swift*, then under charter to the United Arab Emirates, was damaged beyond repair after being hit by what was reported to be a C802/Noor type anti-ship missile. US Navy vessels despatched to safeguard the area were subject to a number of similar attacks – all unsuccessful – later in the month. This resulted in punitive counter-strikes using Tomahawk cruise missiles to disable the Houthi positions. However, other dangers persist. On 30 January 2017, the Saudi frigate *Al-Madinah* was attacked by explosive-carrying boats. One detonated near her stern, damaging the ship and killing two of her crew.

The Indian Navy's underwater flotilla is currently somewhat depleted at a time when many of its neighbours are acquiring Chinese-built submarines. The situation should improve as new Project 75 'Scorpène' type submarines start to commission from the second half of 2017 onwards; the lead Indian boat, *Kalvari*, is pictured here. *(DCNS)*

Table 2.3.1: FLEET STRENGTHS IN THE INDIAN OCEAN, AFRICA AND THE MIDDLE EAST – LARGER NAVIES (MID 2017)

COUNTRY	ALGERIA	EGYPT	INDIA	IRAN	ISRAEL	PAKISTAN	SAUDI ARABIA	SOUTH AFRICA
Aircraft Carrier (CV)	–	–	1	–	–	–	–	–
Strategic Missile Submarine (SSBN)	-	-	1	-	-	-	-	-
Attack Submarine (SSN/SSGN)	–	–	1	–	–	–	–	–
Patrol Submarine (SSK/SS)	4	5	13	3	5	5	–	3
Fleet Escort (DDG/FFG)	2	7	24	–	–	10	7	4
Patrol Escort/Corvette (FFG/FSG/FS)	8	4	10	7	3	–	4	–
Missile Armed Attack Craft (PGG/PTG)	c.12	c.30	10	c.24	8	9	9	3
Mine Countermeasures Vessel (MCMV)	–	c.14	4	–	–	3	3	4
Major Amphibious (LPD)	1	2	1	–	–	–	–	–

Notes:

1 Algerian fast attack craft and Egyptian fast attack craft and mine-countermeasures numbers approximate.

2 Iranian fleet numbers exclude large numbers of indigenously-built midget and coastal submarines, as well as additional missile-armed craft operated by the Revolutionary Guard.

3 The South African attack craft and mine-countermeasures vessels serve in patrol vessel roles.

INDIAN OCEAN NAVIES

Bangladesh: The most important development for the Bangladesh Navy over the past year was undoubtedly the delivery of its two 'new' submarines already discussed in the introduction. The two Type 035G variants of the 'Ming' type were extensively refurbished under a reported c. US$200m contract before being handed-over at Dalian in northern China on 14 November 2016. They subsequently sailed for Bangladesh after their crews had completed initial familiarisation. The submarines have been acquired largely for training purposes and it seems more modern boats will be purchased during the 2020s. A major new naval base will also be completed at Patuakhali in central-southern Bangladesh to support the expanding fleet.

Renewal of surface forces is also largely being placed in China's hands. Acquisitions of larger surface combatants have been focused on three elderly 'Jianghu' series frigates to date, although there have been reports that slightly more modern 'Jiangwei II' types will start to join the fleet in 2018. These are being supplemented by new-built construction of smaller ships. Two C13B derivatives of the domestic Type 056 'Jingdao' class were delivered by the Wuchang shipyard, Wuhan at the end of 2015 and work on a second pair commenced at the same facility in the summer of 2016. Local reports suggest that assembly of a further four will then transfer to a domestic yard. Unlike their Chinese half-sisters, the Bangladesh Navy vessels reportedly carry no anti-submarine sensors or weapons, being largely focused on the anti-surface role. This mission is allocated to the c. 650-ton *Durjoy* class anti-submarine patrol vessels, which are equipped with an ESS-3 bow-mounted sonar and EDS-25A rocket launchers. Two of these units were delivered from Wuchang in 2013, with construction subsequently transferring to the domestic Khulna Shipyard. *Durgam*, the lead ship of a second pair, was launched in December 2016 with *Nishan* following in March 2017. As for the larger C13B corvettes, it appears eight ships are eventually planned.

The Bangladesh Coast Guard is also expanding, taking delivery of two pairs of former Italian Navy *Minerva* class corvettes in August 2016 and mid-2017. The refurbished ships have been stripped of most of their armament and fitted with a small helicopter deck in place of their former surface-to-air missile launcher. The Coast Guard is also acquiring smaller inshore patrol vessels with two variants of the navy's 350-ton *Padma* class currently being completed by Khulna.

The Bangladesh Navy is undertaking an impressive fleet expansion and modernisation programme that is seeing a variety of new vessels commissioned. This picture shows *Shadhinota*, the first of a series of Chinese-built C13B corvettes that are derived from the PLAN's Type 056, at an International Fleet Review held in May 2017 to support the Republic of Singapore Navy's 50th Birthday. *(Singapore Ministry of Defence)*

India: The last year has continued to see the Indian Navy's ambitious plans for enhancement and expansion held back by a number of familiar problems. These have included an inadequate – and declining – budget for naval modernisation; a strong emphasis on local shipbuilding in spite of a poor track record of executing programmes on time and to budget; an over-reliance on defective Russian-supplied equipment; and ongoing mishaps attributable to human error.

India's 2017–18 defence budget estimate amounted to INR262,390 crore (US$41bn) excluding pensions, a year-on-year increase of 5.3 percent.[5] This rate of growth is slower than in recent years. One consequence has been a rise in the proportion of the budget allocated to revenue expenditure to the obvious detriment of modernisation accounts. With much capital consumed by Indian Air Force requirements, the navy has been suffering. The 2017–18 naval modernisation account was INR18,750 crore (US$2.9bn), a twelve percent reduction on the previous year and totally inadequate to support expansion efforts.

A further problem relates to the wisdom with which money is being spent. With the exception of the strategic missile submarine *Arihant*, only one major warship has been commissioned into Indian Navy service in the last twelve months. This is largely a reflection of local industry's inability to deliver new ships in accordance with the navy's expectation. The extent of the problem was revealed in a 'Performance Audit on Construction of Indigenous Aircraft Carrier' published by the Comptroller and Auditor General (CAG) of India in July 2016.[6] This revealed there was complete disagreement between the navy and builders' Cochin Shipyard as to when the new carrier, *Vikrant*, was likely to be delivered. Whilst the Indian Navy is holding to a schedule requiring delivery of the ship by December 2018, the shipyard now believes that 2023 is a more likely estimate. The end result is likely to be India having to rely on just one operational aircraft carrier for an extended period in spite of a long-stated aim to have three in commission.

Whilst the performance of the Indian shipbuilding sector has been far from stellar, reliance on Russian industry has also proved problematic. The

CAG report referred to above makes frequent references to delays in the receipt of Russian design documentation and equipment. It also highlighted a host of problems with respect to the Russian MiG-29K/KUB strike fighters acquired for both *Vikrant* and the existing *Vikramaditya*. These were riddled with defects relating to their airframe, engine and fly-by-wire system, resulting in serviceability as low as sixteen percent.

Meanwhile, on 5 December 2016, the navy suffered the latest in a series of mishaps to impact the fleet when the frigate *Betwa* fell over onto her side whilst the cruiser graving dock in Mumbai Dockyard was being flooded-up. Two sailors were killed in an incident that left the ship with significant damage. *Betwa* was subsequently righted in February 2017. Repairs are expected to take around twelve months to complete.

Table 2.3.2 suggests that there has been very little change in the overall structure of the Indian Navy year-on-year. The formal decommissioning of *Viraat* (the former HMS *Hermes*) on 6 March 2017 – she had already effectively been out of service for around a year – leaves *Vikramaditya* as the navy's sole aircraft carrier. The surface fleet welcomed the final Project 15A *Kolkata* class destroyer, *Chennai*, which

The series of mishaps that have impacted the Indian Navy in recent years saw two further fatalities in December 2016 when the Project 16A *Brahmaputra* class frigate *Betwa* shifted from the blocks and fell on her side whilst the dry dock she was in at Mumbai Dockyard was being flooded. Although she has subsequently been returned to an even keel, the damage she has suffered will have been extensive. As such, Indian Navy estimates that she will be back in service within the year should be regarded with caution. *(Indian Navy via M Mazumdar)*

Table 2.3.2: INDIAN NAVY: PRINCIPAL UNITS AS AT MID 2017

TYPE	CLASS	NUMBER	TONNAGE	DIMENSIONS	PROPULSION	CREW	DATE
Aircraft Carriers							
Aircraft Carrier (CV)	Project 1143.4 **VIKRAMADITYA** (KIEV)	1	45,000 tons	283m x 31/60m x 10m	Steam, 30 knots	1,600	1987
Principal Surface Escorts							
Destroyer – DDG	Project 15A **KOLKATA**	3	7,400 tons	163m x 17m x 7m	COGAG, 30+knots	330	2014
Destroyer – DDG	Project 15 **DELHI**	3	6,700 tons	163m x 17m x 7m	COGAG, 32 knots	350	1997
Destroyer – DDG	Project 61 ME **RAJPUT** ('Kashin')	5	5,000 tons	147m x 16m x 5m	COGAG, 35 knots	320	1980
Frigate – FFG	Project 17 **SHIVALIK**	3	6,200 tons	143m x 17m x 5m	CODOG, 30 knots	265	2010
Frigate – FFG	Project 1135.6 **TALWAR**	6	4,000 tons	125m x 15m x 5m	COGAG, 30 knots	180	2003
Frigate – FFG	Project 16A **BRAHMAPUTRA**	3	4,000 tons	127m x 15m x 5m	Steam, 30 knots	350	2000
Frigate – FFG	Project 16 **GODAVARI**	1	3,850 tons	127m x 15m x 5m	Steam, 30 knots	315	1983
Corvette – FSG	Project 28 **KAMORTA**	2	3,400 tons	109m x 13m x 4m	Diesel, 25 knots	195	2014
Corvette – FSG	Project 25A **KORA**	4	1,400 tons	91m x 11m x 5m	Diesel, 25 knots	125	1998
Corvette – FSG	Project 25 **KHUKRI**	4	1,400 tons	91m x 11m x 5m	Diesel, 25 knots	110	1989
Submarines							
Submarine – SSBN	**ARIHANT**	1	7,500+ tons	112m x 11m x 10m	Nuclear, 25+ knots	100	2016
Submarine – SSN	Project 971 **CHAKRA** ('Akula II')	1	9,500+ tons	110m x 14m x 10m	Nuclear, 30+ knots	100	2012
Submarine – SSK	Project 877 EKM **SINDHUGHOSH** ('Kilo')	9	3,000 tons	73m x 10m x 7m	Diesel-electric, 17 knots	55	1986
Submarine – SSK	**SHISHUMAR** (Type 209)	4	1,900 tons	64m x 7m x 6m	Diesel-electric, 22 knots	40	1986
Major Amphibious Units							
Landing Platform Dock – LPD	**JALASHWA** (AUSTIN)	1	17,000 tons	173m x 26/30m x 7m	Steam, 21 knots	405	1971

The Indian Navy Project 17 frigate *Sahyadri* operating with the Project 28 corvette *Kamorta* in May 2017. Both ships are amongst the newest of their types in the Indian Navy but new designs are in preparation. *(Singapore Ministry of Defence)*

Sluggish Indian construction of domestic designs may be reinforced by orders for new Project 1135.6 frigates of Russian origin. There have previously been reports that some may come from members of the class intended for the Russian Navy but laid-up at the Yantar yard in Kaliningrad because of unavailability of Ukrainian-supplied turbines. However, it now seems more likely that most construction will be licenced to the Goa Shipyard and that any Russian-built ships will be new orders. This picture shows earlier members of the class under construction at Yantar. *(Yantar)*

commissioned on 21 November 2016. Her arrival was balanced by the withdrawal from active service of the Project 16 *Godavari* class frigate *Ganga*, which undertook her last operational voyage on 27 May 2017. She will be formally decommissioned later in the year. *Godavari* ended her service life in December 2015 and *Gomati*, the third member of the class, is also expected to leave the fleet soon.

Construction of major surface combatants encompasses three programmes. The most advanced is that for four Project 15B *Visakhapatnam* class destroyers, which are based on the previous Project 15A design. Builders Mazagon Dock Ltd launched *Mormugao*, second of the class, on 17 September 2016. Deliveries are scheduled to take place every two years from 2018, although this seems ambitious on past performance. Mazagon will also undertake construction of four of the seven Project 17A stealth frigates derived from the *Shivalik* class. The joint programme shared with Kolkata's Garden Reach Shipbuilders & Engineers (GRSE) should enter the production stage in the next twelve months. A third programme involves the licence-built construction of Russian-designed Project 1135.6 *Talwar* class frigates. This project – which was initially reported to include transfer of some of the similar Russian *Admiral Grigorovich* class frigates currently laid up in a half-completed state at the Yantar yard due to non-delivery of their Ukrainian gas turbines – has seemingly been allocated to Goa Shipyard. It seems that the plan is to acquire four frigates, taking the complete class up to ten ships.[7]

There has been little material change to the composition of the balance of the surface fleet. Two Project 28 *Kamorta* class corvettes remain under construction following a decision to redesign their superstructures around a carbon-fibre composite structure to mitigate top-weight issues found with the earlier pair. It was originally envisaged that more ships of the type would be built to a slightly improved Project 28A design. However, this plan appears to have been overtaken by the issue of a request for information in October for seven 'Next Generation Corvettes'. The 120m ships will be larger and more heavily armed than the *Kamorta* class and have a more general-purpose orientation than the anti-submarine focused Project 28 design. There is also a requirement to replace the smaller corvette/fast attack type vessels of the Soviet-designed *Veer* and *Abhay* classes, which are starting to decommission.[8] Discussions have been held with industry around next generation missile ships and smaller shallow water anti-submarine vessels but there has been no confirmation of any orders to date.

There is also an urgent requirement to recapitalise the mine-countermeasures fleet. This has now been reduced to just four vessels following the decommissioning of two of the remaining *Pondicherry* class minesweepers on 5 May 2017. The remaining ships – all around thirty years old – are also expected to retire shortly. They will be replaced by a new class of mine countermeasures vessels being licence-built by Goa Shipyard under a controversial deal with South Korea's Kangnam Corporation. However, construction has yet to begin and there will therefore be a considerable gap before the new ships are delivered. Progress has also been slow with plans to order four new LPD-type amphibious transport docks under a programme estimated to amount to around US$3bn. Two shortlisted local private-sector yards – allied with DCNS and Navantia – have been asked to resubmit bids for the project after plans to involve state-owned Hindustan Shipyard in construction were reversed.[9]

Developments with respect to the submarine force have been somewhat more positive. The unconfirmed entry of *Arihant* into service is a major step forward for the navy's strategic ambitions whilst good progress with trials of the lead Project 75 'Scorpène' type boat *Kalvari* has been followed by the maiden voyage of the second member of the class, *Khanderi*, on 1 June 2017. It has now been decided not to fit an indigenously-developed air-independent propulsion (AIP) system on the last two members of the class, potentially speeding their completion.[10] Current plans envisage submarine production switching to a new Project 75I class boat when the six 'Scorpènes' have been completed. However, a builder has yet to be selected. The 'Scorpène's' designer DCNS had hopes that the delay might result in an interim order for a further batch of boats but the leak of over 22,000 pages of technical information on the design to *The Australian* newspaper has considerably reduced this prospect. The Indian Navy hopes to transition to constructing a class of nuclear-powered attack submarines after the Project 75I boats but this is some distance in the future. In the interim, it appears that a second Russian 'Akula II' type boat will be acquired on lease once the current arrangement with respect to *Chakra* expires. The agreement could see Indian technical experts involved with completing the boat, therefore helping pave the way for its own programme.

The Myanmar Navy is steadily improving the capabilities of its domestically-constructed frigates and corvettes, notably through the incorporation of stealth technology in its latest designs. This picture shows the latest frigate *Sin Phyushin*; the image makes an interesting contrast with the earlier *Aung Zeya* pictured on p.59 of *Seaforth World Naval Review 2017*. *(Singapore Ministry of Defence)*

Myanmar: The arrival of submarines in neighbouring Bangladesh has reinforced the Myanmar Navy's interest in acquiring its own underwater flotilla. In May 2017, Myanmar's Deputy Defence Minister Major General Myint Nwe was widely quoted as saying 'Our neighbours have submarines and we want then as well' before mentioning some of the financial and practical hurdles to such an acquisition.

Meanwhile, the progressive modernisation of Myanmar's surface forces took another step forward with delivery of the modified *Anawrahta* type corvette *Tabinshwheti*, third in the series, in December 2016. In similar fashion to that seen in the evolution of Myanmar's larger frigates of the *Aung Zeya* and *Kyan Sittha* classes, she displays significant differences to earlier ships, not least extensive use of stealth technology. An eclectic range of weapons and sensors appears to include an Italian-designed gun, Russian close-in-weapons systems and Chinese surface-to-surface missiles. Unlike the earlier corvettes, a hangar is provided to support an embarked helicopter.

Pakistan: Pakistan's navy has been steadily losing ground to regional rivals in recent years. The priority given to the army and air force in terms of defence procurement, as well as a reduction in the availability of American-supplied equipment following a deterioration in relations with the United States, has meant there has been little new procurement since the arrival of new frigates around the turn of the decade. The next months should see the delivery of the third Chinese-designed *Azmat* class fast attack craft, which was launched by Karachi Shipbuilding & Engineering Works (KSEW) in September 2016 and is currently undergoing trials. Construction of the fourth and final member of the class began in mid-December. A new fleet tanker being built by Turkey's STM was launched in August 2016 and should also be delivered before the end of the year. China and Turkey have effectively become the suppliers of choice for the Pakistan Navy.

The future looks brighter following the launch of a number of new procurement projects. Dominant amongst these is the construction of eight AIP-equipped variants of China's Type 039A/Type 041 'Yuan' class submarines split equally between China and KSEW under a contract agreed towards the end of 2015. Details of the contract emerging over the last year suggest the Chinese-built quartet will all be delivered by 2023, with the KSEW-assembled boats following by 2028. Some commentators suggest that there is an ambition to give at least some of these submarines the capability to deploy strategic missiles. Meanwhile, STM is working with KSEW to manage the upgrade of the three existing 'Agosta 90-B' class submarines under a contract agreed in June 2016.

STM also looks to play a major part in the renewal of Pakistan's fleet following signature of a long-awaited letter of intent in May 2017 to support the local construction of four 'Milgem' type corvettes, also at KSEW. A firm contract is expected shortly. The importance of KSEW to future programmes was further emphasised the following month with the announcement of a contract for it to build a new 90m offshore patrol vessel to a Damen design. The 1,900-ton vessel will be capable of supporting helicopter operations and may be the first of a number of similar ships.[11]

Sri Lanka: The Sri Lanka Navy is continuing its evolution towards a more balanced force structure with greater capacity for longer-range constabulary missions after the end of the country's lengthy civil war. The two new offshore patrol vessels – *Sayural* and *Sindurala* – ordered from India's Goa Shipyard have now both been launched and both should enter service over the next year. It has also been reported that discussions have been held with Russia's Zelenodolsk over the possible acquisition of at least one Project 1161.1 'Gepard' type frigate.

AFRICAN NAVIES

South Africa's planned naval modernisation continues to creep forward against the backdrop of a sluggish economy and endemic political corruption. Long-awaited selection of preferred bidders for a new survey ship (Project Hotel) and three offshore and three inshore patrol vessels (Project Biro) were finally announced in February 2017, more than two years after tenders first opened. Durban-based Southern African Shipyards has been selected for Project Hotel. If negotiations are successfully concluded, construction of a vessel based on an ice-strengthened Vard Marine 9 105 95m design broadly similar to the British Royal Navy's *Echo* and *Enterprise* will commence in 2018. Damen Shipyards Cape Town has been named preferred bidder for the Project Biro vessels but the contract for these has been deferred to the 2018/19 financial year. Meanwhile, the currently stretched nature of South Africa's maritime patrol capabilities was revealed by a March 2017 admission that none of the Air Force's veteran C-47TP maritime patrol aircraft had been available for a period, leaving the country without any aerial surveillance of its one million km? exclusive economic zone.[12]

The South African Navy frigate *Amatola* undertook a lengthy deployment to West Africa and Europe in the first half of 2017. She is pictured here at Portsmouth, UK on 24 February 2017. Unfortunately, inadequate investment is depriving South Africa's armed forces of sufficient resources to police its own waters effectively. *(Conrad Waters)*

Elsewhere in Sub-Saharan Africa, procurement also continues to be dominated by constabulary vessels as the importance of protecting maritime economic assets gains increasing traction. A wide range of yards in countries ranging from traditional shipbuilding exporters such as France to more recent market entrants such as China, India and even Bangladesh are supplying relatively basic inshore patrol vessels to meet this requirement. Typical of these new ships is the French OCEA type OPV 190 Mk III, an example of which was handed over to **Senegal** at the OCEA facility at Les Stables d'Olonne on 26 October 2016. Named *Fouladou*, the 58m long vessel has a maximum speed of 24 knots and is designed to deploy for up to three weeks on maritime security, search-and-rescue, and anti-pollution activities. There is accommodation for a core crew of twenty-four together with thirty-two additional passengers, for example a military force or survivors of a humanitarian incident. Another interesting development, as in South Africa's case, is increased local construction of such vessels. For example, **Nigeria** has been building a series of locally-developed *Andoni* class seaward defence vessels, with each being larger and more sophisticated than its predecessor. The second, *Karaduwa*, was commissioned on 16 December 2016 and a third is under construction. December 16th also saw the formal entry into service of *Unity*, the second Chinese-built P18N offshore patrol vessel. Like Bangladesh's C13B corvettes, she is a derivative of the Type 056 'Jingdao' class.

Turning to North Africa, Algeria and Egypt continue to dominate naval developments. **Algeria** is seeing the benefits of a massive naval expansion programme encompassing new submarines, frigates and naval helicopters that is now drawing to a close. The last year has seen deliveries of the MEKO A-200AN frigate *El Moudamir*, the latter of a pair ordered from Germany in 2012, and the C28A frigate *Ezzadjer*, last of a three-ship class built by

The economic importance of ensuring the effective policing of national waters is leading many smaller African countries to devote resources to constabulary assets. A good example of the relatively simple but effective ships being acquired is Senegal's *Fouladou*, built by France's OCEA. *(OCEA)*

China's Hudong-Zhhonghua yard. The new mine-countermeasures vessel *El-Kasseh*, which has been built in Italy to a design derived from the Finnish *Katanpaa* class, has also commenced sea trials and the lead boat of two additional Project 636 'Kilo' class submarines was launched in March 2017. At this time, it is uncertain whether options for further vessels will be taken up given the impact of lower energy prices on Algeria's public finances. However, when combined with the hybrid amphibious assault and air defence ship *Kalaat Béni Abbès* that was previously delivered from Italy, these units already provide the Algerian National Navy with a balanced front-line force of six major surface combatants and six submarines.

It seems likely that **Egypt's** own recent substantial naval investment has also been partly spurred by Algeria's example. However, some analysts also point to rumoured Saudi Arabian support as suggesting a Saudi intent to develop a 'proxy' for its own limited naval capabilities. Major deliveries in the past twelve months have included the *Mistral* class amphibious assault ship *Anwar El Sadat* (the former *Sevastopol*), the second of two ships initially ordered from France's DCNS by Russia but never delivered due to Russian actions against Ukraine. DCNS has also commenced trials of the corvette *El Fateh*, the first of four new ships being built to their 'Gowind 2500' design. A light frigate-sized ship of c. 2,600 tons displacement, she has a balanced armament that includes surface-to-air and surface-to-surface missiles, a 76mm gun, anti-submarine torpedo tubes and a helicopter hangar and flight deck. Sensors are integrated into a distinctive mast located just aft of the bridge. The other three will be built locally in Alexandria with French technical assistance. Good progress is also being made by ThyssenKrupp Marine Systems (TKMS) in the supply of four Type 209/1400 submarines. These were ordered in two batches in 2011 and 2014. The first boat – *S41* – was handed over on 12 December 2016, with the second – *S42* – being named on the same day. The investment in new hardware is being supported by improvements to support infrastructure. This includes an upgraded Special Forces facility at Safaga on the Red Sea and a major expansion of the main fleet base at Alexandria.[13]

The Algerian MEKO A-200AN frigate *El Moudamir* is seen here at Kiel in September 2016 shortly before completion. She is similar to the South African MEKO A-200SAN *Amatola* also pictured in this chapter. *(P Winninger under Creative Commons Licence 4.0)*

The lead Egyptian 'Gowind 2500' type corvette *El Fateh* pictured departing Lorient in March 2017 at the start of preliminary sea trials. Three further members of the class will be assembled under licence in Alexandria. *(DCNS)*

MIDDLE EASTERN NAVIES

To the west of Egypt, **Israel** is also in the middle of a major naval upgrade aimed both at maintaining the strategic potential of the country's underwater flotilla and improving its ability to protect its exclusive economic zone. The most significant recent development has been confirmation of a plan to replace the Israeli Navy's three older *Dolphin* class submarines – delivered by TKMS around the turn of the Millennium – with three new German-built boats. Although the earlier submarines are still less than twenty years old, they lack the AIP and, possibly, the weapon-carrying capacity of the three newer boats, the last of which has yet to be delivered. The submarines are widely reported to carry nuclear-tipped cruise missiles to provide Israel with a second-strike capability and are arguably the country's most important asset. Although the deal has become entangled in allegations of political corruption, it appears set to move ahead following approval by Germany's National Security Council.[14]

Modernisation of the surface fleet is also advancing. Four modified A-100 type corvettes – the 'Sa'ar 6' class – were ordered from TKMS in May 2015. Construction of the first is expected to commence shortly to meet a targeted 2019 delivery schedule. Reports on the detailed specification of the new ships vary. However, it seems they will be around 90m in length and displace up to 2,000 tons. A powerful weapons load-out will include between thirty-two and forty-eight vertical launch cells for Barak-8 medium range and C-Dome point defence surface-to-air missiles, both controlled by an EL/M-2248 MF-STAR multi-function radar. The ships will also be capable of embarking an anti-submarine helicopter. Surplus US Navy SH-60 Seahawks are being acquired for this purpose.

The Israeli Navy is also upgrading existing ships to achieve broad commonality with the acquired vessels.[15] The majority of the existing eight 'Sa'ar 4.5' and three 'Sa'ar 5' class ships will be equipped with the new rotating EL/M-2258 ALPHA lightweight phased arrays. Some of the 'Sa'ar 5s' will also be equipped with C-Dome. This is, however, too large for the smaller 'Sa'ar 4.5s'. Other improvements will include combat system software enhancements, improved electronic warfare equipment and replacement of the 76mm Super Rapid guns mounted on some of the 'Sa'ar 4.5s' with more modern variants.

Meanwhile, **Saudi Arabia's** plans to modernise its Persian Gulf-based eastern fleet with American support under a Saudi Naval Expansion Program II

An unnamed 'Sa'ar 4.5' type fast attack craft of the Israeli Navy pictured in 2012. The eight vessels of the class are being modernised as part of a plan to ensure commonality across Israel's surface fleet. *(Israel Defence Forces)*

move slowly onwards. Licences for US$110bn of American defence sales to the country were announced during President Trump's visit in May 2017. Although it is difficult to untangle the rhetoric from the reality, a subsequent announcement by Lockheed Martin made reference to a letter of intent to procure the multi-mission surface combatant variant of its *Freedom* (LCS-1) design as part of a wide package of deals. The on-off purchase is said to amount to c. US$6bn for four ships. This substantial sum may impact plans to acquire other replacement vessels, a contract for which Turkey's 'Milgem' design and a Navantia patrol vessel variant are reportedly in contention. Saudi finances are being severely impacted by the continuation of low oil prices, whilst the cost of the ongoing war in Yemen is not helping the situation. As such, naval programmes may remain a low priority for a while yet. Elsewhere in the Gulf, the **United Arab Emirates** completed an important phase of its own modernisation plans with the commissioning of the sixth and final *Baynunah* class corvette, *Al Hili*, during the IDEX defence exhibition in February 2017. Conversely, there has been little further news about the massive naval contract with Italy's Fincantieri signed by **Qatar**, a country currently suffering from a falling-out with its larger neighbours over alleged terrorist links. However, recent rumours suggest a formal contract will be agreed by the end of 2017.

Iran's alleged support for Yemen's rebels – support that could extend to the supply of missiles that have been used to target warships belonging to the Saudi-led coalition and the US Navy – has led to broader regional tensions at a time when the American Trump presidency's hostility to Iran has also been a destabilising force. Although there have been continued confrontations with Western warships operating in the Persian Gulf, Iran's re-elected president, Hassan Rouhani, appears to want to avoid an out-and-out clash. Both the navy and the independent Iranian Revolutionary Guard Corps Naval Forces continue to focus on developing asymmetric capabilities. These have been supplemented by longer-range deployments by Iran's handful of larger surface combatants.[16]

Notes

1. The commissioning of the two submarines was widely reported, for example, in an unattributed article entitled 'Bangladesh's first 2 submarines commissioned' in *The Daily Star* – 12 March 2017 (Dhaka: Transcom Group, 2017).

2. China's strategy has often been likened to a 'string of pearls' comprising commercial and military investments and alliances extending from mainland China across the Indian Ocean that will be used to gain regional geopolitical dominance. By contrast, China emphasises the commercial nature of its intentions through terminology such as the creation of a new maritime Silk Road trading route. Amongst those seeing the submarine deal with Bangladesh as part of moves to build Chinese regional dominance is Jeff M. Smith in 'Why China's Submarine Deal with Bangladesh Matters' in *The Diplomat* – 20 January 2017 (Tokyo: Trans-Asia Inc., 2017). An alternative perspective – stressing the benefits to India of Bangladesh developing its naval capabilities is provided by Nilanthi Samaranayake in *Bangladesh's Submarines from China: Implications for Bay of Bengal Security* (Singapore: RSiS, 2017)

3. See Richard Sisk's 'China's First Overseas Military Base Nearing Completion' carried on the *Defense Tech* website on 13 March 2017 at: https://www.defensetech.org/2017/03/13/chinas-first-overseas-military-base-nearing-completion/

4. See the press release 'Maritime piracy report sees first Somali hijackings after five-year lull' carried on the International Chamber of Commerce's Commercial Crime Services website on 4 May 2017, available via: https://www.icc-ccs.org/news

5. The inclusion of military pensions of INR85,740 crore (US$13bn) in the total takes total defence spending to around US$54bn or well over two per cent of GDP. A good overall analysis of the latest budget is, again, provided by Laxman K Behera in *India's Defence Budget 2017-18: An Analysis* (New Delhi: Institute for Defence Studies and Analyses, 2017). An online version can currently be found at: http://www.idsa.in/system/files/issuebrief/ib_india-defence-budget-2017-18_lkbehera_030217.pdf

6. The audit is contained within *Union Government (Defence Services) Navy and Coast Guard: Report No. 17 of 2016* (New Delhi: Comptroller and Auditor General of India, 2016).

7. There is some conflict between different sources as to how many of the suspended Russian ships will be included within the agreement. Whilst initial reports suggested all three laid-up *Admiral Grigorovich* class frigates would be transferred to India, there seems to have been opposition to this from the Russian Navy. It now seems that one or two ships will be built by Yantar in Kaliningrad but it is not clear whether these are the incomplete Russian frigates or entirely new ships.

8. Two *Veer* class ships – a variant of the Russian 'Tarantul' class – were decommissioned in April 2016. The *Abhay* class anti-submarine corvette *Agray* became the first of these four 'Pauk' type ships to be taken out of service in January 2017.

9. See Rahul Bedi's 'India invites rebids for US$3.11 billion LPD programme' in *Jane's Defence Weekly* – 10 May 2017 (Coulsdon: IHS Jane's, 2017), p.14.

10. See Vivek Raghuvanshi's 'Indian Navy Not To Mount Homegrown AIP on Scorpene Subs', *Defense News* – 26 October 2016 at: http://www.defensenews.com/ articles/indian-navy-not-to-mount-homegrown-aip-on-scorpene-subs

11. A good overview of the likely course of the Pakistan Navy's future direction is provided in an article by Bilal Khan entitled 'The Pakistan Navy in 2016 (and beyond)' posted on the *Quwa Defence News & Analysis Group* website on 1 January 2017 and currently available at: http://quwa.org/2017/01/01/the-pakistan-navy-in-2016-and-beyond/

12. The naval portal of the *defenceWeb* news site, located at: http://www.defenceweb.co.za/ remains an excellent source of information on ongoing developments with respect to the South African and other African navies.

13. For further information, see Jeremy Binnie's 'Egypt inaugurates upgraded naval base', *Jane's Defence Weekly* – 18 January 2017 (Coulsdon: IHS Jane's, 2017), p.21.

14. The approval was reported by Anna Ahronheim in an article entitled 'Germany approves sale of 3 more submarines to Israel' carried in *The Jerusalem Post* – 30 June 2017 (Jerusalem: The Jerusalem Post Group, 2017).

15. A detailed review of the upgrades is provided by Yaakov Lappin in 'Saar 4.5 upgrade: Israel looks to technologically aligned fleet', *Jane's Defence Weekly* – 7 June 2017 (Coulsdon: IHS Jane's, 2017), pp.22–3.

16. An interesting analysis of the rationale behind longer-range Iranian Navy deployments was provided by James Fargher in '"This presence will continue forever": An assessment of Iranian capabilities in the Red Sea' published on the *Center for International Maritime Security's* website on 5 April 2017 and currently available at: http://cimsec.org/presence-continue-forever-assessment-iranian-naval-capabilities-red-sea/31593

2.4 REGIONAL REVIEW

Author:
Conrad Waters

EUROPE AND RUSSIA

INTRODUCTION

The dominant influence on European naval operations and procurement remains the unforeseen re-emergence of Russian military adventurism, initially in Crimea and the wider Ukraine, more recently in Syria. Russia's renewed appetite for overseas intervention largely caught NATO's European members off guard. However, the post-Cold War understanding that the direct threat to Europe's borders had ended has now been adjusted. In line with changed operational priorities, a significant re-orientation towards rebuilding warfighting capabilities is underway.

The changed reality is reflected in the new German white paper on defence published on 13 July 2016, the first for ten years.[1] The white paper concludes that Russia's willingness to use force to advance its own interests has put the 'peace order' Europe has enjoyed since the Cold War's end into question. Although it recognises that the post-Cold War emphasis on international crisis management remains important, the white paper mandates the provision of national and collective defence as the armed forces' priority task. The defence budget is increasing to reflect the additional burden, which is also resulting in greater naval investment. The fleet of K-130 corvettes, well-suited to defensive operations in the Baltic, is being doubled from five to ten ships. The much-reduced submarine fleet also looks set be expanded. The next programme for major surface combatants – the MKS-180 or F-126 project – will be increased to six units rather than the four plus two options initially envisaged.[2]

The boost to German defence spending is being replicated – to a greater or lesser extent – across Europe, particularly by those countries feeling most exposed to Russia's resurgence. Consequently, there is some irony in the fact that many believe that Russia's own defence budget has now reached its peak given the financial strain caused by lower energy prices and the impact of Western sanctions. Planned expenditure for 2017 amounts to c. RUB2.84 trillion (c. US$48bn). This is significantly down from the headline figure of RUB3.80 trillion (c. US$65bn) for 2016. The seemingly large reduction is distorted by one-off adjustments and funding remains considerably higher than in the earlier years of the decade.[3] However, the period of large increases in defence spending is drawing to a close. Another relevant factor is the emaciated domestic industrial base's continuing inability to support the desired defence build-up. This is evidenced by the long delays in bringing new technology such as the Project 677 'Lada' type submarines and Project 2235.0 *Admiral Gorshkov* class frigates into service.

A clearer understanding of the Russian Navy's future direction will be possible when the updated Russian state armaments programme for the period 2018–25 is published. This is due before the end of 2017. However, there have already been reports suggesting that conventional naval strength will be assigned a comparatively modest priority compared with strategic missile forces – always a focus – and the army. From a practical perspective, this means that ambitious programmes for new 'blue water' ships such as the Project 23000 'Shtorm' type aircraft carrier and Project 23560 'Lider' destroyer are unlikely to enter production within the next decade. The best the ocean-going navy can expect is further upgrades to increasingly elderly Soviet-era warships and continued development of the *Admiral Gorshkov* frigate design. Most investment will be channelled into submarines and littoral surface combatants. These type of forces have already showed themselves well-capable of supporting interventionist operations during Syria's civil war.[4]

The other event likely to have a marked impact on future European naval developments is the UK's narrow vote on 23 June 2016 to leave the European Union (Brexit). Whilst defence and security were not significant factors in the Brexit debate, Britain's increased political distance from Europe is likely to have major security implications. From a British perspective, Brexit will challenge ambitious force modernisation plans. For example, sterling's depreciation makes the cost of defence imports much more expensive than hitherto. More broadly, Brexit could provide a catalyst for European Union defence integration, long held back by British opposition. This possibility is made all the stronger by the American Trump administration's hesitant support for the NATO alliance. Certainly, the German white paper is clear that: 'The long-term goal of German security policy is to create a European Security and Defence Union.' The new French Premier Emmanuel Macron is also on record as a supporter of strengthened European defence co-operation.[5]

The Russian frigate *Admiral Essen* pictured against the backdrop of the Admiralty at Saint Petersburg in August 2016. The most recent Russian major surface combatant in commission, she has subsequently been used for 'Kalibr' cruise missile strikes on anti-Syrian government forces in May and June 2017. It seems unlikely the Russian Navy will commission any vessels larger than a frigate during the life of the next state armaments programme. *(Conrad Waters)*

TABLE 2.4.1: FLEET STRENGTHS IN WESTERN EUROPE – LARGER NAVIES (MID 2017)

COUNTRY	FRANCE	GERMANY	GREECE	ITALY	NETHERLANDS	SPAIN	TURKEY	UK
Aircraft Carrier (CVN/CV)	1	–	–	1	–	–	–	–
Support/Helicopter Carrier (CVS/CVH)	–	–	–	1	–	–	–	–
Strategic Missile Submarine (SSBN)	4	–	–	–	–	–	–	4
Attack Submarine (SSN)	6	–	–	–	–	–	–	7
Patrol Submarine (SSK)	–	6	11[2]	8	4	3	12	–
Fleet Escort (DDG/FFG)	18	9	13	18[2]	6	11	16	19
Patrol Escort/Corvette (FFG/FSG/FS)	15	5	–	2	–	–	8	–
Missile Armed Attack Craft (PGG/PTG)	–	–	17	–	–	–	19	–
Mine Countermeasures Vessel (MCMV)	14	10[1]	4	10	6	6	11	15
Major Amphibious (LHD/LPD/LPH/LSD)	3	–	–	3	2[3]	3	–	6

Notes:

1 Two further units used as support vessels.

2 Headline figures overstate the actual position, as old ships will be withdrawn once new units are fully operational.

3 Also one joint support ship with amphibious capabilities.

MAJOR REGIONAL POWERS – FRANCE

The future direction of travel for France's *Marine Nationale* will be heavily influenced by a strategic review announced by the new Macron administration in June 2017. The review will be concluded before the end of 2017 and used to set priorities for the forthcoming 2019–25 military programme. Whilst the review is unlikely to go into the detail of future equipment priorities, its conclusions on required capabilities will inevitably have a significant impact on planned naval programmes. These include the BATSIMAR offshore patrol vessel programme, FLOTLOG replenishment vessels and – in the longer term – the construction of new strategic submarines and a replacement for the carrier *Charles de Gaulle.*

In the meantime, Table 2.4.2 summaries the major units of the current French fleet. Major developments with principal ship types are highlighted below:

France's upgrade of its force of strategic missile submarines with the new M51 ballistic missile is reaching its conclusion. *Le Téméraire*, the final boat to be modernised, is pictured being towed to Brest naval dockyard in December 2016 for the work to begin. Preliminary design work is already underway on a new class of strategic submarines that will be needed to replace the existing *Le Triomphant* class from the late 2030s. *(DCNS)*

Aircraft Carriers and Amphibious Ships: The aircraft carrier *Charles de Gaulle* was docked for a scheduled major refit in February 2017, leaving France without an operational strike carrier. The eighteen-month refuelling and modernisation programme will include a number of improvements to sensors and systems. These include installation of a Thales SMART-S surveillance radar and the Artemis electro-optical surveillance system. The *Mistral* amphibious assault ship class provides a substantial helicopter-operating capability whilst *Charles de Gaulle* is absent.

Meanwhile the withdrawal of the BATRAL type landing ships from the amphibious flotilla is almost complete. *La Grandière* was decommissioned as previously planned in July 2016. She will soon be followed by the final member of the class, *Dumont d'Urville*, which is currently returning to France from Martinique. *Dumont d'Urville* will be replaced in the course of 2018 by the last of four B2M *bâtiment multi-mission* type vessels of the *D'Entrecasteaux* class. This fourth unit was formally ordered in January 2017 and will assume the previous ship's name. The other three B2Ms contracted at the end of 2013 are now all in service.

TABLE 2.4.2: FRENCH NAVY: PRINCIPAL UNITS AS AT MID 2017

TYPE	CLASS	NUMBER	TONNAGE	DIMENSIONS	PROPULSION	CREW	DATE
Aircraft Carriers							
Aircraft Carrier – CVN	**CHARLES DE GAULLE**	1	42,000 tons	262m x 33/64m x 9m	Nuclear, 27 knots	1,950	2001
Principal Surface Escorts							
Frigate – FFG	**AQUITAINE** (FREMM)	4	6,000 tons	142m x 20m x 5m	CODLOG, 27 knots	110	2012
Frigate – FFG	**FORBIN** ('Horizon')	2	7,100 tons	153m x 20m x 5m	CODOG, 29+ knots	195	2008
Frigate – FFG	**CASSARD** (FAA-70)	2	5,000 tons	139m x 15m x 5m	CODAD, 30 knots	250	1988
Frigate – FFG	**GEORGES LEYGUES** (FASM-70)	5	4,800 tons	139m x 15m x 5m	CODOG, 30 knots	240	1979
Frigate – FFG	**LA FAYETTE**	5	3,600 tons	125m x 15m x 5m	CODAD, 25 knots	150	1996
Frigate – FSG	**FLORÉAL**	6	3,000 tons	94m x 14m x 4m	CODAD, 20 knots	90	1992
Frigate – FS[2]	**D'ESTIENNE D'ORVES** (A-69)	9[1]	1,300 tons	80m x 10m x 3m	Diesel, 24 knots	90	1976
Submarines							
Submarine – SSBN	**LE TRIOMPHANT**	4	14,400 tons	138m x 13m x 11m	Nuclear, 25 knots	110	1997
Submarine – SSN	**RUBIS**	6[2]	2,700 tons	74m x 8m x 6m	Nuclear, 25+ knots	70	1983
Major Amphibious Units							
Amph Assault Ship – LHD	**MISTRAL**	3	21,500 tons	199m x 32m x 6m	Diesel-electric, 19 knots	160	2006

Note:

1 Now officially reclassified as offshore patrol vessels.

2. One scheduled to be withdrawn from service before the end of 2017.

Submarines: Submarine construction continues to be dominated by the 'Barracuda' type nuclear-powered attack submarines of the *Suffren* class. The first boat's assembly has been delayed, although launch is still expected during the course of 2017 for delivery during 2019. 2017 should also see an order placed for *Casablanca* (formerly *Duquesne*), the fifth member of the planned six-boat class.

The 'Barracudas' are likely to be followed into production by a new class of third-generation strategic submarine. These will need to start entering service during the late 2030s given the likely forty-year maximum lifespan of the existing *Le Triomphant* class, which were delivered from 1997 onwards. Preliminary design studies for the new boats are already underway. Meanwhile, mid-life modernisation of the current fleet is coming to a close with the transfer of *Le Téméraire* into dockyard hands in December 2016 for the modifications needed to handle the new M51 ballistic missile. Two of the four-boat class have already completed the modification process – the fourth submarine, *Le Terrible*, was equipped with the missile from build.

Surface Combatants: Current construction continues to be dominated by the ongoing FREMM programme. This reached its midway point in April 2017 with delivery of *Auvergne*, the fourth ship in the truncated eight-ship French programme. The fifth ship, *Bretagne*, was launched in September 2016 and all three remaining ships will be in the course of construction before the end of 2017. The programme will end with the delivery of *Lorraine*, one of two specialised air-defence variants in 2022.

DCNS – renamed Naval Group at the end of June 2017 – announced the award of a development and construction contract for five follow-on *frégates de taille intermédiaire* (FTI) on 21 April 2017. These 'frigates of intermediate size' fit between the larger FREMM and the smaller 'Gowind' series. Principal characteristics include a length of 122m and beam of 17.7m. Full load displacement is in the order of 4,000 tons. This is sufficient to accommodate sixteen Sylver A50 vertical launch cells for Aster 15 and 30 surface-to-air missiles, eight Exocet surface-to-surface missiles, two twin torpedo tubes, a 76mm gun and space for an embarked helicopter. Other key equipment includes a four-faced Thales Sea Fire multifunction radar, and both bow-mounted and towed sonar systems. Sensors and weapons are controlled by a SETIS combat management system. Diesel propulsion provides a maximum speed of 27 knots whilst endurance is in excess of 5,000 nautical miles. The basic design can be flexed to meet the needs of export customers; the export variant being marketed under the name 'Belh@rra' (*sic*). The French ships will replace the existing *La Fayette* class on a one-for-one basis from 2023 onwards. They provide a significantly enhanced anti-submarine capability compared with the existing ships. However, the projected c. €4bn (US$4.6bn) cost of the five ship programme has not been without its critics.[6]

Minor Warships and Auxiliaries: To date, investment in the high-priority surface combatant and submarine programmes already discussed has delayed renewal of France's ageing patrol fleet. This is currently a disparate collection of vessels of

The lead French FREMM frigate *Aquitaine* pictured leaving Brest harbour at speed on 23 June 2017. Four of the eight ships destined for French naval service have been delivered to date. *(Bruno Huriet)*

The French FREMM frigates will be supplemented by a new design of lighter and cheaper frigate optimised for export success. The new ship will be marketed under the 'Belh@rra' name in export markets. *(DCNS)*

L'Astrolabe – pictured here fitting-out in Brittany in April 2017 – is a new French ice-patrol vessel designed for Antarctic service. She will be operated by a navy crew but will carry out a wide range of both military and civilian tasks, many of the latter for the French Polar Institute. *(Bruno Huriet)*

An Italian submarine of the Type 212A *Todaro* class. The Italian Navy has taken delivery of an additional pair of these German-designed boats in the past year and hopes to acquire more under a memorandum of understanding signed with the German Ministry of Defence in March 2017. *(Fincantieri)*

varying origins and capabilities. However, the navy hopes to launch its BATSIMAR (*bâtiments de surveillance et d'intervention maritime*) project before the end of the decade. As many as eighteen offshore patrol vessels are planned, largely for service in the overseas territories. A potential prototype already exists in the form of the DCNS 'Gowind' type patrol vessel *L'Adrioit*. Built with company funds, this has been on loan to the *Marine Nationale* since completion in 2012. However, other competitors are likely to emerge. In the meantime, orders for a number of specialised ships such as the B2Ms and two light patrol vessels specifically designed for operations in the shallow waters of French Guiana have helped to balance withdrawals of older ships.

A somewhat unusual ship scheduled for delivery over the next twelve months is the new polar patrol and logistic support ship *L'Astrolabe*. Constructed in Poland and outfitted at the Piriou yard at Concarneau in Brittany to a Finnish design, she is owned by the French Southern & Antarctic Lands Administration and operated by the navy in conjunction with the French Polar Institute. The 72m, ice-capable vessel can accommodate up to sixty personnel for thirty-five days and is equipped with a helicopter platform and crane to assist logistic operations. Up to 1,200 tonnes of cargo can be shipped in support of the French Antarctic bases. Piriou is also working on four BSAH (*bâtiment de surveillance et d'assistance hauturier*) type ocean-going tugs of the *Loire* class ordered in two batches in August 2015 and October 2016.

The next major programme for auxiliaries will be FLOTLOG project for three new replenishment tankers to replace the remaining members of the *Durance* class. The previous plan to use the DCNS' BRAVE design appears to have been overtaken by Fincantieri's acquisition of the former STX-owned yard at Saint-Nazaire. It now appears that the new ships will be built to the Italian Navy's new *Vulcano* design at the newly-acquired facility.[7]

MAJOR REGIONAL POWERS – ITALY

The Italian Navy's current composition is set out in Table 2.4.3. The most significant development in the last year has been the arrival of the second pair of German-designed but Italian-built Type 212A air-independent propulsion (AIP)-equipped submarines. *Pietro Venuti* was delivered on 6 July 2016, with *Romeo Romei* following on 11 May 2017. Although this takes the current underwater flotilla

Table 2.4.3: ITALIAN NAVY: PRINCIPAL UNITS AS AT MID 2017

TYPE	CLASS	NUMBER	TONNAGE	DIMENSIONS	PROPULSION	CREW	DATE
Aircraft Carriers							
Aircraft Carrier – CV	**CAVOUR**	1	27,100 tons	244m x 30/39m x 9m	COGAG, 29 knots	800	2008
Aircraft Carrier – CVS	**GIUSEPPE GARIBALDI**[1]	1	13,900 tons	180m x 23/31m x 7m	COGAG, 30 knots	825	1985
Principal Surface Escorts							
Frigate – FFG	**CARLO BERGAMINI** (FREMM)[2]	6	6,700 tons	144m x 20m x 5m	CODLOG, 27 knots	145	2013
Frigate – FFG	**ANDREA DORIA** ('Horizon')	2	7,100 tons	153m x 20m x 5m	CODOG, 29+ knots	190	2007
Destroyer – DDG	**DE LA PENNE**	2	5,400 tons	148m x 16m x 5m	CODOG, 31 knots	375	1993
Frigate – FFG	**MAESTRALE**	6	3,100 tons	123m x 13m x 4m	CODOG, 30+ knots	225	1982
Frigate – FFG	**ARTIGLIERE**	2	2,500 tons	114m x 12m x 4m	CODOG, 35 knots	185	1994
Frigate – FS	**MINERVA**	2	1,300 tons	87m x 11m x 3m	Diesel, 25 knots	120	1987
Submarines							
Submarine – SSK	**TODARO** (Type 212A)	4	1,800 tons	56m x 7m x 6m	AIP, 20+ knots	30	2006
Submarine – SSK	**PELOSI**	4	1,700 tons	64m x 7m x 6m	Diesel-electric, 20 knots	50	1988
Major Amphibious Units							
Landing Platform Dock – LPD	**SAN GIORGIO**	3	8,000 tons	133m x 21m x 5m	Diesel, 20 knots	165	1987

Note:
1 Now operates largely as a LPH.
2 Class includes BERGAMINI (GP) and FASAN (ASW) variants.

to as many as eight submarines, the four existing *Pelosi* class boats are scheduled to retire between 2018 and 2022. The navy would like to acquire additional German-designed submarines to help fill the gap. To this end, a memorandum of understanding on future underwater collaboration was signed between the German and Italian ministries of defence in March 2017. With Germany and Norway already working together to acquire a total of six new submarines derived from the Type 212 design, it would seem Italian participation in this programme is likely.

Developments in the surface fleet continue to be dominated by the FREMM frigate programme. Deliveries continue at the rate of one ship p.a. with the sixth member of the class, *Luigi Rizzo*, entering service on 20 April 2017. The seventh unit, *Federico Martinengo*, was launched on 4 March. Although all ten planned Italian members of the class have now been ordered, the design is also one of three short-listed for Australia's SEA 5000 future frigate programme. The fourth member of the class, *Carabiniere*, departed on a voyage to the Indian Ocean and Australia to support Italy's marketing efforts at the end of 2016. She undertook several port visits and a number of exercises with friendly navies in the course of a four-month deployment.

The arrival of new ships is being balanced by the

The Italian FREMM frigate variant is one of three contenders for Australia's next generation of frigates. *Carabiniere* – the fourth member of the class – departed from Italy in December 2016 on a tour intended to support local builder Fincantieri's efforts to win the contract. She is pictured here against the backdrop of Sydney Harbour in February 2017. *(Australian Department of Defence)*

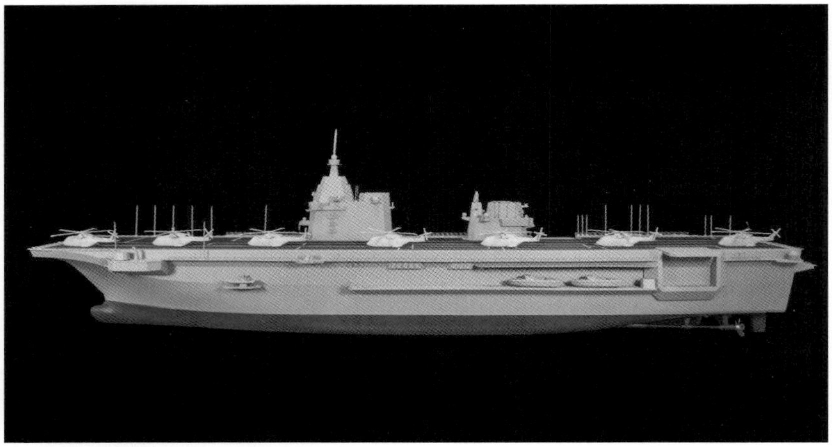

More details of Italy's future multi-role 'patrol vessels' and amphibious assault ship, *Trieste*, both authorised under the 2014 naval funding law, are emerging. The 'PPA' patrol vessels will be acquired in light, light plus and full variants dependent on the extent of weapons and sensors installed – the image shows the more heavily-armed 'full' version. The amphibious assault ship's twin-island design seems to owe much to the new British *Queen Elizabeth* class aircraft carrier. *(Rolls-Royce Group)*

withdrawal of older vessels. The *Maestrale* class frigate *Aliseo* reduced to reserve in September 2016 prior to final withdrawal planned for 2017. Other units leaving the fleet included the *Minerva* class corvettes *Sfinge* and *Fenice*, which were retired at a ceremony at the Augusta naval base on 29 May 2017. Their decommissioning leaves just two ships of an original eight-ship class in service with the Italian fleet. Both of these are likely to be retired in 2018, along with the pair of remaining 'Soldati' class frigates.

The longer-term modernisation of the fleet is secured by the €5.4bn (c. US$6.2bn) fleet renewal programme approved under the 'Naval Law' of 2014. This provided funding for a new amphibious assault ship, a logistic support ship, a new class of frigate-like patrol vessels and two high-speed Special Forces insertion craft.[8] The major component of this programme is the PPA (*Pattugliatore Polivalente d'Altura*) patrol vessels. Seven of these vessels have been ordered from an industrial consortium headed by Fincantieri and there are options for three more. They will be delivered in three configurations – light (two vessels), light plus (three vessels) and full (two vessels) – of increasing warfighting potential and sophistication on a common hull. The first PPA, being constructed in light configuration, entered production at Fincantieri's Muggiano yard near La Spezia in February 2017. This was followed by a formal keel-laying ceremony on 9 May. She will be launched in the middle of 2019 for delivery in 2021. The remaining vessels will be delivered at roughly annual intervals through to 2026, with the first full 'combat'-specified variant an additional delivery during 2024.

Good progress is also being achieved with the rest of the renewal programme. The 94m bow section of the new logistic support ship *Vulcano* was launched from Fincantieri's Castellammare di Stabia yard in Naples on 10 April 2017, subsequently being transported to Muggiano for integration with the Riva Trigoso-built 86m aft section. Sea trials should commence towards the end of 2018 prior to delivery in 2019. Fabrication of the 'flagship' of the new construction programme, the 245m LHD-type amphibious assault ship *Trieste*, is also likely to commence imminently under an ambitious build schedule that envisages launch from Castellammare di Stabia in 2019 and the commencement of sea trials from Muggiano before the end of 2020. Work has already started at Sarzana on the final element of the programme, the two 40m high-speed interceptors entrusted to the Intermarine group.

Meanwhile Italy has returned to the front line of the Mediterranean refugee crisis after the broadly successful implementation of measures to close off the Aegean migrant routes. Whilst statistics vary dependent on source, Italy has seen the vast majority of the c. 80,000 migrants entering Europe during the first half of 2017 land on its shores. Lack of help from other EU countries has seen Italy threaten to close its borders to new entrants amongst claims that non-governmental humanitarian agencies have been colluding with people-smugglers and making the crisis worse. A definitive solution to a crisis that has claimed thousands of lives seems as far away as ever at the current time.

MAJOR REGIONAL POWERS – SPAIN

There has been no change to the principal units of the *Armada Española* listed in Table 2.4.4 for a fourth consecutive year. This reflects two main factors: the effective halt to all but the most essential procurement activity during the depths of the Eurozone's financial crisis and the significant design problems that have impacted the flagship S-80 submarine project.

It will be some years before the fleet benefits from the arrival of any new major units as both these factors are only being slowly overcome. The S-80 programme – previously effectively suspended due to a weight problem that had a critical impact on buoyancy – passed a critical design review in mid-2016. The revised design – known as the S-81 Plus – has been extended by insertion of a c. 10m hull plug and the first boat will now go forward to completion. This is scheduled for 2021, some nine years behind the original plan. If all goes well, the other boats will follow during the 2020s. However, it seems that the problems impacting the indigenously-developed AIP system to be installed in the class have yet to be fully overcome and the system will not – at least initially – be installed in the first submarine. To date, over 2.1bn (US$2.4bn) has been allocated to what has been a disastrous programme for the Spanish Navy. Moreover, recent reports suggest this will be sufficient to complete only the first boat.

Meanwhile, future procurement is likely to be hindered by a massive legacy of unpaid bills relating to so-called 'special defence projects' that amounted to as much as €21bn across the three branches of the armed forces.[9] Information provided to the defence committee of the Spanish Chamber of Deputies in March 2017 is particularly interesting in so far as it provides both the capital cost and amounts outstanding of the bulk of recent naval procurement projects, viz.

- Amphibious Assault Ship; *Juan Carlos I:* €462m (US$525m); €340m still to be paid.
- F-100 Frigate Programme; First Four Units: €1,998m (US$2,280bn); €1,510m still to be paid.
- F-100 Frigate Programme; F105 *Cristóbal Colón:* €827m (US$945m); €751m still to be paid.
- BAM Programme; First Four Units: €626m (US$710m); €626m still to be paid.
- BAM Programme; *Audaz* and *Furor*: €334m (US$380m); programme ongoing.
- Replenishment Vessel; *Cantabria*: €255m (US$290m); €234m still to be paid.

At present, construction for the Spanish Navy is focused on the two additional BAM-type *Meteoro* class offshore patrol vessels ordered in December 2014. The programme is running a little ahead of schedule, with the first ship – *Audaz* – launched from the San Fernando yard at Cadiz on 30 March 2017. The final vessel *Furor* – allocated to the yard at Ferrol in Galicia – will follow in the second half of the year. Both ships should be delivered during 2018. The plan is that construction will shift to the new F-110 multi-mission frigates that will replace the existing FFG-7 class during the 2020s. Design work on the new ships commenced in 2015.

Meanwhile, in similar fashion to Italy's *Carabiniere*, the F-100 class frigate *Cristóbal Colón* deployed to the Far East early in 2017 for a four-month attachment with the Royal Australian Navy. The deployment was ostensibly intended to help Australia prepare for the imminent arrival of *Hobart*, a near-sister of the Spanish ship, scheduled in the second half of 2017. However, Spain's Navantia is another company shortlisted for the SEA 5000 future frigate programme and the voyage can also be seem as another marketing effort to win the lucrative contract.

Spain is another country vying for Australia's new frigate programme and has the advantage that its existing F-100 design was used for the preceding Australian *Hobart* class. *Cristóbal Colón* – seen here with the frigate *Darwin* in 2017 – has spent much of the first half of this year in Australian waters helping the RAN prepare for the new ships whilst doubtless also promoting the merits of Spanish technological excellence. *(Australian Department of Defence)*

MAJOR REGIONAL POWERS – UNITED KINGDOM

A detailed review of the British Royal Navy is contained in Chapter 2.4A whilst Table 2.4.5 shows an unchanged fleet structure year-on-year. The British government is giving some prominence to 2017 as 'The Year of the Navy', reflecting a number of expected positive developments. These include planned deliveries of no fewer than three new classes of new surface vessel, with the lead *Queen Elizabeth* class aircraft carrier, the first 'Tide' class replenishment *Tidespring* and the new Batch II 'River' class patrol vessel *Forth* all expected to be in the Royal

Table 2.4.4: SPANISH NAVY: PRINCIPAL UNITS AS AT MID 2017

TYPE	CLASS	NUMBER	TONNAGE	DIMENSIONS	PROPULSION	CREW	DATE
Principal Surface Escorts							
Frigate – FFG	**ÁLVARO DE BAZÁN** (F-100)	5	6,300 tons	147m x 19m x 5m	CODOG, 28 knots	200	2002
Frigate – FFG	**SANTA MARIA** (FFG-7)	6	4,100 tons	138m x 14m x 5m	COGAG, 30 knots	225	1986
Submarines							
Submarine – SSK	**GALERNA** (S-70/AGOSTA)	3	1,800 tons	68m x 7m x 6m	Diesel-electric, 21 knots	60	1983
Major Amphibious Units							
Amph Assault Ship – LHD	**JUAN CARLOS I**	1	27,100 tons	231m x 32m x 7m	IEP, 21 knots	245	2010
Landing Platform Dock – LPD	**GALICIA**	2	13,000 tons	160m x 25m x 6m	Diesel, 20 knots	185	1998

Navy's hands before the end of the year. Other welcome news is expected to include the naming and floating out of *Queen Elizabeth*'s sister-ship, *Prince of Wales*, in the second half of 2017 and the long-awaited order for the first three Type 26 frigates before Parliament breaks for its summer holidays in mid-July.

The actual position is more nuanced.[10] The last year has seen the navy facing ongoing overstretch in the face of commitments that included involvement in as many as twenty-two operations. At times, over 8,000 of its nearly 30,000 trained personnel have been deployed. A continued shortage of surface vessels reflecting well-publicised problems with the Type 45 destroyers' integrated propulsion systems and the loss of Type 23 frigates to a programme of rolling refits to install the new Sea Ceptor missile has been exacerbated by difficulties in recruiting sufficient personnel. The underwater flotilla has also faced similar manning difficulties and has also suffered from a number of technical defects and operational mishaps. Most significantly, the nuclear-powered attack submarine *Ambush* needed significant repairs to her fin and passive intercept array after colliding with a merchant vessel whilst surfacing in the Strait of Gibraltar on 20 July 2016. It is likely to take a number of years – and no further adverse developments – before the Royal Navy recovers from a period of austerity.

Although 2017 has been promoted by the British Ministry of Defence as 'The Year of the Navy', the reality has been more nuanced. For example, the need to rectify the defective propulsion system design of the new Type 45 destroyers – the final member of the class, *Duncan*, is pictured here – has been one of a number of pressures in fleet availability. *(Derek Fox)*

MID-SIZED REGIONAL FLEETS

Germany: The last year has been generally positive for the *Deutsche Marine*. New ships are in the course of delivery and important programmes are moving ahead. Looking first at the underwater flotilla, the problems that affected the delivery of the additional pair of improved Type 212A submarines ordered in 2006 have now seemingly been overcome. The final boat in the batch, *U36*, entered service on 10 October 2016, taking the fleet's number of submarines up to six. The agreement of a strategic partnership with Norway on future submarine production should see an additional two boats based on the Type 212 design ordered in 2019 as part of a joint acquisition totalling six submarines.

Table 2.4.5: BRITISH ROYAL NAVY: PRINCIPAL UNITS AS AT MID 2017

TYPE	CLASS	NUMBER	TONNAGE	DIMENSIONS	PROPULSION	CREW	DATE
Principal Surface Escorts							
Destroyer – DDG	**DARING** (Type 45)	6[1]	7,500 tons	152m x 21m x 5m	IEP, 30 knots	190	2008
Frigate – FFG	**NORFOLK** (Type 23)	13	4,900 tons	133m x 16m x 5m	CODLAG, 30 knots	185	1990
Submarines							
Submarine – SSBN	**VANGUARD**	4	16,000 tons	150m x 13m x 12m	Nuclear, 25+ knots	135	1993
Submarine – SSN	**ASTUTE**	3	7,800 tons	93m x 11m x 10m	Nuclear, 30+ knots	100	2010
Submarine – SSN	**TRAFALGAR**	4[2]	5,200 tons	85m x 10m x 10m	Nuclear, 30+ knots	130	1983
Major Amphibious Units							
Helicopter Carrier – LPH	**OCEAN**	1	22,500 tons	203m x 35m x 7m	Diesel, 18 knots	490	1998
Landing Platform Dock – LPD	**ALBION**	2	18,500 tons	176m x 29m x 7m	IEP, 18 knots	325	2003
Landing Ship Dock – LSD (A)	**LARGS BAY**	3	16,200 tons	176m x 26m x 6m	Diesel-electric, 18 knots	60	2006

Notes:

1. One at extended readiness as a harbour training ship pending refit.
2. One scheduled to be withdrawn from service before the end of 2017.

Royal Navy submarine availability has also been under strain due to both technical problems and operational incidents. These pictures show the *Astute* class attack submarine *Ambush* limping into Gibraltar with damage to her fin and intercept array after a collision with the merchant vessel *Andreas* whilst surfacing in the Strait of Gibraltar on 20 July 2016. *(Moshi Anahory)*

The new boats – currently referred to as the Type 212NG – will enter service in the second half of the next decade, increasing the underwater flotilla to eight units.

Meanwhile, good progress is being made with construction of the new F125 class stabilisation frigates that form the current surface warship construction programme.[11] All four ships have now been christened and the lead vessel, *Baden-Württemberg*, should be commissioned before the end of 2017 after completion of an extensive set of trials. The next major surface combatant programme will be for the MKS-180 multi-purpose combat ships, large destroyer-like ships also referred to as the F126 type. It is expected that six ships will be ordered in a contract expected before the end of 2017 at a unit cost in the order of €1bn. Three consortia – TKMS/Lürssen; Blohm & Voss/Damen; and German Naval Yards/BAE Systems – have submitted proposals for the new class. However, Lürssen's subsequent acquisition of the Blohm & Voss yard in Hamburg will doubtless impact this process. In practice, it would seem likely that an industry-wide partnership similar to those that have built previous German warship classes will eventually undertake the project.

Given it will take time for the detailed design of the MKS-180 to be drawn up, a decision has been taken to maintain continuity of production in the shipyards by ordering a second batch of five K-130 corvettes to join the five existing ships commissioned between 2008 and 2013. This will also ensure that at least four of the class are available to support NATO operations at all times. The German ministry of defence attempted to circumvent normal competition processes by awarding construction to the TKMS/Lürssen consortium that built the initial batch but a challenge from German Naval Yards means that they have also been brought in on the new deal.

Departures during the year have included the

The German Navy has decided to double the number of K-130 *Braunschweig* class corvettes to ten ships. They are particularly well suited for operation in the Baltic and other littoral waters – the second member of the class, *Magdeburg*, is seen operating in support of the UNIFIL mission off Lebanon in this 2011 image. *(German Armed Forces)*

F122 type frigate *Karlsruhe*, leaving only two of the class in service. The last four Type 143A *Gepard* class fast attack craft were also decommissioned at a ceremony on 16 November 2016, bringing to an end sixty years of post-Second World War German Navy operations with the type.

Greece: The significant modernisation achieved by the Hellenic Navy during the first decade of the Millennium has largely ground to a halt under the burden of economic austerity. With the commissioning of the third and fourth Type 214 submarines – *Matrozos* and *Katsonis* – on 23 June 2016, the only remaining major new-build project still in hand is that for the completion of the sixth and seventh *Roussen* (Super Vita) class fast attack craft. A restructuring of this much-delayed contract agreed with the Elefsis yard in 2016 provides for the delivery of these two vessels before the end of 2018. The yard completed the repair and modernisation of the 'La Combattante III' type fast attack craft *Mykonios* – damaged by fire in 2010 – towards the end of 2016. Unlike her three sisters, she is equipped with Harpoon rather than Exocet surface-to-surface missiles.

The Hellenic Navy remains numerically large. Whilst the operational status of some of its older ships must now be questionable, it remains active in support both of NATO operations and to deter any possible Turkish encroachment in the Aegean Sea. A major fleet re-orientation announced in mid-2016 will see much greater emphasis placed on the naval base at Souda Bay in Crete, which will support a rotating presence of surface vessels and submarines. This will allow a stronger presence to be maintained in Eastern Mediterranean and south-eastern Aegean waters, both focal points for refugee activity in recent years.

The Netherlands and Belgium: The Dutch and Belgian navies have become increasingly integrated in recent years under the operational leadership of the Admiral Benelux function.[12] They have also steadily progressed to operating largely identical equipment. This collaboration is now being reinforced in the area of future procurement by a letter of intent signed by the countries' respective defence ministries in November 2016 that commits to the joint acquisition of a new generation of warships. The arrangement will encompass joint procurement of two new frigates and six new mine-countermeasures vessels for each fleet. The aim is to have all the new ships in service by 2030.[13]

The frigates are required to replace the existing pairs of 'M' or *Karel Doorman* class ships that were first commissioned in the early 1990s. It has been agreed that the Netherlands will lead this part of the project and the ships are likely to be built by Damen and incorporate Thales sensors. The existing frigates have an emphasis on anti-submarine missions and it appears that this remains an important area of focus for the Dutch. However, it appears that, perhaps surprisingly, the Belgian Naval Component is interested in acquiring a ballistic missile defence (BMD) capability for the new vessels. The EWC (early warning capability) of Thales SMART-L radar used in a number of existing Dutch and European surface vessels certainly has the capacity to track such targets but it is a moot point whether or not Belgium could afford a full BMD capability.

Although Belgium has been given leadership of the mine-countermeasures vessel (MCMV) element of the procurement plan, it seems likely that the ships will also be assembled in the Netherlands given the lack of suitable Belgian infrastructure. The new vessels will effectively act as motherships for a range of unmanned and autonomous systems and will therefore not incorporate the expensive signature-reduction measures built into the existing 'Tripartite' class.

The Netherlands also has a requirement to replace its existing submarine fleet in roughly the same timescale as the frigate and MCMV procurement. There appears to be some hesitancy in committing to this expensive programme in parallel with the surface ship projects and further studies of replacement options, including the possible use of unmanned vessels, is being explored. The existing *Walrus* class boats are being refitted to allow them to serve into the mid-2020s. *Zeeleeuw*, the first submarine to undergo the upgrade, completed post-refit trials in the autumn of 2016 and is now back with the fleet. Also back at sea is the joint support ship *Karel Doorman*. The almost brand-new ship was taken out of service after the emergence of defects in one of her main electric motors in March 2016 that required both units to be repaired.

Turkey: Turkey has experienced a turbulent year dominated by the failed military coup of 15 July 2016 in which at least one major warship, the frigate *Yavuz*, was involved. A subsequent widespread purge initiated by President Recep Erdogan to strengthen his administration's control of Turkish politics has seen the arrest of numerous senior officers across the military in a move that is unlikely to improve the armed forces' long-term efficiency.

The navy has seen a marginal diminution in is strength over the past twelve months, due largely to the decommissioning of the elderly lead Type 209/1200 submarine *Atilay* in November 2016. Commencement of the programme for replacement Type 214 submarines was significantly delayed. Although work on the first boat, *Piri Reis*, is now underway it is possible that the underwater flotilla will suffer further erosion before the new boats arrive. One major positive development, however, is the arrival of the submarine rescue vessel *Alemdar*, which commissioned on 28 January 2017. She provides an important capability for what is still Europe's largest submarine fleet. *Akın* and *I?ın*, two rescue and towage vessels, ordered at the same time as *Alemdar*, are expected to be commissioned in the second half of 2017.

Surface construction continues to be dominated by the indigenous 'Milgem' corvette/light frigate programme. A transition from production of the initial 'Ada' class corvette to the enlarged 'I' frigate is currently underway. *Kinalida*, the fourth and final member of the 'Ada' class to be built at Istanbul Naval Shipyard, is expected to be launched at the start of July 2017, clearing the slipway for assembly work to start on the lead 'I' class vessel, *Istanbul*. The new class is about 600 tons heavier and 14m longer than the original design and accommodates a sixteen-cell Mk 41 vertical launch system. Four ships – *Istanbul*, *Izmir*, *Izmit* and *Içel* – are planned, with the first to be commissioned in 2021.

Modernisation of existing surface combatants is also progressing, with a contract signed between the navy and an Aselsan-Havelsan joint venture in June 2017 for a mid-life upgrade of the four MEKO 200 Track IIA/B class frigates of the *Barbaros* class. The legacy combat-management systems installed in the ships will be replaced by a new *Barbaros* Combat Management System (BI-SYS). This will be a derivative of the existing GENESIS CMS made locally and used on 'Ada' class corvettes and 'Gabya' (FFG-7) class frigates.

Meanwhile, the last year has seen progress made with a number of other important indigenous programmes. Notably, the first of a new class of 7,250-ton tank landing ships, *Bayraktar*, was delivered on 14 April 2017. She is able to transport 350 troops, around twenty tanks and other supporting vehicles at distances of up to 5,000 miles and has a platform but no hangar for a large military helicopter. A sister-ship, *Sancaktar*, was launched on 16 July 2016 – the day after the abortive coup – and will be delivered in the course of the next year. Work continues at the Sedef Shipyard on the navy's 'flagship' project, the *Juan Carlos I*-type amphibious assault ship *Anadolu*.[14]

The new Turkish tank landing ship *Bayraktar* is pictured at anchor whilst in the course of sea trials in this February 2017 image, A second member of the class was launched in 2016. A target float for gunnery tests can be seen on deck forward of the twin cranes. *(Devrim Yaylali)*

OTHER REGIONAL FLEETS

Black Sea and Mediterranean: Although tensions remain high in the Black Sea due to NATO's response to events in the Ukraine, procurement of new vessels by the region's smaller navies has yet to gain traction. In November 2016, the **Romanian** defence ministry announced it had selected Damen's 'Sigma' 10514 light frigate design to meet a requirement for four new corvettes in a deal variously reported to be worth between €1.1bn and €1.6bn. However, the Romanian senate's refusal to ratify the decision and subsequent elections have put the programme on hold. The navy remains keen to pursue a project that seems entirely logical given Damen's significant 'in-country' shipbuilding capacity at its Galati facility. Meanwhile, there has been little tangible progress with **Bulgaria's** €420m plan to buy two new offshore patrol ships. It seems likely that these ships will also be constructed at a domestic yard, with Varna-based shipbuilder MTG Dolphin presenting a K-90 multi-purpose corvette concept at the October 2016 Euronaval exhibition in Paris.[15] The ship appears to have much in common with the 'Sigma' design.

Elsewhere in the south of Europe, new construction is largely focused on smaller constabulary vessels, sometimes equipped with a secondary warfighting role in mind. Typical of the type of ships being acquired is the first of a new class of five coastal patrol vessels for **Croatia's** coast guard. Work on the new 43.5m vessel began at the Brodosplit yard in 2015 and she was launched on 3 June 2017.

The Caspian Sea-based Kazakh Naval Force was established in 2003 and operates a growing force of patrol vessels of steadily increasing power. These 2017 images show *Mangistau*, the fourth of a series of 250-ton *Kazakhstan* class vessels that are equipped with an AK-630 gun mount forward of the bridge and a multiple rocket launcher position on the aft deck. The Caspian Sea is home to a range of vessels, the largest being Russian and Iranian light frigates. *(Kazakhstan Ministry of Defence via M Mazumdar)*

A main armament of a Turkish Aselsan 30mm remotely-operated and stabilised gun is supplemented by 12.7mm machine guns and portable surface-to-air missile launchers. Maximum speed is 28 knots, whilst the ship is capable of remaining at sea for up to fourteen days.[16] She should be delivered by the end of 2017. Meanwhile, on Europe's borders, land-locked **Kazakhstan** is taking delivery of a new class of broadly similar 250-ton, 46m patrol vessels of the *Kazakhstan* (Project 0250) class for service in the Caspian Sea. The fourth of these, *Mangistau*, was delivered by the local Zenit shipyard in Orál in April 2017. She is armed with a six-barrelled 30mm AK-630 gun, a 'Grad' multiple rocket launcher and lighter weapons.

Atlantic: The smaller Atlantic-facing navies also have a heavy current emphasis on constabulary asset procurement, albeit of much larger proportions due to a much harsher operating environment. **Ireland** took delivery of its third 90m *Samuel Beckett* (OPV90) class offshore patrol vessel, *William Butler Yeats*, in July 2016 prior to an official commissioning ceremony held on 17 October. It was announced that the unexpected fourth – and final – vessel of the class ordered in June 2016 will be named *George Bernard Shaw* during a ceremony to mark her keel laying held at Babcock's Appledore yard on 28 February 2017.

Also in the middle of OPV construction is **Portugal**, which ordered a second pair of *Viana do Castello* class vessels from the West Sea Viana Shipyard under a €77m order announced in July 2015. The design shares many features with the Irish ships, not least a comparable maximum speed and a similar propulsion system. However, the Portuguese vessels are somewhat smaller and more lightly armed, possibly reflecting the availability of front-line frigates should a more coercive capability be required. *Sines*, the lead ship of the new batch, was launched on 3 May 2017 and should be delivered during 2018. The new class are steadily replacing the navy's 'colonial corvettes' on Atlantic patrol duties, with only three of the old ships left in service. A similar process of renewal is seeing the remaining elderly *Cacine* class 44m patrol ships being replaced by the four decommissioned *Flyvefisken* (StanFlex 300) class vessels acquired from Denmark. These are being refurbished and converted for their new role in Portugal. *Tejo*, the first to complete this process, entered the fleet in May 2016. She has recently been joined by *Duoro*, with the others following at annual intervals.

The third Irish *Samuel Beckett* class offshore patrol vessel, *William Butler Yeats*, pictured shortly before being floated out of her building dock in March 2016. A fourth member of the class, *George Bernard Shaw*, is currently under construction. *(Irish Defence Forces)*

The Baltic and Scandinavia: The Baltic's status as part of the new 'front line' between Russia and the rest of Europe has resulted in a new focus on defence that shows no signs of abating. Efforts to strengthen military posture have been particularly evident in **Sweden**, which has announced it will reintroduce military conscription from 1 January 2018. The measure is aimed at resolving a recruitment shortage and – unlike the previous measure abolished in 2010 – will extend to both men and women. Other significant developments have included the reintroduction of the RBS-15 land-based surface-to-surface missile system, some seven years after the coastal artillery network was abandoned. The missiles have been taken from stocks released from decommissioned surface vessels, whilst some of the launch vehicles have reportedly been recovered from museums.[17] Significant efforts are also being made to extend the lives of existing surface vessels, including a SEK1.2bn (c. US$150m) contract announced on 30 June 2017 for the modernisation of the two *Gävle* class corvettes. They will join the five *Visby* class vessels to form a seven-strong force of front-line surface combatants, supported by two older *Stockholm* class ships acting in a second-line surveillance role. Some of the *Koster* class MCMVs are also receiving further modernisation, whilst a comprehensive recapitalisation of the submarine fleet was announced in 2015.

Amongst the other Baltic navies, **Poland** has plans for a major programme of naval investment encompassing submarines, corvette-like patrol vessels and MCMVs. However, progress finalising many of these projects has been slow. Accordingly, it seems likely to be well into the next decade before new warships arrive in significant numbers. Most advanced is the programme for three 'Kormoran II' minehunters, the first of which commenced sea trials in 2016. The MEKO-100-based patrol vessel *?l?zak* should also start trials soon. There have been reports that Poland is examining the acquisition of two of the Royal Australian Navy's modernised FFG-7 *Adelaide* class frigates to replace its own unmodernised ships of the type to maintain fleet numbers until the new programmes come good.

Elsewhere in Scandinavia, **Finland** is currently evaluating contractors for its 'Squadron 2020' programme. This will see four corvette-like patrol vessels replace the existing *Rauma* class fast attack craft (FAC) and *Hämeenma* class minelayers and supplement upgraded *Hamina* class FACS at the core of the fleet. Meanwhile, the third and final *Katanpää* class MCMV, *Vahterpää*, was delivered by Intermarine at the end of 2016 and introduction of 'Jehu' class assault vessels is progressing well.[18] To the west, **Denmark** should shortly see its third *Knud Rasmussen* class ice-strengthened oceanic patrol vessel, *Lauge Koch*, enter operational service. **Norway** also has plans for new, large ice-strengthened Arctic patrol vessels for its coast guard. Orders for three ships – to be built in a domestic yard – are expected by 2018. The following year should see orders for the new Type 212-NG submarines to be jointly procured with Germany under the partnership agreement reached in February 2017. Meanwhile – as for the British 'Tide' class – the new logistics support vessel *Maud* has been delayed whilst under construction at DSME in Korea and is not now expected until the end of the year.

RUSSIA

The Russian Navy has spent another active year supporting President Putin's overseas adventures in Syria and elsewhere. The October 2016 deployment of a strike group headed by Russia's sole aircraft carrier, *Admiral Kuznetsov*, for operations off Syria gained much attention. However, the loss of two of the limited pool of carrier aircraft to technical incidents and reports that much of the air group were subsequently dispatched to land-based airfields for the duration of the deployment suggest something of the fragility of Russian naval aviation. More extensive use has been made of 3M-54 series 'Kalibr' (NATO SS-N-27 Sizzler) cruise missiles. These have been launched to attack Islamic State targets from a number of the navy's most modern platforms, including Project 636.3 'Kilo' class submarines, Project 1135.6 *Admiral Grigorovich* class frigates and even Project 2163.1 'Buyan M' light corvettes. The success of these strikes may be a factor influencing the delays to larger surface combatant programmes expected in the new state armaments plan.[19]

Although there is much political rhetoric surrounding the buoyant state of Russian naval construction, many of the problems referenced in previous editions of *Seaforth World Naval Review* – including weaknesses in domestic industrial infrastructure and technology; lack of access to imported equipment following Western and Ukrainian sanctions on the supply of defence equipment; and dispersal of effort amongst an almost bewildering array of designs – remain. Some progress is being achieved. For example, the NPO Saturn aircraft engine manufacturer is reportedly ready to put a new range of marine gas turbines into production to overcome previous reliance on Ukrainian-supplied models. Equally, the priority given to submarine construction is starting to achieve a marked upgrade in the capabilities of the underwater flotilla. However, this progress remains patchy, with – for example – only modest progress achieved with surface fleet renewal. The status of major projects is summarised below.

Submarines: The strategic submarine force has received the highest priority amongst Russian naval modernisation programmes and a good degree of success has been achieved after considerable effort. Three new Project 955 'Borey' class submarines are

The Norwegian submarine *Utsira* pictured in February 2017. A decision has been taken to replace he six members of the existing *Ula* class with four Type 212-NG submarines of German origin. *(Synne Emilie Svee Solberg/Norwegian Armed Forces)*

Sweden's re-armament in the face of increasing concern over Russia's assertive steps against neighbouring countries has extended to re-establishing mobile coastal defence missile systems disbanded after the end of the Cold War. Local press reports suggest that some land-based launch equipment for the potent RBS-15 Mk 3 missile – pictured here – has had to be recovered from museums to get the systems back up-and-running. *(Copyright Saab AB)*

now in service with the Northern and Pacific Fleets. A further five improved Project 955A variants are now under construction following the start of work on *Knyaz Pozharskiy* on 23 December 2016. This reflects the originally planned total of eight boats. The new state armaments plan will provide clarity on future production. Some reports suggest transition to a new fifth generation 'Husky' type is planned. Meanwhile, most of the problems that plagued the RSM-56 Bulava strategic missile that equip the class appear to have been resolved. One test-firing of two missiles in September 2016 was a partial success in so far as one missile hit the target and one self-destructed; a subsequent test in June 2017 involving just one missile was a total success.

The counterpart of the 'Borey' class in the nuclear-powered attack submarine flotilla is the improved Project 855M variant of the 'Yasen' type. A modification of the original Project 855 submarine *Severodvinsk*, the first new boat – *Kazan* – was launched in March 2017 and has now started trials. A further four have been laid down at approximately annual intervals since 2013; work on a sixth is expected to start in the summer of 2017.[20]

Conventional submarine production continues to be focused on assembly of the Project 636.3 variant of the 'Kilo' class submarine. The six-strong batch intended for the Black Sea fleet were completed with the delivery of *Velikiy Novgorod* and *Kolpino* in October and November 2016. A further six are earmarked for the Pacific fleet, with construction expected to commence in the second half of 2017. It also appears that protracted trials of the lead Project 677 boat *Sankt Peterburg* – which was first commissioned in 2010 – have now progressed sufficiently to allow delivery of the other two members of the class during 2018/19. There are conflicting reports in the Russian press as to whether further construction is envisaged. However, the evident disappointment with the type suggests the opportunity might be taken to move on to a new, AIP-equipped design.

Aircraft Carriers and Amphibious Vessels: The much-delayed major overhaul of *Admiral Kuznetsov* is now scheduled for 2018. The refit has been delayed several times in the past, possibly due to the decision to modernise the battlecruiser *Admrial Nakhimov* for further service and lack of shipyard capacity to carry out both projects simultaneously. Although there are longstanding plans for a replacement carrier, tangible progress is unlikely in the short term.

The amphibious fleet should shortly be bolstered by the arrival of the new Project 1171.1 landing ship *Ivan Gren*. The completion and commissioning process relating to this ship has also proved to be protracted. Although a second ship, *Petr Morgunov*, is also under construction, it seems the focus will then turn to the proposed 'Priboy' amphibious assault ships. It seems that these ships – essentially replacements for the cancelled *Mistral* class order – may be one of the few programmes for major warships approved in the 2018–25 armaments plan.

TABLE 2.4.6: RUSSIAN NAVY: SELECTED PRINCIPAL UNITS AS AT MID 2017

TYPE	CLASS	NUMBER[1]	TONNAGE	DIMENSIONS	PROPULSION	CREW	DATE
Aircraft carriers							
Aircraft Carrier – CV	Project 1143.5 **KUZNETSOV**	1	60,000 tons	306m x 35/73m x 10m	Steam, 32 knots	2,600	1991
Principal Surface Escorts							
Battlecruiser – BCGN	Project 1144.2 **KIROV**	1 (1)	25,000 tons	252m x 29m x 9m	CONAS, 32 knots	740	1980
Cruiser – CG	Project 1164 **MOSKVA** ('Slava')	3	12,500 tons	186m x 21m x 8m	COGAG, 32 knots	530	1982
Destroyer – DDG	Project 956/956A **SOVREMENNY**	c.5	8,000 tons	156m x 17m x 6m	Steam, 32 knots	300	1980
Destroyer – DDG	Project 1155.1 **CHABANENKO** ('Udaloy II')	1	9,000 tons	163m x 19m x 6m	COGAG, 29 knots	250	1999
Destroyer – DDG	Project 1155 **UDALOY**	c.8	8.500 tons	163m x 19m x 6m	COGAG, 30 knots	300	1980
Frigate – FFG	Project 1136.6M **GRIGOROVICH**	2	4,000 tons	125m x 15m x 4m	COGAG, 30 knots	200	2016
Frigate – FFG	Project 1154 **NEUSTRASHIMY**	2	4,400 tons	139m x 16m x 6m	COGAG, 30 knots	210	1993
Frigate – FFG	Project 1135 **BDITELNNY** ('Krivak I/II')	c.2	3,700 tons	123m x 14m x 5m	COGAG, 32 knots	180	1970
Frigate – FFG	Project 2038.0 **STEREGUSHCHY**	5	2,200 tons	105m x 11m x 4m	CODAD, 27 knots	100	2008
Frigate – FFG	Project 1161.1 **TATARSTAN** ('Gepard')	2	2,000 tons	102m x 13m x 4m	CODOG, 27 knots	100	2002
Submarines							
Submarine – SSBN	Project 955 **YURY DOLGORUKY** ('Borey')	3	17,000+ tons	170m x 13m x 10m	Nuclear, 25+ knots	110	2010
Submarine – SSBN	Project 941 **DONSKOY** ('Typhoon')	1	33,000 tons	173m x 23m x 12m	Nuclear, 26 knots	150	1981
Submarine – SSBN	Project 677BDRM **VERKHOTURYE** ('Delta IV')	6	18,000 tons	167m x 12m x 9m	Nuclear, 24 knots	130	1985
Submarine – SSBN	Project 677BDR **ZVEZDA** ('Delta III')	3	12,000 tons	160m x 12m x 9m	Nuclear, 24 knots	130	1976
Submarine – SSGN	Project 855 **SEVERODVINSK** ('Yasen')	1	13,500 tons+	120m x 14m x 9m	Nuclear, 30+ knots	90	2013
Submarine – SSGN	Project 949A ('Oscar II')	c.5	17,500 tons	154m x 8m x 9m	Nuclear, 30+ knots	100	1986
Submarine – SSN	Project 971 ('Akula I/II')	c.10	9,500 tons	110m x 14m x 10m	Nuclear, 30+ knots	60	1986
Submarine – SSK	Project 677 **ST PETERSBURG** ('Lada')	1	2,700 tons	72m x 7m x 7m	Diesel-electric, 21 knots	40	2010
Submarine – SSK	Project 877/636 ('Kilo')	c.20	3,000 tons	73m x 10m x 7m	Diesel-electric, 20 knots	55	1981

Notes:
1 Table only includes main types and focuses on operational units: bracketed figures are ships being refurbished or in maintained reserve.

The new Russian Project 636.3 'Kilo' class submarine *Krasnodar* seen transiting the English Channel on 8 May 2017. One of six new members of the class delivered between 2014 and 2016, she was undertaking cruise missile strikes against anti-government forces in Syria before the end of the month. *(Crown Copyright 2017)*

Surface Combatants: Construction of major surface combatants is currently focused on two major classes. The larger of these are the Project 2235.0 *Admiral Gorshkov* class frigates. The lead member of the class commenced sea trials in November 2014 but there have been problems both with the platform management system and the Poliment-Redut surface-to-air missile complex. Current reports suggest that these have been largely overcome and that the ship will join the fleet before the end of 2017. However, several deadlines have been missed previously. A second ship, *Admiral Kasatonov*, was launched in 2014 and two other ships are under construction. It was originally planned to build up to twenty members of the class but recent reports suggest production will transition to an improved Project 2235.0M class that will be considerably larger than the c. 5,500-ton full load displacement of the current design. This improved variant may also well supersede construction of Project 1135.6 *Admiral Grigorovich* class frigates. The third of these, *Admiral Marakov*, will shortly commission after completion of sea trials but work on three sister-ships has been suspended due to lack of Ukrainian-built turbines. There have been suggestions that these ships might be transferred to India – which operates a variant of the type as the *Talwar* class – as a means of getting round the problem. However, NPO Saturn's success in developing an alternative production plant suggests it is now likely they will be completed for Russia.

Construction of smaller, light frigate and corvette-sized vessels is split into three major streams. The largest-sized and longest-established is that for Project 2038.0/2038.1 *Steregushchy* class light frigates. Five of a projected class of at least ten of these 2,200-ton ships has been completed to date. The initial plan was for production to transition to an improved Project 2038.5 *Gremyashchy* variant but significant use of now-embargoed foreign equipment means that this type has been terminated at just two units. Instead, the first of a new Project 2038.6 *Derzkiy* variant – larger and making much greater use of stealth technology – was laid down on 28 October 2016. At the other end of the scale are the Project 2163.1 'Buyan M' light corvettes. Twelve

The post-Cold War Russian Navy remains dominated by submarines and smaller surface vessels; the East German-built Project 1331M 'Parchim II' class anti-submarine corvette *Kazanets* is seen here in August 2016. Whilst numerous projects are underway for new ships of similar size, progress in replacing larger Soviet-era ships is progressing extremely slowly. *(Conrad Waters)*

of these c. 950-ton ships have either been completed or are on order, supplementing the three older and smaller vessels of the original Project 2163.0 'Buyan' type. The latest development has been construction of the new Project 2280.0 'Karakurt' design, which are of roughly similar size to the 'Buyan M' class but better-suited for high seas deployment. At least six have been started to date and the first is scheduled for delivery within the next year.

These major construction programmes are supplemented by numerous projects for second-line warships that include new classes of Arctic and offshore patrol vessel, MCMVs and auxiliaries. Refurbishment work also continues to extend the lifespan of Soviet-era designs but this is becoming progressively more difficult. A number of types, such as Project 1155 *Udaloy* class destroyers, have been seemingly less active in recent years. This may also reflect Ukrainian supply problems given that country's domination of gas turbine supply in Soviet times.

Notes

1. See White Paper 2016 on *German Security Policy and the Future of the Bundeswehr* (Berlin: Federal Ministry of Defence, 2016).

2. German Navy investments were highlighted by Dr Lee Willett in: 'German Navy bolsters fleet to address high-end threats', *Jane's Defence Weekly* – 7 June 2017 (Coulsdon: IHS Jane's, 2017).

3. Although the headline reduction looks to be in the order of twenty-five percent, the nominal 2016 budget figure was boosted by a RUB700bn adjustment writing-off debts owed by defence suppliers. Additionally, the 2017 budget could well gain from windfall allocations. In practice, the real reduction is likely to be closer to five percent. For analysis, see Mark Galeotti's post 'The truth about Russia's defence budget' of 24 March 2017 on the *European Council on Foreign Relations'* website at: http://www.ecfr.eu/article/commentary_the_truth_about_russias_defence_budget_7255

4. The use of the successful integration of long-range cruise missiles with Russian Navy light frigates and corvettes as a justification against heavy investment in larger, blue water combatants was heralded by James Bosbotinis in his review of The Russian Navy in the editor's *Navies in the 21st Century* (Barnsley: Seaforth Publishing, 2016), pp.83–90.

5. An interesting overview of the impact of Brexit on European defence has been provided by a series of publications produced by the European arm of the RAND Corporation consultancy. These are summarised by James Black, Alexandria Hall, Kate Cox, Marta Kepe and Erik Silfversten in *Defence and Security after Brexit: Overview Report* (Santa Monica, CA & Cambridge UK: RAND Corporation, 2017). For a balanced appraisal on the likely impact of the Macron Presidency on European defence see Andrea Frontini's 'The 'Macron effect' on European defence: En Marche, at last?' published by the *European Policy Centre* on 1 June 2017 and available by searching: http://www.epc.eu/

6. A good overview of the FTI's characteristics is provided by Christopher P Cavas and Pierre Tran in 'France Unveils New FTI Frigate Designed for the French Navy and Export' in an article posted to the *Defense News* website on 18 October 2016. A French view on the motivation for the programme – which revolves around the desirability of creating a warship more attractive to foreign buyers and the need to keep DCNS and Thales design teams occupied – is provided by Michel Cabriol in 'Les frégates de taille intermédiaire se jettent à l'eau', *La Tribune* – 17 October 2016 (Paris: La Tribune Nouvelle, 2016).

7. Fincantieri's acquisition of a sixty-seven percent stake in STX France was announced on 19 May 2017. The French state holds the other thirty-three percent and had two months to block the deal. A conclusion to discussions between Fincantieri and the French government had not been announced as of 30 June 2017.

8. A comprehensive update on the vessels being constructed under the Naval Law was provided by Luca Peruzzi in 'The Italian Navy's New PPA and LSS: First Design Details Emerge', *European Security & Defence* – 5/2016 (Bonn: Mittler Report Verlag GmbH, 2016), pp.56–62.

9. The large amounts owing from previous Spanish defence programmes have been highlighted by the Spanish-language *infodefensa.com* website by B Carrasco.

10. An interesting commentary on the nine positive milestones set out by the Ministry of Defence in celebrating the 'Year of the Royal Navy' was provided on 3 January 2017 by the *Save the Royal Navy* website under the title 'Will 2017 be "the year of the Royal Navy"? It is currently available at: http://www.savetheroyalnavy.org/will-2017-be-the-year-of-the-royal-navy/

11. A detailed review of the F125 design is provided in Chapter 3.2.

12. A review of 'The Royal Netherlands Navy: Recent developments and current status was provided by Theodore Hughes-Riley in *Seaforth World Naval Review 2016* (Barnsley: Seaforth Publishing, 2016), pp.92–103.

13. A detailed overview of the joint Belgian-Netherlands naval procurement agreement is provided by Jamie Karreman in 'Belgian Modernisation Plans in Cooperation with The Netherlands', *European Security & Defence* – 2/2017 (Bonn: Mittler Report Verlag GmbH, 2016), pp.30–3. Mr Karreman is editor-in-chief of the Dutch naval website marineschepen.nl, an excellent source of information on current Dutch naval developments.

14. Devrim Yaylali's Bosphorus Naval news blog at https://turkishnavy.net/ remains the premier English language source for Turkish Navy developments.

15. See 'Euronaval 2016: Bulgarian shipbuilder MTG Dolphin unveils its K-90 multi-purpose corvette' posted to the ever informative *Naval Recognition* website on 20 October 2016 and available by searching: http://www.navyrecognition.com/

16. Further details of the Croatian programme can be found in an article entitled 'Croatia launches its first inshore patrol vessel' posted to *the NavalToday.com* website – another invaluable source of naval news – on 7 June 2017 and available by searching: http://navaltoday.com/

17. See 'Sweden rebuilds Cold War missile system from museums' published on *The Local.se* website on 20 November 2016, available at: https://www.thelocal.se/20161120/sweden-brings-cold-war-missile-defence-system-out-of-museums

18. A comprehensive review of Finland's naval procurement plans was provided by Peter Felstead in Briefing: Finnish procurement – Northern composure in *Jane's Defence Weekly* – 19 April 2017 (Coulsdon: IHS Jane's, 2017), pp.24–32.

19. If true, this arguably reflects flawed thinking. A carrier strike group is able to conduct sustained strikes against multiple targets whilst the cruise-missile capacity of surface combatants is limited – just eight cruise-missile cells on the Project 1135.6 and Project 2163.1 classes.

20. For further analysis of the improved 'Yasen' design see Matthew Bodner's 'Russia adds "Kazan" to its nuclear attack submarine fleet' posted to the *Defense News* website on 31 March 2017 and currently available at: http://www.defensenews.com/articles/as-rd-180-ban-looms-space-companies-make-steady-progress-on-new-launch-technologies

2.4A Fleet Review

Author:
Richard Beedall

ROYAL NAVY The start of a new era?

The last twelve months have seen two prominent speeches seemingly setting out a bright future for Britain's Royal Navy. On 1 January 2017, the British Defence Secretary, Sir Michael Fallon, said:

> We are investing billions in growing the Royal Navy for the first time in a generation with new aircraft carriers, submarines, frigates, patrol vessels and aircraft all on their way. 2017 is the start of a new era of maritime power, projecting Britain's influence globally and delivering security at home.

A few weeks earlier Admiral Sir Philip Jones, First Sea Lord and professional head of the naval services,[1] issued a message saying:

> ... the Government has repeatedly stated its ambition to grow the size of the Royal Navy by the 2030s ... For most of my thirty-eight year career, the story of the Royal Navy has been one of gradual, managed contraction. Now, at long last, we have an opportunity to reverse this trend, I intend to work with the Government in the coming months and years to deliver their ambition for a larger navy. Only this will ensure the Royal Navy can continue to deter our enemies, protect our people and promote our prosperity in these uncertain times.[2]

This chapter examines the extent to which this vision is matched by reality.

FUNDING

The root cause of the contraction referred to by Sir Philip is underfunding. The United Kingdom's (UK) defence spending has been essentially static in recent years: £35.1bn in 2015/16, changing to £36.0bn in 2017/18 (planned). The government has committed to continuing to meet the NATO target of spending two percent of Gross Domestic Product (GDP) on defence. However, it now includes previously excluded items such as military pensions and contributions to UN peacekeeping missions as part of this total. Spending is projected to slowly increase to £39.7bn in 2020/21 – but for comparison the figure in 2010/11 was £39.0bn (2.5 per cent of GDP in that year).[3]

A significant problem that emerged in late 2016 was the affordability of the Ministry of Defence's (MOD) ten-year Equipment Plan. Cost overruns, a fall in the sterling/US dollar exchange rate and other unexpected problems led the National Audit Office to conclude in January 2017 that risks to the £178bn plan were '... greater than at any point since reporting began in 2012'.[4] British newspapers have reported that the MOD is seeking economies of at least £1bn a year, including £500m from the RN. Many commentators expect major projects to be delayed, scaled back or even cancelled.

PERSONNEL

For several years, a critical challenge facing the RN has been a shortage of personnel. The *Strategic Defence and Security Review* in 2010 (SDSR 2010)

The new Royal Navy aircraft carrier *Queen Elizabeth* pictured on sea trials on 30 June 2017, the day on which the first helicopter operations were conducted from the ship. The new aircraft carrier's imminent delivery is being used to justify government assertions of a rebirth in British naval capability but the reality is more complex. *(Crown Copyright 2017)*

The Type 23 frigate *Sutherland* pictured escorting the Russian frigate *Yaroslav Mudry* through the English Channel in December 2016. The strategic environment is more unsettled than previously and the Royal Navy is struggling to meet the resulting challenges. *(Crown Copyright 2016)*

An image of the new *Dreadnought* strategic missile submarine, previously known as 'Successor'. Construction of the first boat is now underway. *(BAE Systems)*

reduced the number of trained regular personnel in the Royal Navy and Royal Marines (RM) by 5,000, to 30,000. The cuts fell almost entirely on the sailors manning the surface and submarine flotillas and by 2013, with redundancies still being made, it was clear that there were no longer enough sailors left to man the remaining ships. SDSR 2015 cancelled the planned reduction of another 1,000 personnel and restored 400 posts at the rate of ten a month. Internally, the RN had already decided that 300 officer appointments would be replaced by 600 rating billets. Two hundred RM posts will also be reallocated to the general service (i.e. from marines to sailors).

On 1 May 2017, the RN/RM had 29,430 trained personnel (22,410 RN, 6,690 RM, 330 full-time reservists); a 2.6 percent undershoot compared to an approved strength of 30,220. The RN is actively recruiting, with the goal of again reaching 23,000 trained sailors (last seen in 2014) as soon as possible. Any further increase will depend upon the outcome of SDSR 2020.

A CHANGING STRATEGIC ENVIRONMENT

Defence spending wasn't a major issue in the UK's General Election of 8 June 2017, with all three major national parties pledging to meet the two percent of GDP spending commitment. However, the ruling Conservative Party's unexpected poor showing in the polls resulted in the formation of a minority government with uncertain long-term prospects. The likelihood that the UK will leave the European Union in 2019 (Brexit) and the possibility – now more remote – that Scotland and even Northern Ireland could, in turn, leave the UK add further uncertainties for the MOD and RN.

Since the end of the Cold War in 1990 the UK has faced no serious maritime threats. This, combined with the continuing global dominance of the US Navy, has allowed a significant reduction to the size of the RN. This era seems to be ending, SDSR 2015 arguably underestimating the situation when it said 'International Military Conflict: The risk is growing. Although it is unlikely that there will be a direct military threat to the UK itself, there is a greater possibility of international military crises drawing in the UK, including through our treaty obligations.'[5]

In Europe, Russia is re-asserting itself militarily and the RN and Royal Air Force (RAF) are struggling to cover the frequent appearance of Russian warships and aircraft in the North Sea and English Channel. On multiple occasions since 2014 the MOD has been forced to ask for American, French and Canadian assistance to track Russian submarines operating near or even in UK waters. Gibraltar has also become a problem, with Spanish vessels regularly testing the UK's resolve to defend the outpost by entering British Gibraltar Territorial Waters, including three incursions by warships in April 2017. The Royal Navy's impressively-named Gibraltar Squadron actually consists of just two small (24-ton) patrol craft – *Scimitar* and her sister-ship *Sabre*. Until about 1999 a frigate or destroyer was assigned as a Gibraltar guard ship but that is no longer possible. The number of RN and RFA ships making port visits to Gibraltar has been maximised but still only averages two ships a month. Basing an Offshore Patrol Vessel (OPV) at Gibraltar (in a manner similar to *Clyde* for the Falkland Islands) is being considered by the MOD.

In the Middle East, the UK continues to conduct military operations in Syria and Iraq, whilst maintaining the general stability of the region is considered a vital national interest. A symbol of the UK's commitment was the official opening of a UK Mina Salman Support Facility in the Kingdom of Bahrain by Prince Charles on 10 November 2016. Bahrain has paid most of the £30m cost, with the UK contributing around £7.5m. Construction work has slipped, but it is hoped that the base will be complete by November 2017. It will then be commissioned as HMS *Jufair*, replacing a naval base of the same name that closed in 1971. With about 200 headquarters and base staff, it will be the home port for four mine-countermeasures vessels (in April 2017 these were *Penzance*, *Middleton*, *Chiddingfold* and *Bangor*) and the forward operating base for a frigate or destroyer.

In the Far East, the Royal Navy maintains a small logistics base (with just six naval and civilian staff – termed Naval Party 1022) at Sembawang Wharf in Singapore. This has gained renewed importance because of its strategic location on a major maritime trade route, and relative proximity to potential flash points such as the South China Sea and the Korean peninsula. It supports visiting RN warships as well as those from Australia and New Zealand.

Post-Brexit the UK government wants to demonstrate that the country remains a serious player on the world stage. The first operational deployment of the reformed UK Carrier Task Group, centred on the new aircraft carrier *Queen Elizabeth*, is planned for 2021. This is likely to be an impressive six-to-nine month flag waving tour that includes the Far East.

DREADNOUGHT AND THE UK'S NUCLEAR DETERRENT

The highest defence priority of the current UK government is renewal of the country's nuclear deterrent, which was approved by Parliament on 18 July 2016. Total expenditure on the £41bn programme (including a contingency) amounted to about £2.6bn by late 2016, and will exceed £1bn in 2017/18 alone.

A key element of the programme is the construction of four new *Dreadnought* class nuclear-powered ballistic missile submarines (SSBN). Previously known as 'Successor', they will replace the *Vanguard* class submarines completed in the 1990s. One submarine will always be on patrol to provide 'Continuous At Sea Deterrence' (CASD). To avoid a re-occurrence of the problems that have plagued the *Astute* attack submarine project, a Submarine Delivery Authority is being created to ensure that the *Dreadnought* programme meets its targets.

Secrecy surrounding the design has slightly dissipated and we now know that they will be 152.9 metres long and displace 17,200 tons. Although slightly larger than the *Vanguard* class they will have twelve rather than sixteen missile tubes, and only eight will carry Trident II D-5 missiles. There is the interesting potential to use the four spare tubes for other purposes, e.g. to carry cruise missiles.

On 9 September 2016, the programme entered the Demonstration and Manufacture stage (or Delivery Phase 1). The MOD then awarded BAE Systems contracts worth £1.3bn (US$1.7bn) to fund early manufacturing work as well as '…furthering the design of the submarine, purchasing materials and long lead items, and investing in facilities at the BAE Systems yard in Barrow-in-Furness where the submarines will be built'. On 6 October, the shipyard cut the first steel for the auxiliary machine space that will contain switchboards and control panels for *Dreadnought*'s reactor. Each *Dreadnought* submarine will be built in sixteen units that are then grouped into three mega-units – aft, mid and forward – and joined together.

Items already ordered for *Dreadnought* include the PWR3 nuclear propulsion system from Rolls-Royce and the missile compartment from General

A picture of the fourth *Astute* class submarine *Audacious* being transported out of BAE Systems' submarine manufacturing hall at Barrow-in-Furness in April 2017. To help avoid a re-occurrence of the problems that plagued the *Astute* project, a Submarine Delivery Authority is being created to support the *Dreadnought* programme meeting its targets. *(BAE Systems)*

Dynamics Electric Boat in the USA. The latter will be largely common with the US Navy's new *Columbia* class, formerly known as the *Ohio* Replacement Submarine.[6]

Dreadnought should enter service in 2028, with three more boats following by 2037. The existing *Vanguard* class will continue to be primarily responsible for providing CASD until the early/mid 2030s; this will push them about ten years over their original 25-year design life. The failure of a Trident missile test launch by *Vengeance* in 2016 led to considerable but poorly-informed concern in the British media about the reliability of CASD now, let alone in the 2020s.

The uncertain result of the UK general election also produces some uncertainty about the *Dreadnought* programme. Whilst the largest opposition party, Labour, upported the renewal of the Trident nuclear deterrent in its manifesto, its leader, Jeremy Corbyn, is a long-standing opponent of British nuclear weapons.

HUNTER-KILLER SUBMARINES

The RN's attack submarine flotilla is going through a difficult period. It entered 2017 with just seven submarines – three 'A' or *Astute* class and four 'T' or *Trafalgar* class. A combination of accidents (e.g.

Table 2.4A.1: ROYAL NAVY FLEET COMPOSITION MID 2017

TYPE	CLASS	NUMBER	DATE[1]	NOTES
Principal Surface Escorts				
Destroyer – DDG	**DARING** (Type 45)	6	2008-2013	*Daring* (D32), *Dauntless* (D33), *Diamond* (D34), *Dragon* (D35), *Defender* (D36) , *Duncan* (D37). Of these, *Dauntless* is being used as a harbour training ship pending refit.
Frigate – FFG	**NORFOLK** (Type 23)	13	1991-2002	*Argyll* (F231), *Lancaster* (F229), *Iron Duke* (F234), *Monmouth* (F235), *Montrose* (F236), *Westminster* (F237), *Northumberland* (F238), *Richmond* (F239), *Somerset* (F82), *Sutherland* (F81), *Kent* (F78), *Portland* (F79), *St Albans* (F83). Three additional ships disposed of early to Chile.
Submarines				
Submarine – SSBN	**VANGUARD**	4	1993-1999	*Vanguard* (S28), *Victorious* (S29), *Vigilant* (S30), *Vengeance* (S31). To be replaced by *Dreadnought* class.
Submarine – SSN	**ASTUTE**	3	2009 onwards	*Astute* (S119), *Ambush* (S120), *Artful* (S121). Four further boats under construction.
Submarine – SSN	**TRAFALGAR**	4	1987-1991	*Torbay* (S90), *Trenchant* (S91), *Talent* (S92), *Triumph* (S93). Survivors of an original class of seven, being replaced by the new *Astute* class submarines. *Torbay* to decommission in 2017.
Amphibious Ships				
Helicopter Carrier – LPH	**OCEAN**	1	1998	*Ocean* (L12). To decommission 2018.
Landing Platform Dock – LPD	**ALBION**	2	2003-2004	*Albion* (L14), *Bulwark* (L15). Only one ship kept operational post-SDSR 2010. *Bulwark* is currently in service, with *Albion* replacing her in the front-line fleet on completion of a refit.
Landing Ship Dock – LSD(A)	**LARGS BAY**	3	2006-2007	*Lyme Bay* (L3007), *Mounts Bay* (L3008), *Cardigan Bay* (L3009). *Largs Bay* sold to Australia post-SDSR 2010 in 2011. Operated by RFA.
Minor War Vessels				
Patrol Vessel – OPV	**TYNE** ('River')	3	2003	*Tyne* (P281), *Severn* (P282), *Mersey* (P283). Being replaced by Batch II 'River' class. *Severn* to decommission 2017, *Mersey* in 2019.
Patrol Vessel – OPV	**CLYDE** (Modified 'River')	1	2007	*Clyde* (P257). An upgraded 'River' used as Falkland Islands Patrol Vessel. To be replaced by a Batch II 'River' class vessel in 2019.
Ice Patrol Ship - OPV	**PROTECTOR**	1	2011	*Protector* (A173). Former Norwegian commercial ice breaker built in 2001. Antarctic Patrol Vessel.
Minehunter – MCMV	**SANDOWN**	7	1997-2001	*Penzance* (M106), *Pembroke* (M107), *Grimsby* (M108), *Bangor* (M109), *Ramsey* (M110), *Blyth* (M111), *Shoreham* (M112). Five earlier vessels sold to Estonia, laid up or used for static training.
Minehunter – MCMV	**BRECON** ('Hunt')	8	1980-1989	*Ledbury* (M30), *Cattistock* (M31), *Brocklesbury* (M33), *Middleton* (M34), *Chiddingfold* (M37), *Atherstone* (M38), *Hurworth* (M39), *Quorn* (M41). Five other vessels sold or used for static training.

Notes:

1 Date refers to the delivery date(s) of ships in the class remaining in service.

2 Other vessels include three survey ships of the *Scott* and *Echo* classes; the survey launch *Gleaner*; and eighteen inshore patrol vessels of the *Archer* and *Scimitar* classes. The Royal Fleet Auxiliary service also includes two 'Wave' class fleet tankers; the auxiliary oiler and replenishment vessel *Fort Victoria*; two *Fort Rosalie* class stores vessels; and the casualty receiving ship *Argus*.

3 Ships under construction or undergoing trials include the aircraft carriers *Queen Elizabeth* and *Prince of Wales*; the strategic submarine *Dreadnought*, the attack submarines *Audacious*, *Anson*, and *Agamemnon* and an unnamed seventh boat; five Batch II 'River' class offshore patrol vessels – *Forth*, *Medway*, *Trent*, *Tamar* and *Spey* – and four 'Tide' class tankers – *Tidespring*, *Tiderace*, *Tidesurge* and *Tideforce*. Long-lead items for three Type 26 frigates have also been ordered.

Ambush collided with a merchant ship on 20 July 2016) and technical problems (e.g. the discovery of reactor damage to *Trenchant* that could also affect her sister boats) meant that it was probable that none of these were operational at that time. As such, the service has been unable to maintain the recent norm of one submarine deployed 'East of Suez' whilst another operates in home waters with a primary responsibility of protecting *Vanguard* class SSBNs. Furthermore, the service will end 2017 with just six boats, as the thirty-year-old *Torbay* is scheduled to decommission during the year.

The fourth 'A' class, *Audacious*, was first floated on 29 April 2017 and should begin sea trials in 2018. A £1.4bn order to complete the construction of the sixth boat, *Anson*, was announced on 19 April 2017. A seventh unnamed boat is planned and assembly of materials for her began in January 2014. No more are planned, despite eight being the accepted operational requirement.

AIRCRAFT CARRIERS

In December 2010, the aircraft carrier *Ark Royal* was unexpectedly decommissioned as a result of SDSR 2010, leaving no such ship in RN service – a gap that has since been frequently regretted. The British Government has often responded to criticism of this and other naval cuts by pointing to the construction of two new aircraft carriers – the 65,000-ton *Queen Elizabeth* class (QEC).

The first of the new ships, *Queen Elizabeth*, was expected to begin sea trials in March 2017. Unfortunately these were delayed by technical problems, including the discovery that some paintwork had not properly adhered to the undercoat and needed to be replaced. Under the command of Captain Jerry Kyd she finally sailed for the first time from Rosyth Dockyard – where she had been assembled – on 26 June 2017. On conclusion of initial trials, she will move to Portsmouth Naval Base, where £130m has been spent on upgrades to accommodate her. The historic North Corner has been

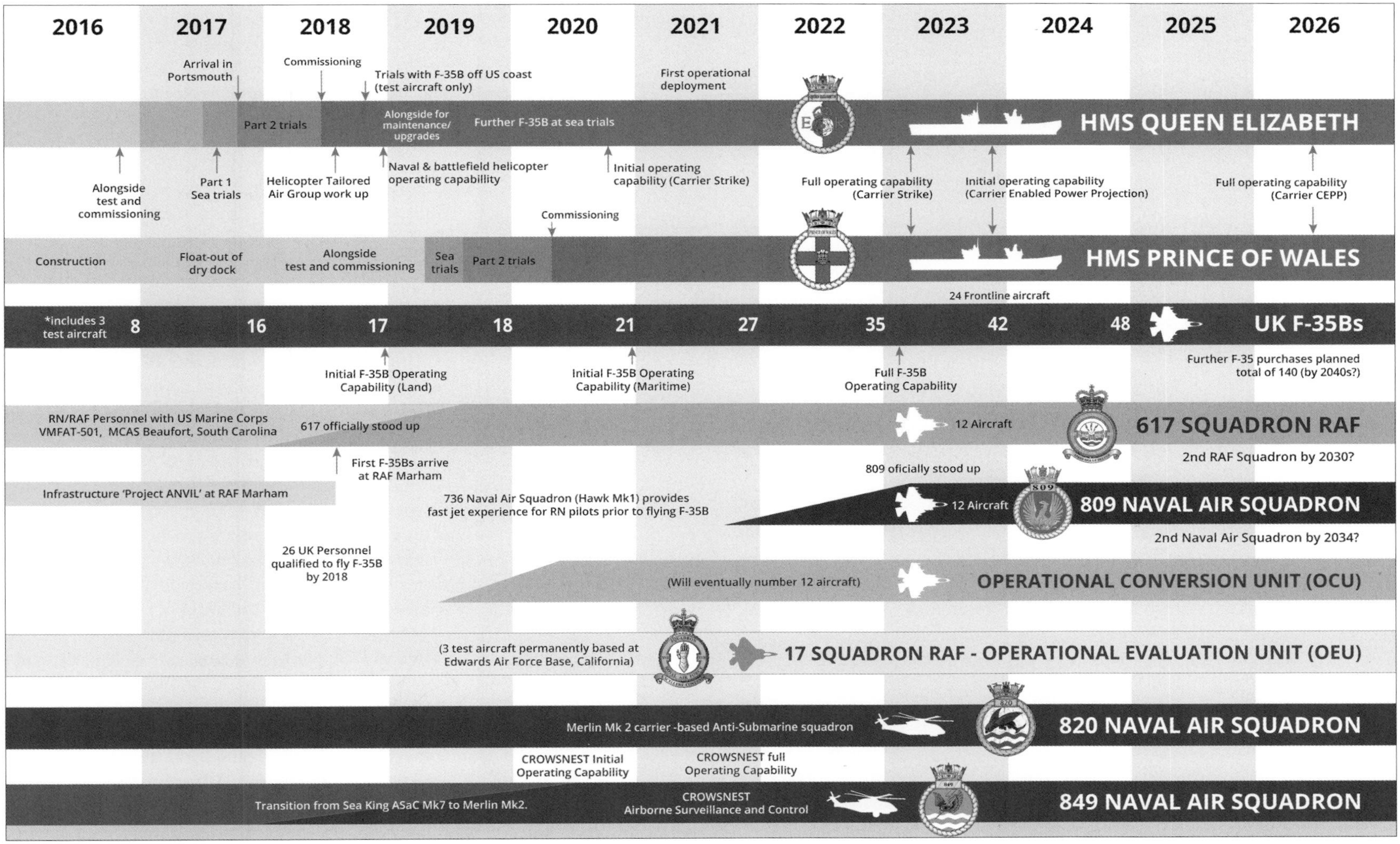

A graphic of the timeline involved in getting the Royal Navy's new carrier strike capability to full operational capability. This gives a good indication of the complexity involved in re-establishing a capacity once it has been lost. *(Courtesy Save the Royal Navy.org)*

The Type 45 destroyer *Defender* pictured operating in company with the French FREMM type frigate *Provence* in the Indian Ocean in early 2016. The Type 45 class's integrated electric propulsion has proved unreliable, particularly in hot climates. An expensive rectification programme should get underway in 2018. *(Crown Copyright 2016)*

The Type 26 frigate has experienced a protracted gestation period but construction should finally get underway during 2017. The class will replace some of the existing Type 23 frigates that are currently going through a programme of capability upgrades to keep them effective. *(BAE Systems)*

rebuilt and renamed the Princess Royal Jetty, whilst three million cubic metres of material have been dredged from the harbour and approaching channel to accommodate the draft of the two vessels. *Queen Elizabeth* will undertake intensive flight trials with helicopters during the summer of 2018, before sailing to the east coast of the United States in late 2018 for further trials with F-35B Lightning II strike fighters – a major milestone on her way towards becoming fully operational by December 2020. No date has yet been set for the commissioning of the ship, although the ceremony is expected to be attended by Her Majesty Queen Elizabeth II.

Clarity has emerged on the nature of the air group of a QEC carrier operating in the carrier strike role. It is envisaged this will consist of two squadrons of twelve F-35B strike fighters (617 Squadron RAF and 809 Naval Air Squadron), one squadron (820 NAS) of six Merlin HM 2 anti-submarine helicopters, and one flight (from 849 NAS) of three Merlin HM 2 helicopters fitted with the Crowsnest airborne early-warning radar system. One or two utility helicopters may also be carried. An additional six F-35Bs can be embarked in an emergency. The RN is considering the acquisition of a small number of Bell-Boeing V-22 Osprey aircraft for Carrier Onboard Delivery (COD) and the air-to-air refuelling of F-35Bs.

Although 617 Squadron should be fully operational in time for the maiden deployment of *Queen Elizabeth* in 2021, 809 NAS will not form until 2023. On 15 December 2016 UK Defence Secretary Sir Michael Fallon and then US Defense Secretary Ash Carter signed an agreement to allow US Marine Corps aircraft to embark on the QEC. A squadron of their F-35Bs and, possibly, a number of MV-22 Ospreys will join *Queen Elizabeth*'s air group for her first deployment. The agreement has resulted in some interesting legal questions.

Meanwhile, it is anticipated that second carrier, *Prince of Wales*, will be floated out of her construction dock at Rosyth in the summer of 2017. She should commission in 2020 and will incorporate various design improvements. The enhancements will make her more flexible and able to deliver all aspects of what is termed Carrier Enabled Power Projection (CEPP) – Carrier Strike, Littoral Manoeuvre, Humanitarian Assistance and Defence Diplomacy.

Currently the RN does not have enough money,

The prospects of orders for a new a new Type 31 light frigate, designed for both RN service and export, has attracted the attention of a number of design houses. These computer graphics show BMT Defence Services' Venator 110 design and an alternative concept from Steller Systems. *(BMT Defence Services / Steller Systems Ltd)*

manpower, aircraft, escorts or support ships to keep both carriers operational. Whilst arrangements will probably not be finalised until SDSR2020, it seems likely that around 2022 *Prince of Wales* will take over from *Queen Elizabeth* as the UK's on-call carrier. The latter will then be reduced to extended readiness (i.e. in an emergency she could re-enter service in three to six months) before starting in 2025 a refit to bring her up to the same standard as *Prince of Wales*.

ESCORT FORCE (DESTROYERS AND FRIGATES)

The Royal Navy is struggling to maintain a viable escort force. As recently as 2005 it had thirty-one frigates and destroyers in service whereas it now nominally has nineteen – described by the House of Commons Defence Select Committee as a 'woefully inadequate total'.[7]

The six Type 45 *Daring* class destroyers suffer from serious engineering problems; their complex WR-21 gas turbines and sophisticated Integrated Full Electric Propulsion (IFEP) have proven to be unreliable (particularly in hot weather) and on multiple occasions ships have lost all power.[8] The worst-affected ship – *Dauntless* – was temporarily reduced to the role of a harbour training ship in 2016, just six years after entering service. The agreed solution or Power Improvement Plan (PIP) is to install powerful new diesel generators and relegate the WR-21s to a boost role. However, shoehorning the diesels in to the ship's hull and integrating them with the IFEP will be a substantial undertaking. A contract for the PIP, which might approach £300m in value, is expected to be awarded in early 2018. The work will be added to the scheduled programme of refits, so it will be the early 2020s before all of the class are fixed.

The thirteen Type 23 frigates are going through a programme of refits to keep them effective to end-of-service dates that extend out into the future as far as 2035. The most notable change is replacement of the Sea Wolf GWS.26 anti-air missile system with the longer-range Sea Ceptor Common Anti-Air Modular Missile (CAAM), *Westminster* was the first to be fitted, emerging from refit in January 2017. *Argyll* followed early in February and both ships have been undergoing post-refit trials. One unmodernised unit, *Lancaster*, was placed in reserve in late 2015, apparently due to an inability to man her. There was speculation that she would never again become operational before her scheduled decommissioning in 2024 but she commenced refit at Devonport in April 2017 and should return to the fleet during 2018.

The Type 23 frigates and three of the Type 45 destroyers are armed with Harpoon Block 1C surface-to-surface missiles. Most of the RN's stock of missiles are reaching the end of their thirty-year service life and there is no funding for replacements. The system will thus be taken out of service by the end of 2018. Until a replacement can be afforded, the primary surface weapon will be their single Mk 8 gun, inferior in some respects to the twin Mk 6 mounted on most RN escorts in the 1960s.

Currently RN escorts largely operate alone, with four or five typically dispersed on various tasks around the world. The formation in 2020 of a UK Carrier Task Group will be a game-changer as at least one Type 45 destroyer and one Type 23 frigate will be assigned to this, possibly two of each if allied navies fail to make hoped-for contributions. The result will be a significant reduction in the ability of the RN to regularly assign an escort to NATO maritime groups, the North and South Atlantic patrol tasks, and other duties around the world.

UK NAVAL SHIPBUILDING

In 2005, the MOD published a *Defence Industrial Strategy* (DIS).[9] Subsequent naval orders have born little resemblance to that projected by DIS. One of the few substantial outcomes of DIS was a fifteen-year Terms of Business Agreement made in 2009 between the MOD and BAE Systems; this guaranteed that the company would receive a core naval work load worth at least £230m a year. Since 2013 the MOD has partially met this by ordering five *Forth* class Offshore Patrol Vessels (OPVs) at about £120m each – double their probable cost if competitively tendered. Three or four will replace existing OPVs that are still relatively young; the balance will apparently be a much-needed addition to a disproportionately small patrol force.

For the last twenty years the RN has been trying to build a new class of frigates. The design has been recast many times but it seemed that construction by BAE Systems Naval Ships of the first of thirteen Type 26 frigates would finally start in 2015. However, due to escalating costs, SDSR 2015 reduced the class to eight (optimised for anti-submarine warfare) and announced vague plans to build at least five of a new general-purpose frigate, since designated the Type 31.

The ongoing changes and delays have adversely impacted the unit cost of a Type 26, which has been reported as reaching up to £1bn per ship. Between 2010 and the end of 2016, the MOD spent or committed nearly £2bn to the project. This was mostly on design development but also encompassed lead items for the first three Type 26 frigates, including the ordering of 5in Mk 45 guns manufactured by BAE Systems in the United States. The MOD has repeatedly said that the construction of the first Type 26 will begin during 2017, with entry in to service by the 'mid-2020s'.[10] A serious problem is that by then the Type 23s will be decommissioning at the rate of one per year – far faster than the MOD can afford to build Type 26s. As a result, one option being considered by the MOD is to build the Type 26s in two batches of three and five ships, separated by a batch of five (or possibly more) of the lower-cost Type 31s. This approach would allow more ships to be ordered whilst reducing expenditure during the life of the current ten-year Equipment Plan.

The capability requirements for the Type 31 light frigate are still being defined but several companies have already optimistically submitted preliminary designs of varying size, capability and cost to the MOD. Official briefings indicate that a maximum unit cost of £350m has been set. Some industry sources do not believe that it will be possible to deliver a warship that would be useful as a member of a task group for that price and that capabilities will need to be limited to those required for lower-tempo operations.

Given all the uncertainty, another attempt at a Naval Shipbuilding Strategy seemed justified, and Sir John Parker was commissioned to produce an independent report that would inform this. His report was published on 29 November 2016.[11] It made thirty-four recommendations and a covering letter summarised these as:

- Govern the design and specification of Royal Navy ships to a target cost within an assured capital budget and inject pace to contract on time.
- Design ships suitable for both RN and export.
- Build via a Regional Industrial Strategy to achieve competitive cost and reduced build cycle time.
- Maintain RN Fleet numbers over the next decade via urgent and early build of Type 31e ('e' for exportable).

Sir John's report resulted in a largely deafening silence from official sources. It was anticipated that the Chancellor's March 2017 budget would give at least an update on the shipbuilding strategy, but no mention was made. As a result, it appeared that both the report and the strategy were moribund. However, a commitment to publishing the strategy was included in both the Conservative and Labour election manifestos. Considerable scepticism remains about the likelihood of Type 31 frigates being assembled anywhere other than at the BAE Systems shipyards on the Clyde, although there may be some regional fabrication.

The dock landing ship *Mounts Bay*, amphibious transport dock *Bulwark* and helicopter carrier *Ocean* pictured in company. A sustained and expensive series of investments in amphibious capability from the mid-1990s onwards gave the Royal Navy an impressive amphibious flotilla but this is being steadily run down. *(Crown Copyright 2016)*

AMPHIBIOUS SHIPS

The only large permanent task group maintained by the Royal Navy is the Joint Expeditionary Force (Maritime), or JEF(M), previously known as the Response Force Task Group (RFTG). This is commanded at sea by Commander Amphibious Task Group (COMATG). Every year since 2011 the JEF(M) has been deployed to the Mediterranean and often further afield for about six months, the deployment being called Exercise Cougar.

For the Cougar 2015 and 2016 deployments the flagship of the force was the 21,500-ton helicopter or 'Commando' carrier *Ocean*. She cost about £300m to build in 2017 money and can carry an Embarked Military Force (EMF) of Royal Marines supported by aviation and landing craft assets – 690 personnel in total. If *Ocean* joins Cougar 2017 it may be her last operational tasking. In 2014, she completed a major £65m refit which extended her life until 2025. However, in 2015 the MOD decided that it only had the money and personnel to operate her until March 2018. *Ocean* will then pay off without replacement after less than twenty years' service. She is expected to be sold to a foreign navy, possibly Brazil

The QEC will provide a limited replacement for *Ocean. Queen Elizabeth* will be able to accommodate an EMF of about 250 personnel; *Prince of Wales* will be completed with more extensive amphibious capabilities and will be able to accommodate an EMF of up to 900 personnel (presumably to a lower standard of accommodation). Unlike *Ocean*, neither ship will have a vehicle deck or be able to carry landing craft on davits, but they will be able to operate numerous troop-carrying helicopters such as RAF Chinooks instead of Lightning II fighters. It seems very unlikely that the RN will ever risk using aircraft carriers costing over £3bn each as amphibious ships in anything but the most benign threat environment.

In June 2017, the amphibious assault and command ship *Albion* began sea trials after completing an extensive £90m mid-life technical upgrade and refit. She will take over from her sister ship *Bulwark* before the end of the year. The latter will go in to extended readiness before herself entering a refit in, perhaps, 2021.

Pictured during a maiden 'flag-waving' deployment to the UK in the summer of 2016, two F-35B Lightning II Joint Strike Fighters are seen in company with a Royal Navy Hawk advanced training aircraft of 736 Naval Air Squadron. The FAA's helicopter squadrons are nearing the end of a major re-equipment programme but the costly Lightning II is only being ordered in small batches. *(Crown Copyright 2016)*

ROYAL MARINES

Since the 1970s, the primary combat formation of the Royal Marines has been 3 Commando Brigade, consisting of three Commandos (essentially infantry battalions) plus supporting formations. The brigade has become an important adjunct to the British Army, and as such has largely escaped any major cuts since the 1970s.

However, on 11 April 2017, the MOD announced that the 'Royal Marines are to be restructured in line with a growing Royal Navy'. This rather disingenuous statement hid the reallocation of 200 RM posts to the RN's surface and submarine flotillas to ease their manning shortage. The main impact is that 42 Commando will be reduced in size and cease to a fully-capable combat formation, instead it will become a specialised maritime operations unit, providing personnel for ship force protection teams, small-boat teams, counter-piracy operations and other detachments. The two other Commandos – 40 and 45 – will alternate annually as the 'Lead Commando', ready for deployment anywhere in the world with supporting engineers, artillery, and reconnaissance and logistics units – about 1,800 men in total.

FLEET AIR ARM (FAA)

The FAA is nearing the end of a major re-equipment programme. Recent developments include:

- The veteran Westland Lynx HMA 8 helicopter left service in March 2017, along with its Sea Skua anti-surface guided weapon.
- Deliveries of twenty-eight replacement Leonardo Helicopters (formerly AgustaWestland) Wildcat HMA 2 multi-role helicopters completed in October 2016, although the Sea Venom lightweight anti-ship missile that will be a primary weapon is not expected to enter service until late 2020.
- Deliveries of thirty upgraded Leonardo Helicopters Merlin HM 2 anti-submarine helicopters were completed on 11 July 2016, replacing forty-four HM 1s.
- Seven former RAF operated Merlin HC 3 transport helicopters were given an interim navalisation (e.g. folding rotor heads) and re-delivered to the FAA by April 2016
- Between September 2017 and 2020, twenty-five ex-RAF Merlin HC 3/3A transport helicopters will be fully navalised to the Merlin HC 4 standard and re-delivered at a cost of £454m. They will replace a once 42-strong force of Sea King HC4 medium-lift helicopters, the last of which left service in March 2016

On 16 January 2016, Lockheed Martin UK received a £269m (c. U$350m) contract to manufacture ten

A Merlin Mk 3 helicopter of 846 Naval Air Squadron (NAS) seen training at Royal Naval Air Station Yeovilton in the spring of 2016. Formerly operated by the RAF but now transferred to the navy, twenty-five of this transport variant of the type are being upgraded and 'navalised' to the new HC 4 standard. *(Crown Copyright 2016)*

sets of the Crowsnest airborne surveillance and control (ASaC) system. These can be fitted to Merlin HM 2s as required. The system is due to enter service in 2019, allowing the last seven (all ASaC 7s) of the 112 Sea King helicopters that have served with the RN to be retired.

The FAA is an enthusiastic advocate of the UK's purchase of the carrier capable Lockheed-Martin F-35B Lightning II strike fighter, and a third of the personnel of the 17(R) Test and Evaluation Squadron RAF are actually RN. The MOD is only ordering the aircraft in small batches due to its high cost but enough aircraft should be available by April 2023 to stand-up 809 Naval Air Squadron.

In December 2016, the MOD announced the purchase from Boeing of nine P-8A Poseidon Maritime Patrol Aircraft (MPA) for the RAF at a cost of US$3.87bn (c. £3bn). These badly needed aircraft will enter service in 2020, taking over missions that were performed by the Nimrod MR 2 until it was grounded in 2010.

SUPPORT SHIPS

The Royal Fleet Auxiliary (RFA) service is at a low point in terms of both ships (nine) and personnel (1,800). In June 2016, it was announced that the only forward repair ship, *Diligence*, would be disposed of immediately, leaving the RN without a vital support capability. She had only recently – in February 2015 – completed a refit to extend her service life to 2020. Plans for a replacement ship, termed the Operational Maintenance and Repair (OMAR) capability, have also been abandoned.

The tanker force has been run down significantly in recent years. The sole remaining 'Leaf' class support tanker – *Orangeleaf* – started a major refit (arguably a rebuild) in early 2014, but the discovery of serious corrosion problems caused the refit to be cancelled and she was sold for scrap in February

A Wildcat HMA 2 helicopter pictured operating from the flight deck of the amphibious helicopter carrier *Ocean* during NATO Exercise Trident Junction in October 2015. Deliveries of the new helicopter – a replacement for the Lynx – were completed in October 2016. *(Crown Copyright 2015)*

2016. The last of the hard-worked 'Rover' class – *Gold Rover* – left service in February 2017. This leaves just two tankers (*Wave Knight* and *Wave Ruler*), pending the arrival of the new 'Tide' class. Because they are reasonably fast and well-equipped, they are often diverted to duties that, until 2010, would have been performed by a RN ship. For example, in July 2016 *Wave Knight* was assigned to Atlantic Patrol Task (North) and then substituted as a Royal Yacht for Prince Harry's tour of the Caribbean in November 2016.[12]

Four 39,000-ton 'Tide' class tankers were ordered in 2012 from Daewoo Shipbuilding & Marine Engineering (DMSE) in South Korea at a cost of £452m. Delivery of the first – *Tidespring* – was delayed by ten months due to problems with the design and insulation of electrical systems. She finally arrived in UK waters on 31 March 2017. A&P Falmouth will conduct a £15m final fit-out, installing sensitive equipment such as self-defence weapons, ballistic protection and communications systems before she enters service in late 2017. Three sister-ships should follow at about six-month intervals. A proposed additional tanker optimised to supporting the QEC was never funded.

The RFA has three stores ships which were completed in 1978, 1979 and 1994. The newest, *Fort Victoria*, also has tanker capabilities. Her unexpected departure from the Arabian Gulf in March 2016 to meet a higher priority requirement in the Mediterranean marked the suspension of the (Arabian) Gulf Readiness Tanker, although she later returned. For over thirty years RN and allied warships in the region had been able to rely on the presence of an RFA tanker for refuelling; that is no longer certain.

SDSR 2015 approved three new Military Afloat Reach and Sustainability (MARS) Fleet Solid Support Ships (FSS), but no order will be placed until March 2020. The three existing ships will thus remain in service until at least 2023. The Naval Design Partnership – a consortium of the Ministry of Defence (MOD) and UK defence companies – is refining the FSS design. Early concepts incorporated features to enable the ships to support amphibious operations (e.g. a vehicle deck and stern loading ramp) but these may be dropped to save money. The final design is expected to focus on the requirements of a carrier task group.

Delays to the 'Tide' class has fuelled a debate as to whether the MOD should seek the lowest price for building FSS, or pay a premium for construction in the UK. Assembly of the ships at Rosyth from modules built at locations around the UK has been suggested but this would face problems, not least that the critical Goliath crane is for sale. The Cammell Laird yard in Birkenhead has recently re-entered the ship building market and is another possibility.

The RFA operates three 'Bay' class landing ships. One – currently *Lyme Bay* – is used as an afloat forward support base (aka 'mother ship') for British and allied (typically two US Navy *Avenger* class) minehunters in the Arabian Gulf. This role may disappear when HMS *Jufair* commissions in November 2017 but current plans are to retain all three ships in service for at least another ten years.

The final member of the RFA fleet is *Argus*. Commissioned in 1988, she has a dual role as an aviation training ship and primary casualty treatment ship. She is due to start a major refit in early 2018 to extend her life beyond 2020.

The MOD also has access to up to four (originally six) 'Point' class strategic sealift ships. Operated by Foreland Shipping Ltd, they are made available for commercial use when not required by the MOD.

Finally, since 2003 the chartered MV *Maersk Rapier* has been used by the MOD for the bulk transport of fuel.

The four new 'Tide' class tankers are being built by DSME in Korea before outfitting with sensitive equipment by A&P Group's yard in Falmouth. Construction has been delayed but the first, *Tidespring*, arrived in the River Fal early in April 2017. *(Crown Copyright 2017)*

SDSR 2015 confirmed that three FSS fleet solid support ships would be built in the early 2020s to supply task groups built around the new *Queen Elizabeth* class. This image shows an early conceptual image of the type. There is some debate as to whether these should be built overseas or in the UK. *(Naval Design Partnership)*

MINE COUNTERMEASURES

Progress is slowly being made on the Mine-countermeasures and Hydrographic Capability (MHC) project. A joint Maritime Mine Countermeasures (MMCM) programme with France has resulted in Thales receiving a contract to build two prototype systems – including unmanned surface and underwater vehicles – for trials and evaluation by the RN and the French Navy in 2019. The RN will initially operate any resulting production system from its 'Hunt' class vessels. Replacement of the 'Hunt' and the *Sandown* classes is not now expected to start before 2030.

UNMANNED SYSTEMS

Since 2014 the RN has used – with considerable success – the Boeing Insitu ScanEagle UAV from escorts and RFA support ships. However, the MOD could not afford to extend the £30m contract and operations will cease by November 2017. Procurement of a replacement system is high on the RN's wish list, as demonstrated by running the Unmanned Warrior 2016 exercise, at which companies demonstrated their latest unmanned autonomous systems – on, above and below the waves.[13]

FINAL COMMENTS

Despite the quotes at the start of this article, recent developments have not provided assurance that 2017 will be seen by future historians as the start of a new era for the Royal Navy.

For example, since the early 1960s the RN's possession of a formidable force of 'hunter killer' or attack nuclear submarines has been a source of pride, prestige and raw naval power. Numbering fifteen submarines in 1995 and twelve as recently as 2004, these modern capital ships are often counted above aircraft carriers in importance. The apparent inability of the submarine flotilla to deploy even one operational attack submarine for parts of 2017 should worry both senior naval officers and the British government.

A sustained and expensive effort to rebuild amphibious warfare capabilities between 1990 and 2005 resulted in the UK becoming the only nation other than the USA able to mount a brigade-level landing anywhere around the globe. This capability has been run down since 2010 and will be finally lost in 2018 with changes to the Royal Marines and the decommissioning of the RN's only helicopter carrier, *Ocean*.

Just how hard it is to recover a lost capability is demonstrated by carrier aviation. The UK government decided in November 2010 that this capability would not be required for ten years. Whilst recreating a carrier strike capability soon became a government priority – and the completion of *Queen Elizabeth* is undoubtedly a triumph for the RN and British industry – the reality is that the end of 2020 remains the best estimate as to when she will be operational. Unfortunately, it takes time – a lot of time – to rebuild neglected naval capabilities.

The future of the RN is heavily dependent on investment decisions about the Type 26 and Type 31 frigate projects, as well as fixing the engineering problems of the Type 45 destroyers. It is currently not clear that RN can establish a carrier task group whilst still meeting vital commitments such as providing a 'Fleet Ready Escort' in UK waters.

Finally, there is no point in having ships if they can't be effectively manned. Comparing the number of personnel in the RN (22,400 sailors plus Royal Marines) with, for example, the Indian Navy (67,000) or the Chinese PLAN (255,000) is an interesting but largely academic exercise. However, the RN's efforts to increase its head-count by just hundreds are indicative of a crisis.

The Royal Navy is an enthusiastic advocate of unmanned technology, sponsoring the Unmanned Warrior series of exercises and demonstrations in the autumn of 2016 to assess how a wide range of unmanned vehicles might best be used in an operational context. This image shows just some of the systems deployed in the exercise. *(Crown Copyright 2016)*

The last *Invincible* class aircraft carrier *Illustrious* departs for the breakers in Turkey in December 2016. Although the Ministry of Defence and Royal Navy is 'talking up' the service's future prospects, significant challenges lie ahead to remedy the cuts of the past. *(Derek Fox)*

The nuclear-powered attack submarine *Torbay* flying her decommissioning pennant on departure from HM Naval Base Gibraltar on 6 June 2017. The availability of the *Trafalgar* class boats remaining in service has been patchy and it is possible no RN attack submarines have been available for operations at times. This is one of a number of areas of significant concern with respect to Royal Navy capabilities in spite of positive developments elsewhere. *(Moshi Anahory)*

Notes

1. UK naval services include the Royal Navy (surface flotilla and submarine flotilla), Royal Marines, Fleet Air Arm, Royal Naval Reserve and Royal Fleet Auxiliary Service. The online portal to these is www.royalnavy.mod.uk. For convenience, this article sometimes uses the term 'Royal Navy' or RN as being synonymous with all the services.

2. Sir Michael's comments were included in a MOD press release entitled '2017 is the Year of the Navy' published on 1 January 2017. The First Sea Lord's comments were contained in the RN press release 'A message from the First Sea Lord' issued on 25 November 2016.

3. Figures on planned and actual defence spending are complex and opaque, with changing accounting conventions and the need to account for inflation making comparisons difficult over time. The figure for 2015–16 expenditure is from the National Audit Office's *Departmental Overview 2015-16: Ministry of Defence* (London: National Audit Office, 2017) at https://www.nao.org.uk/wp-content/uploads/2016/10/Departmental-Overview-2015-16-Ministry-of-Defence.pdf. Figures for planned spending are from a MOD press release entitled 'Defence budget increases for the first time in six years' issued on 1 April 2016. There has been scepticism around the two percent commitment: the International Institute for Strategic Studies (IISS) assessed in its annual publication *The Military Balance* (London: IISS, 2017) that UK military spending was 1.98 percent of GDP in 2016. The IISS has a website at: www.iiss.org

4. *The Equipment Plan 2016-2026* (London: National Audit Office, 2017). This can be found at www.nao.org.uk/report/the-equipment-plan-2016-2026.

5. *National Security Strategy and Strategic Defence and Security Review 2015* (London: Ministry of Defence, 2015), currently available at www.gov.uk/government/publications/national-security-strategy-and-strategic-defence-and-security-review-2015

6. The US Congressional Research Service regularly publishes a report on *Navy Columbia Class (Ohio Replacement) Ballistic Missile Submarine (SSBN[X]) Program: Background and Issues for Congress – R41129* (Washington DC, CRS). This contains information about cooperation with the UK not otherwise released. The latest edition can be found on the Federation of American Scientists website at: https://fas.org/sgp/crs/weapons/R41129.pdf

7. The House of Commons Defence Select Committee has published a report *Restoring the Fleet: Naval Procurement and the National Shipbuilding Strategy* (London: House of Commons, 15 November 2016), accessible via www.publications.parliament.uk.

8. The first official confirmation of the problems was a letter dated 3 March 2016 from Secretary of State for Defence Michael Fallon to Dr Julian Lewis MP, chair of the House of Commons Defence Committee. In subsequent remarks to the committee, the First Sea Lord acknowledged that the WR-21 was unable to operate effectively in hot temperatures and that, instead of a 'graceful degradation', the engines were 'degrading catastrophically'.

9. *Defence Industrial Strategy* (London: Ministry of Defence, 2005). This can be found at www.gov.uk/government/publications/defence-industrial-strategy-defence-white-paper

10. A £3.7bn order for the first three Type 26s was announced on 2 July 2017, just after this book's official cut-off date.

11. *An Independent Report to inform the UK National Shipbuilding Strategy* by Sir John Parker (London; Ministry of Defence, 29 November 2016), available at: www.gov.uk/government/publications/uk-national-shipbuilding-strategy-an-independent-report

12. There have been regular campaigns for a new Royal Yacht since the last, *Britannia*, left service in 1997. Neither the Government nor RN consider this a priority and will not contribute towards an estimated £100m construction cost and £10m annual running cost.

13. For more on Unmanned Warrior see the editor's 'Unmanned Warrior 16'in *European Security & Defence – 3/2017* (Bonn, Mittler Report Verlag GmbH, 2017), pp.55–6.

81

3.1 SIGNIFICANT SHIPS

Author:
Norman Friedman

ARLEIGH BURKE (DDG-51) CLASS DESTROYERS

A product of the 1980s that remains valuable today

They are so ubiquitous, so fixed a feature of the United States' naval programme, that it seems difficult to remember that the *Arleigh Burke* (DDG-51) class was a product of the 1980s which, due to a remarkably flexible design, remains modern and extremely valuable more than thirty years later. The class is, moreover, likely to remain in production for many more years. Its putative successor, the *Zumwalt* (DDG-1000), proved far too expensive and hardly worthwhile. Why have the *Burke*s done so well? Why, moreover, have they endured when many of their NATO contemporaries did not survive the end of the Cold War?

THE STRATEGIC BACKGROUND

A large part of the reason is that the *Burke*s were and are genuinely multi-function ships. They reflect the US Navy's strategic ideas of the late Cold War, which differed radically from those of most NATO navies of the time (the French navy was the main exception). The United States' naval view of the time prioritised power projection as the way both to overcome the Soviet naval challenge and to affect the war ashore. The key insight was that powerful carrier battle groups which threatened the one naval asset the Soviets truly valued, the strategic submarines in bastions, could attract and destroy Soviet anti-ship forces which otherwise were likely to overwhelm any defensive NATO naval strength.[1] That applied to the Soviet naval bomber force, which could have destroyed the primarily anti-submarine assets the other NATO navies planned to deploy. It also applied to much of the Soviet attack submarine force. Once the Soviet sea-denial forces had been drawn down or, better, annihilated, the carrier battle groups could threaten the flanks of any Soviet advance into Europe. They could also threaten attacks on Soviet held-territory, even in the Baltic. To deal with this threat, the Soviets would have had to hold substantial land forces in reserve. Those forces would not have been available for the offensive into Europe. That was even more true of the powerful Pacific Fleet, whose presence might keep large Soviet land and tactical air forces tied down in the Far East.

All of this made the Soviet land-based naval bombers a primary target, to be destroyed mainly in the air as they flew out to attack the carriers. To free the carriers' fighters to deal with the bombers, the fleet had to have as much anti-missile firepower as possible, otherwise the fighters would have had to chase the bombers' missiles once they had been fired. Given sufficient firepower, they could concentrate on the bombers even after they released their weapons. New destroyers conceived in the 1980s therefore needed as much anti-aircraft firepower as possible. A decade earlier, anti-submarine warfare had been the priority, as reflected in the *Spruance* (DD-963) class.

This new US Maritime Strategy began to crystallise late in the 1970s, as new digital technology finally seemed reliable enough to promise an effective counter to the mass of Soviet missiles, bombers and submarines. For example, given the new

Flight IIA *Arleigh Burke* class destroyer *Winston S. Churchill* (DDG-81) pictured operating in the Atlantic Ocean in 2015. Although initially a programme of the late-Cold War era, the design has been sufficiently flexible to go through several iterations and is set to remain in production for many more years. *(US Navy)*

A series of images of the lead *Arleigh Burke* (DDG-51) class destroyer taken in February 1994, some three years after she had first commissioned. Her design reflected the US Navy's late-Cold War focus on maximising air defence firepower in its surface warships as part of an offensively-orientated Maritime Strategy that involved threatening the sea areas or bastions where the Soviet strategic submarine fleet operated. The advent of the new Aegis combat system, including the SPY-1 multi-function radar, offered an effective counter to the large Soviet maritime bomber force. *(US Navy)*

Phoenix missile on board fighters and the new Aegis missile control system on board surface ships, it seemed that the US fleet could seize maritime air supremacy – destroy the Soviet naval bomber force – rather than simply survive its attacks. The new towed arrays promised much longer detection ranges against Soviet submarines, and the new Mk 48 torpedo could deal with them.

Another factor in the evolving naval strategy was an understanding that, at least early in a war, the Soviets would be far more interested in defending their strategic submarines in bastions covered by their fleet than in what had been imagined as a vast anti-shipping campaign. That was consistent with Soviet doctrine, which regarded the nuclear balance as crucial, and which envisaged a campaign at the outset of a war to destroy enemy nuclear forces using non-nuclear weapons. That the bastions would have priority was a very controversial view within the US Navy. After the Cold War, US naval intelligence officers wrote that it was accepted as mainstream only after very sensitive (undescribed) sources had confirmed Soviet thinking.[2]

At the time, those in the US Navy looking towards an offensive strategy were running against the current. The Carter administration, which entered office in January 1977, distinguished the Cold War from what it saw as neo-colonial wars in the Third World, such as Vietnam. It associated carriers with the latter, which it wanted to avoid. In its view, the only legitimate naval mission in a war against the Soviets would be to ensure the resupply of Europe in the face of Soviet anti-shipping forces, which were identified as submarines. As for Soviet anti-ship bombers, routing ships sufficiently far south would evade them. This was much the view taken by NATO navies. It included plans for the Pacific Fleet to swing most of its strength to the Atlantic, because that would be the only important theatre of operations.

An early graphic of the DDG-51 design. The design was intended to replace life-expired destroyers of the early Cold War era and be more affordable than the complex and expensive *Ticonderoga* (CG-47) class cruisers that had just introduced Aegis to the fleet. An 8,500-ton, gas turbine-powered design was chosen to meet this aim in February 1982 but went through a number of detailed revisions. This early graphic, believed to date from the second half of 1982, shows the class's 5in gun in the aft position. *(US Navy)*

A later, 1984 graphic of the DDG-51 design is much closer to the *Arleigh Burke* as built. High design priorities were overall combat capability and limiting acquisition cost. Interestingly, the use of automation to reduce crew size and therefore through life costs was considered risky; as such the *Arleigh Burke*s are relatively heavily manned compared to many European ships. *(US Navy)*

The navy's reaction to such ideas began with a study called 'Seaplan 2000', which discussed the impact of the new digital technology.[3] Officers commanding the deployed fleets had definite ideas about an alternative offensively-oriented US naval strategy. President Reagan's administration, which began in January 1981, was much more disposed to a forward naval strategy in place of the earlier defensive one. His Secretary of the Navy, John Lehman, found that the operations envisaged by senior fleet leaders all fitted together into such a strategy.

The *Arleigh Burke* class reflected the new thinking. When the design was conceived, no one imagined that within a few years the Cold War would be over and the Soviet threat to NATO shipping would be gone. The new post-Cold War world would require not escort of shipping but rather the ability to project force into a very unstable world. For NATO that shift began even before the Cold War ended, with deployments to the Gulf in 1990 during the run-up to the first Gulf War.[4]

DESIGN DEVELOPMENT

The design of the *Arleigh Burke* was prompted by the coming need to replace the *Mitscher* (DL-2), *Forrest Sherman* (DD-931) and *Farragut* (DDG-37) class destroyers, which would reach replacement age (at thirty years old) during the 1980s. Of these thirty-two ships, the eighteen *Forrest Sherman*s and four *Mitscher*s were conventional destroyers (although four *Forrest Sherman*s and two *Mitscher*s has been converted to missile destroyers). The *Farragut*s were missile air-defence ships. In the 1980s, the impact of the new US Maritime Strategy was to emphasise air defence, because the fleet would face heavy air attacks in places like the Norwegian Sea. It was a considerable leap to replace all of these ageing ships with highly capable and, hence, expensive air defence ships. Another factor in the new design was an increasing understanding that the Soviets felt compelled to track US nuclear-capable ships (as part of their doctrine of tipping the nuclear balance early in a war). However, their capacity for such tracking was limited. During the 1980s the US Navy deliberately deployed nuclear-capable Tomahawk cruise missiles aboard surface warships as a way to force the Soviets to track many more potential targets simultaneously. This factor had a significant impact on the *Arleigh Burke* design.

When the Reagan administration began to consider building replacement destroyers in 1981,

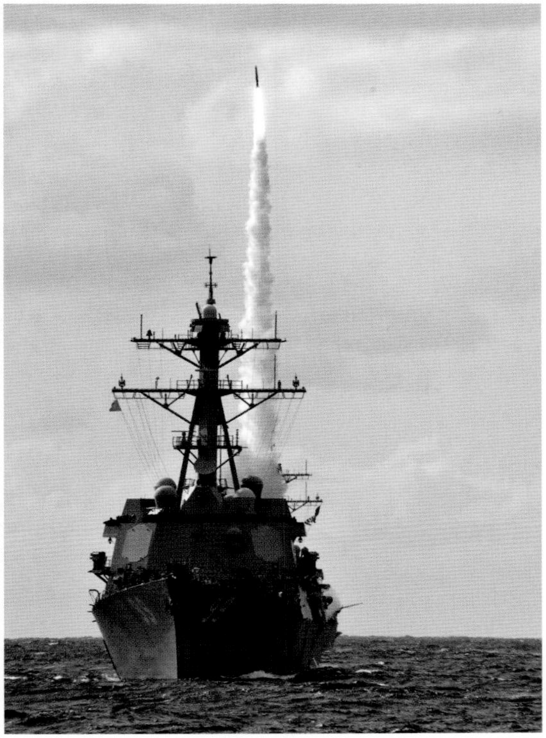

A Standard SM-2 missile is launched from the Flight IIA DDG-51 class destroyer *Farragut* (DDG-99) in March 2012. Missiles fired by Aegis-equipped ships, such as the *Ticonderoga*s and *Burke*s, have programmable autopilots that can be periodically updated by data from the SPY-1 radar to bring them closer to their targets. *(US Navy)*

A detailed view of the forward Mk 41 VLS system and superstructure of the Flight IIA DDG-51 class destroyer *Wayne E. Meyer* (DDG-108) taken during a transit of the South China Sea in April 2017. Points of interest include the SPY-1 radar panel installed below bridge level and the SPG-62 illuminator dish just in front of the mast. The mid-course guidance provided by SPY-1 means the illuminators only need to be used in the final stages of an engagement, significantly increasing the number of targets that can be engaged at any one time. *(US Navy)*

the US Navy was beginning to build *Ticonderoga* (CG-47) class Aegis cruisers.[5] They were widely considered extraordinarily complex and expensive. The most recent air-defence destroyers were the four *Kidd* (DDG-993) class, originally built for Iran on *Spruance* class hulls. The *Kidd*s were armed with the Tartar-D digital version of the Tartar (or Standard) missile system. As they were only equipped with two SPG-51D guidance radars, they could only engage two targets at a time.[6] The Aegis system on board the *Ticonderoga*s had far greater capacity. In Tartar-D a guidance beam had to be pointed at a target throughout a missile's flight. Missiles fired by Aegis ships, however, had programmable autopilots. The SPY-1 radar and the associated combat system could track targets (and the ship's missiles) well enough so as periodically to update missiles in the air to bring them close to their targets. Only at the end of the missile's flight did an illuminator on the ship have to point at the target. Actual target-handling capacity depended on the nature of the targets and the performance of the missiles, but it was clearly far better than that of earlier systems. This type of operation is now becoming widespread (for example in the European PAAMS system) but in the 1980s Aegis was unique. It was the only way to prevent a mass enemy air attack from saturating the defence.

Perhaps the most important difference between the 1981 project and its immediate predecessors was that with the Reagan Administration, determined to grow the navy, in office, immediate construction was likely. Previous studies on new escorts had emphasised new technology and were therefore likely to be risky; these made sense only if the resulting designs were to be built in low numbers, or not at all. Now they were dropped in favour of a ship which could be built in quantity. A senior review panel appointed in 1981 by the Chief of Naval Operations (CNO) called for as many air-defence missiles as possible and continued use of the known powerplant of the *Spruance* and *Ticonderoga* class designs, viz. four LM2500 gas turbines. The panel emphasised dependability. Anything new would be evaluated in terms of the risk it would pose for the new ship. Destroyer operators wanted a towed array and a more powerful version of the SPY-1 radar installed on board the new *Ticonderoga* (which was just being completed at this time).

The Office of the Chief of Naval Operations (OpNav) issued a formal Top Level Requirement laying out design priorities on 30 June 1981. The highest went to increasing combat capability and to limiting acquisition cost. Medium-high (second highest) priorities went to operability, energy conservation, passive survivability (later upgraded to highest priority), and future growth potential.[7] Medium-priority goals were active survivability and limiting displacement. Habitability received a low priority. Here active survivability meant a combination of electronic countermeasures and anti-missile weapons such as Phalanx. It was already known that manning was a major contributor to operating cost, but any automation to limit it represented a risk (the later *Zumwalt* project took an opposite stance). Thus manning reduction was only a medium priority.

By November 1981 Naval Sea Systems Command (NAVSEA) could offer four alternatives.[8] These were an austere 8,000-ton design, two 8,500-ton designs that did not meet speed and range requirements, and a 9,100-tonner which did meet all requirements but which was expected to be too expensive. Unfortunately it was impossible to be certain of what the ship would cost, so it was also impossible to be sure that particular cuts in a sketch design would ensure either that the ship would be affordable or that it would be as effective as desired.[9] It was assumed that displacement determined cost, although in reality that is often not the case. An 8,500-ton gas turbine design presented in February 1982 was finally chosen. At this time OpNav still wanted greater speed and range and reduced beam (beam as well as displacement was considered a measure of likely cost).

DESIGN CHARACTERISTICS

The most visible feature of the NAVSEA design was a 'seakeeping' hull shaped by US observations of Soviet ships. Existing US surface warships were optimised for high speed in smooth water. However, Americans were unhappily aware that foreign ships often behaved much better in rough weather, as in the Atlantic. A fleet intended to fight in the Norwegian Sea needed better seakeeping. The new hull form had a fuller water-plane forward and aft. That increased resistance at high speed, but was considered an advantage in terms of survivability. Worse, as part of the effort to control cost, length was limited (greater

Table 3.1.1A.

ARLEIGH BURKE (DDG-51) PRINCIPAL PARTICULARS[1]

Building Information:	
Laid Down:	6 December 1988
Launched:	16 September 1989
Commissioned:	4 July 1991
Builders:	Bath Iron Works (now part of General Dynamic Inc.), Bath, Maine
Dimensions:	
Displacement:	8,315 tons full load displacement.
Overall Hull Dimensions:	153.9m x 20.1m x 9.3m (sonar). Length at waterline is 142.0m.
Equipment:	
Missiles:	90 x Mk 41 VLS cells for Standard surface-to-air, ASROC anti-submarine and Tomahawk land-attack cruise missiles 2 x quadruple launchers for Harpoon surface-to-surface missiles.
Main Guns:	1 x 127mm Mk 45 Mod 1 gun. 2 x 20mm Phalanx CIWS.
Torpedoes:	2 x triple Mk 32 324mm anti-submarine torpedo tubes.
Countermeasures:	SLQ-32(V) 2 ES. Mk 36 SRBOC decoy launchers. SLQ-25 Nixie torpedo decoy.
Aircraft:	Helicopter landing platform only
Principal Sensors:	SPY-1D multi-function radar (4 panels). 3 x SPG-62 illuminators. Surface search and navigation radar. SQS-53C hull-mounted sonar. SQR-19 towed array.
Combat System:	Aegis. Comprehensive radio and satellite communications including NATO links.
Propulsion Systems:	
Machinery:	COGAG. 4 x LM2500 gas turbines producing 100,000shp. Twin shafts.
Speed:	30+ knots. Endurance is over 4,000 nautical miles at 20 knots.
Other Details:	
Complement:	Normal crew is 303 (23 officers). Accommodation is provided for 356 personnel.
Class:	One Flight I ship (DDG-51) followed by twenty Flight 1A ships (DDG-52 to DDG-71) and seven Flight II ships (DDG-72 to DDG-78).

Notes:

1. Data refers to ship as originally constructed.

length makes it easier to attain higher speed). A stern wedge, inspired by Italian practice, was added to make the hull more efficient. Beam was reduced partly to help reduce resistance, but that raised stability problems. The ship had to be compartmented more tightly, sheer had to be increased to gain reserve buoyancy, and the weight of the superstructure cut. Even so, when the lead ship was named after '31 knot' Arleigh Burke, NAVSEA had to boost gas turbine power to insure that the design would make at least that speed.[10] Fortunately LM2500 output had increased, so the same four gas turbines which drove a *Spruance* with 80,000shp could now supply 100,000shp. Fortunately it turned out that the propellers already used in the *Spruance* and *Ticonderoga* classes could handle the increased power.

Not long after the 8,500-ton design was chosen, the British fought the Falklands War, losing four warships. Survivability was already important in the design; the draft Top Level Requirement was unusual among Cold War US surface combatant specifications in listing shock, air blast, electromagnetic pulse (EMP) and thermal requirements. However, Secretary of the Navy Lehman ordered special studies to derive lessons from the British experience in the Falklands, and a Survivability Planning Group was established. It was claimed (incorrectly) that British ships in the Falklands had suffered because of their aluminium structures; consequently, in October 1982, it was decided that the new destroyer would be all-steel. Probably for the first time in a US warship, measures to improve passive survivability included reduced radar cross-section, in hopes of making missile decoys more effective. The most visible sign of this effort is that all of the ship's exterior surfaces are sloped. The ship was also designed with protection against chemical and biological weapons.

The CNO and Secretary of the Navy decided that

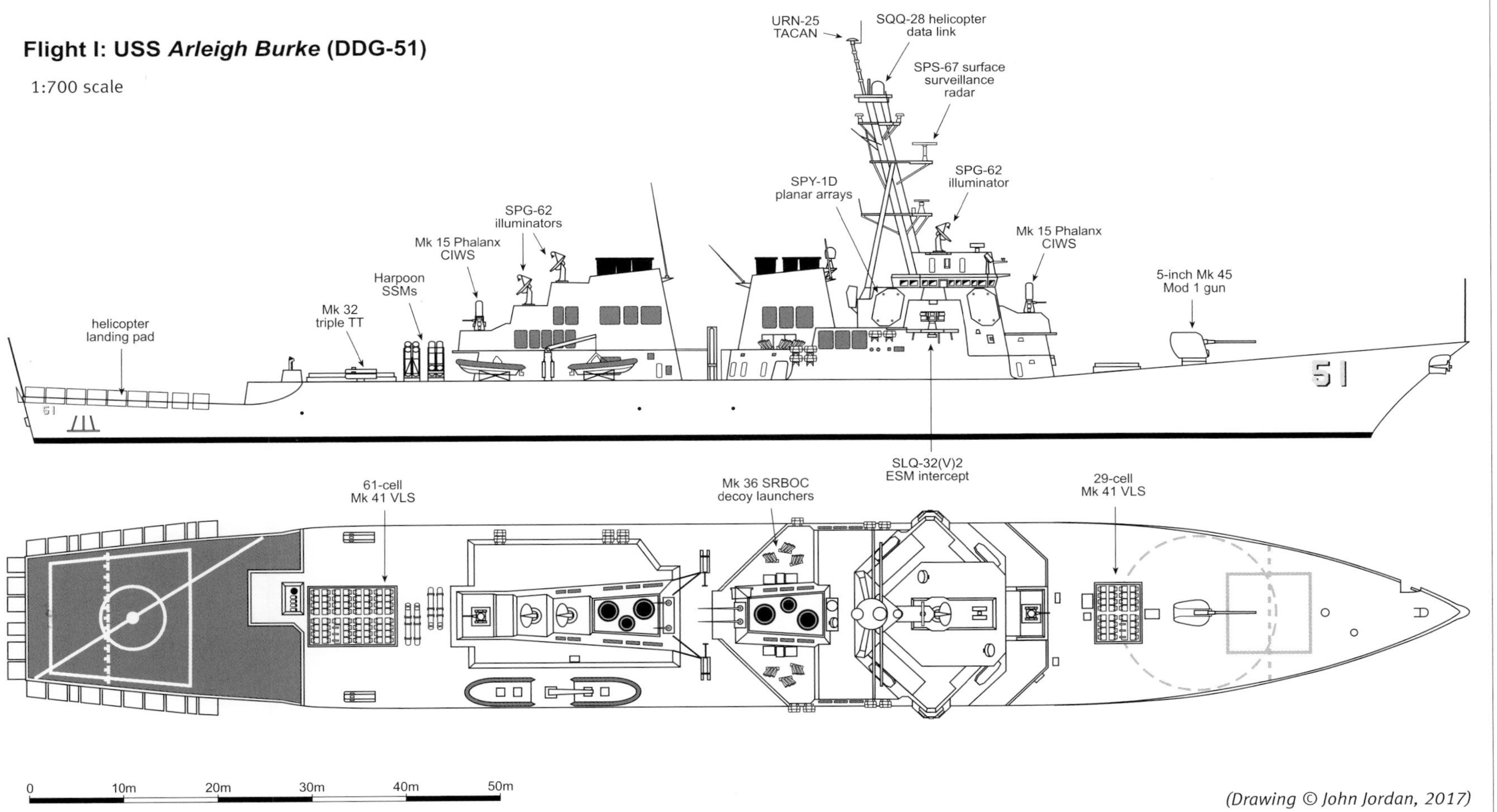

(Drawing © John Jordan, 2017)

the new destroyer should cost no more than seventy-five percent as much as a *Ticonderoga*; otherwise she would seem to be a degraded cruiser rather than a mass-production destroyer. That translated into seventy-five percent of the capabilities of the cruiser, which was soon defined as having three rather than four illuminators. This meant that the ship could provide simultaneous terminal guidance to three rather than four missiles. Internally, the destroyer lacked the second pair of large-screen (command) displays found on board the cruiser. Because the ships nominally to be replaced all lacked helicopter facilities, the new destroyer (unlike the cruiser) also lacked them. The justification was that the destroyer would always operate as part of a battlegroup, other ships in which would have helicopter hangars. However, the ship did have a helicopter pad.

In appearance the new ship was a considerable departure from past practice; it was far more rakish than, say, a *Ticonderoga*. Like the seakeeping hull, that owed a great deal to internal navy studies of the emerging Soviet surface fleet. One of those studies pointed to the impact of appearance on roles such as naval presence. The *Arleigh Burke*s were consciously designed to look more impressive than the boxy warships the United States had been building.

NAVSEA produced what it hoped was a final baseline design late in December 1982. It was judged to be too expensive. In January 1983 NAVSEA was ordered to examine two alternatives, a downgraded *Ticonderoga* and an upgraded *Kidd* with the New Threat Upgrade weapon system.[11] After three weeks NAVSEA deemed the baseline ship best, and in February 1983 Secretary Lehman approved it. Work on the Contract Design, which would be the basis for shipyards' bids, began on 2 May 1983. Meanwhile Litton, which had built the *Spruance* class and *Ticonderoga*, was allowed to propose a modified *Spruance* hull as an alternative. For a time its representatives thought the US Navy would buy that design, but that was never very likely; the existence of an alternative was more a way to impose fiscal discipline on NAVSEA.[12]

When the ship was announced, opponents in Congress pressed for the *Kidd* design as an affordable alternative. The Aegis system had just gone to sea on board *Ticonderoga*, and had experienced teething problems. Opponents claimed that it was fatally flawed. The Secretary of the Navy and the CNO showed otherwise. A *Kidd* would cost at least as much as the new ship, but would be far less capable.

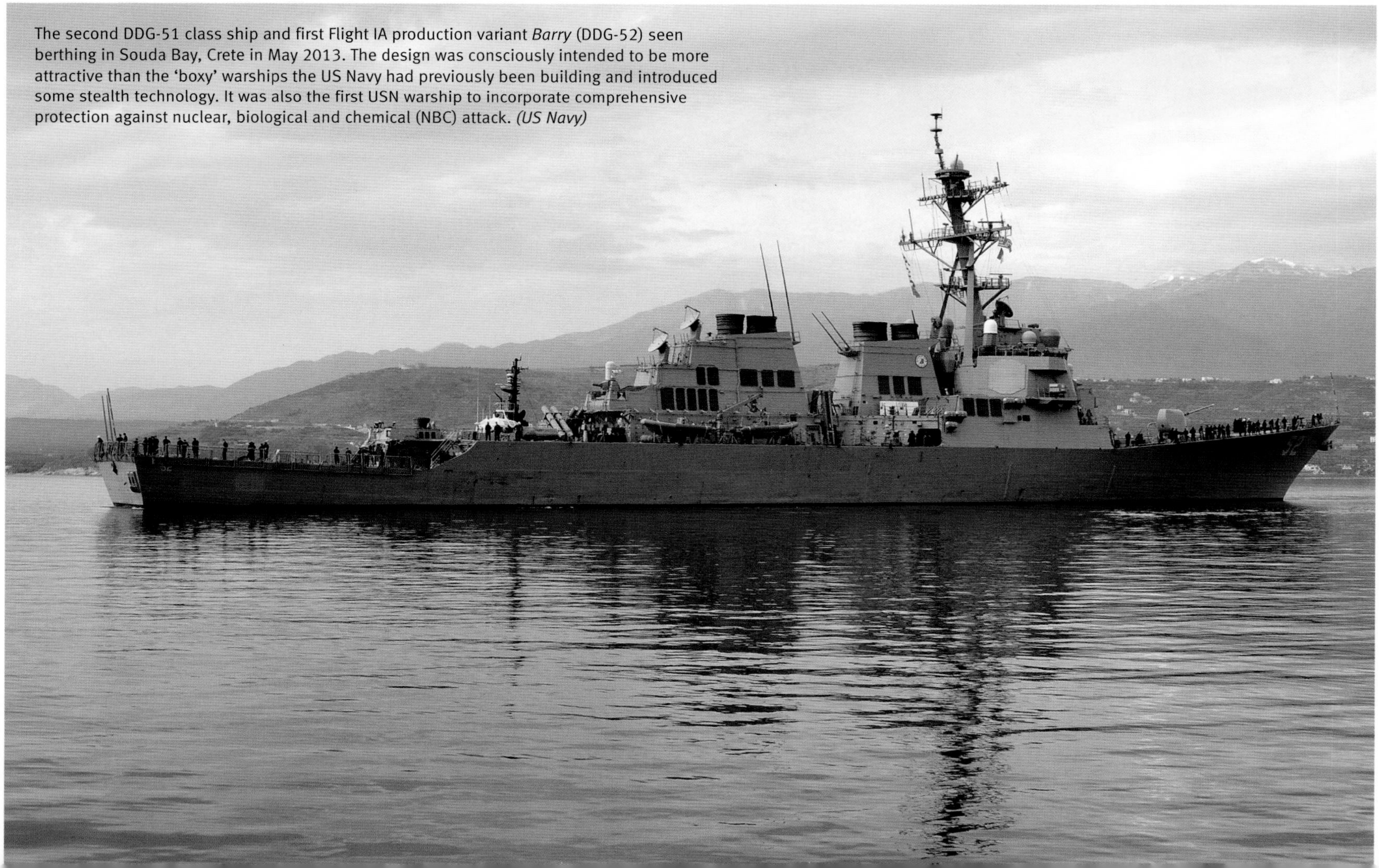

The second DDG-51 class ship and first Flight IA production variant *Barry* (DDG-52) seen berthing in Souda Bay, Crete in May 2013. The design was consciously intended to be more attractive than the 'boxy' warships the US Navy had previously been building and introduced some stealth technology. It was also the first USN warship to incorporate comprehensive protection against nuclear, biological and chemical (NBC) attack. *(US Navy)*

The new ship was described as two-thirds as expensive as a *Ticonderoga* with three-quarters of the capability. As for *Ticonderoga* herself, the teething problems were cured by a software upgrade. In the past such problems would have led to a lengthy stay in a shipyard. Now they were quickly resolved. That happy experience hinted at the extraordinary flexibility inherent in a software-controlled combat system. This flexibility in turn has kept the *Burke* class viable over nearly four decades.

An important difference from both *Ticonderoga* and *Kidd* was the missile launching system. Both of the earlier ships had mechanical launchers with limited flexibility.[13] They were, moreover, a single point of failure in a missile system; if the one mechanical launcher atop a magazine failed, the entire system failed. Launchers were subject to all sorts of sea damage. *Arleigh Burke* incorporated vertical launchers, also now commonly known as a vertical launch system (VLS). They were far more reliable than mechanical launchers, because there was no single critical item which could jam and thus put the ship's missiles out of service. Moreover, because all the missiles are aimed and launched independently, the ship loses no time slewing a launcher into the direction of a target; she is much better adapted to dealing with saturation raids. A very important decision taken during the design was to install launcher cells long enough to take not only anti-aircraft missiles but also Tomahawk cruise missiles. That fit well with ongoing policy to arm many surface combatants with nuclear Tomahawks to complicate Soviet targeting. No one simply looking at a nest of vertical launchers could be sure of what they contained.[14]

The initial design showed a single 5in (127mm) gun aft, but in the course of design it was moved forward, leaving space aft (and wider arcs) for a second Phalanx close-in weapons system (CIWS). To make the ship more compact than a *Ticonderoga*, the two separate radar power units (in two separate parts of the superstructure) of the earlier ship were

A view aft from the forecastle of the Flight IA *Arleigh Burke* class destroyer *Cole* (DDG-67) showing the wash-down system installed as part of the ship's protection against nuclear, biological and chemical (NBC) attack being tested. The 54-calibre Mk 45 Mod 1 5in gun was moved to the forward position in the later stages of the DDG-51 design process, providing space for a second Phalanx CIWS aft. *(US Navy)*

The *Arleigh Burke* class guided-missile destroyer Jason *Dunham* (DDG-109) is moved from the Land Level Transfer Facility at Bath Iron Works (BIW) into the floating dry dock after its christening ceremony in August 2009. BIW was the lead yard for DDG-51 construction, sharing orders with the Ingalls facility at Pascagoula, Mississippi. Early ships of the class built by BIW were constructed on a traditional slipway but *Mason* (DDG-87) was the last to be launched in this way. *(General Dynamics)*

replaced by a single power unit in the forward superstructure, all four faces of the phased-array SPY-1D radar being located there. That was acknowledged as an acceptable loss of combat survivability.

CONSTRUCTION AND VARIANTS

The contract for detail design and construction of the lead ship, *Arleigh Burke* (DDG-51), was awarded in April 1985, and the ship was included in the US Navy's FY85 programme. The naval architectural firm of Gibbs & Cox was lead ship design agent and Bath Iron Works (BIW) in Maine was allocated the construction contract. In accord with standard practice, no ship was included in the next year's programme, to allow time to develop the design. Two ships were included in the FY87 programme (DDG-52 to DDG-53), but none in FY88. Then the tempo shifted to upwards, often to as many as five ships each year: DDG-54 to DDG-58 in FY89, DDG-59 to DDG-63 in FY90, DDG-64 to DDG-67 in FY91, DDG-68 to DDG-72 in FY92, DDG-73 to DDG-76 in FY93 and DDG-77 to DDG-79 in FY94. Construction of these ships was split between BIW and the then Litton Industries' Ingalls Shipbuilding at Pascagoula, Mississippi, a practice which has continued to this day.

FLIGHT IA

Barry (DDG-52) was designated Flight IA, the class subsequently developing in further Flights (or production blocks). She and later ships had upgraded helicopter pads on which helicopters could be serviced and rearmed with torpedoes and sonobuoys.

Table 3.1.1B.

CHUNG-HOON (DDG-93) PRINCIPAL PARTICULARS[1]

Building Information:	
Laid Down:	14 January 2002
Launched:	15 December 2002
Commissioned:	18 September 2004
Builders:	Ingalls Shipbuilding (now part of Huntington Ingalls Industries), Pascagoula, Mississippi.
Dimensions:	
Displacement:	9,200 tons full load displacement.
Overall Hull Dimensions:	155.3m x 20.1m x 9.8m (sonar). Length at waterline is 143.6m.
Equipment:	
Missiles:	96 x Mk 41 VLS cells for Standard and ESSM surface-to-air, ASROC anti-submarine and Tomahawk land-attack cruise missiles
Main Guns:	1 x 127mm Mk 45 Mod 4 gun.[2]
Torpedoes:	2 x triple Mk 32 324mm anti-submarine torpedo tubes.
Countermeasures:	SLQ-32(V) 3 ECM/ESM. Mk 137 Nulka decoy launchers. Mk 36 SRBOC decoy launchers. SLQ-25A Nixie torpedo decoy.
Aircraft:	2 x SH-60 series helicopters (flight deck and hangars).
Principal Sensors:	SPY-1D (V) multi-function radar (4 panels). 3 x SPG-62 illuminators. Surface search and navigation radar. SQS-53 series hull-mounted sonar.
Combat System:	Aegis. Comprehensive radio and satellite communications including NATO links.
Other:	Fitted for the WLD-1 remote minehunting system, now no longer used in destroyers.
Propulsion Systems:	
Machinery:	COGAG. 4 x LM2500 gas turbines producing 100,000+ shp. Twin shafts.
Speed:	30+ knots. Endurance is over 4,000 nautical miles at 20 knots.
Other Details:	
Complement:	Normal crew is c. 320. Accommodation is provided for 369 personnel.
Class:	Twenty-nine Flight IIA ships (DDG-79 to DDG-112) constructed up to 2012. To be followed by three Flight IIA Restart ships (DDG-113 to DDG-115) and subsequent (DDG-116 onwards) Flight IIA Technology Insertion Variants pending transition to the new Flight III.

Notes:

1. Data refers to ship as originally constructed.
2. 1 x 20mm Phalanx CIWS fitted by 2008.

FLIGHT II

On 31 October 1986 the CNO approved a Flight II version with electronic upgrades. These encompassed the new JTIDS (Link 16) data link, the SLQ-32(V)3 jammer in addition to the previous reliance on passive electronic warfare, a receiver for an over-the-horizon link to target the anti-ship version of Tomahawk (TADIXS B), a radio direction finder to help target the missile (Combat DF), and an upgraded Aegis system. Contract design was conducted in FY90-91. Some of the upgrades applied to *The Sullivans* (DDG-68), but the first full Flight II ship was *Mahan* (DDG-72).

FLIGHT IIA

The next version was Flight IIA, which reflected post-Cold War conditions and some of the lessons of the Gulf War. Open-ocean anti-submarine warfare was no longer so important, but destroyers would often operate alone. Given this, omitting a helicopter hangar from the *Arleigh Burke* class design no longer made much sense. Moreover, helicopters were proving invaluable in the new post-Cold War naval roles, such as enforcing the embargo of shipping to Iraq. Armed helicopters also seemed to be the ideal counter to fast attack craft, as British Lynxes had done during the Gulf War. The US Navy therefore developed an armed surface attack capability for its standard SH-60 helicopter. Another lesson from the Gulf War was the experience of Iraqi minefields, which justified installation of the Kingfisher mine evasion attachment to the ship's sonar.

The path to the Flight IIA design was marked by

Flight IIA: USS *Chung-Hoon* (DDG-93)

1:700 scale

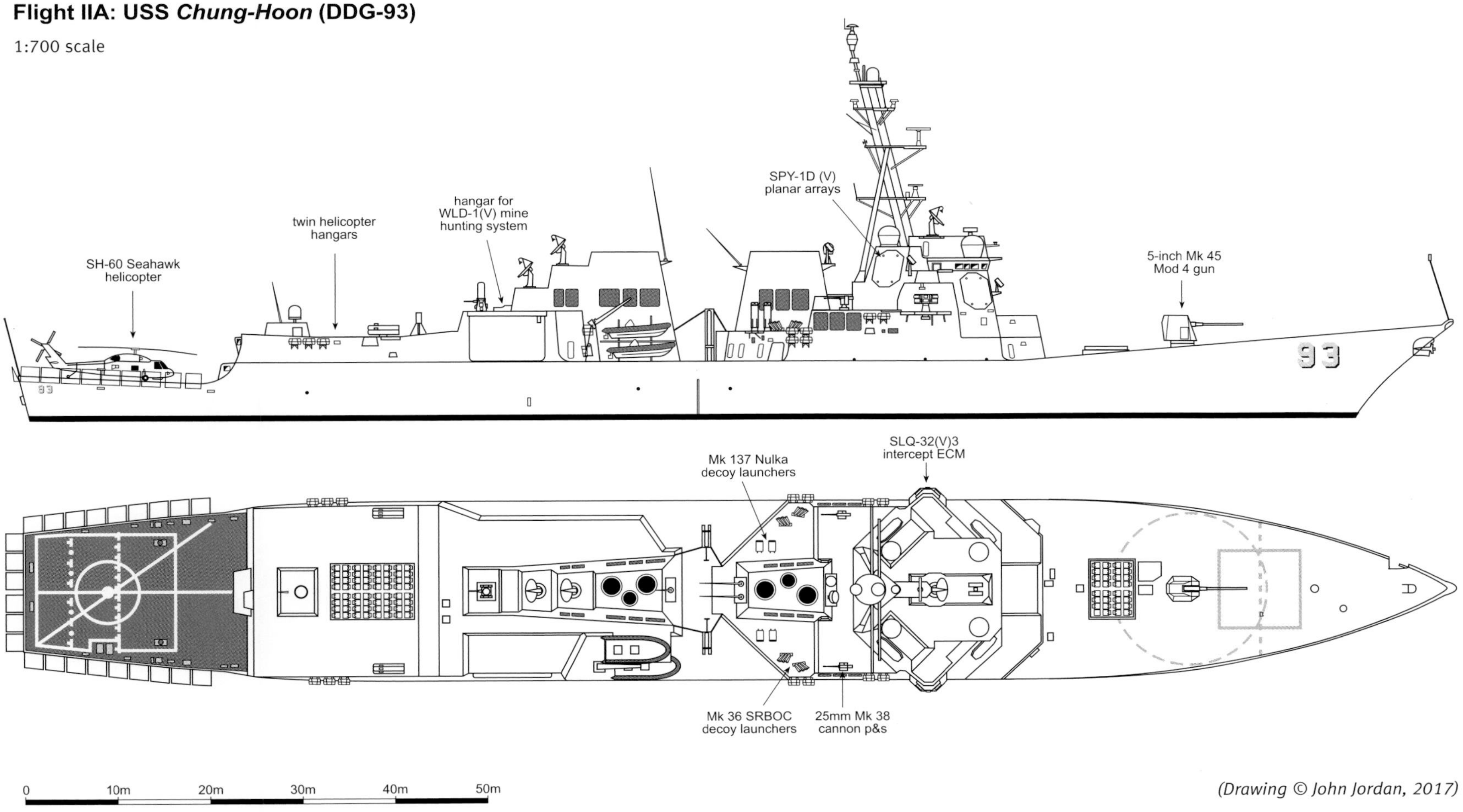

(Drawing © John Jordan, 2017)

a Destroyer Variant Study ordered by the CNO in mid-1991.[15] In theory the aim was to seek a more affordable destroyer but the underlying objective was more to incorporate these new requirements in the current *Arleigh Burke* design already in production. In December 1991 the study concluded by recommending the new Flight IIA version, the most visible change being a two-helicopter hangar. To provide space, the two lightweight torpedo launchers mounted amidships in earlier ships were moved to external positions aft. The helicopter deck was extended aft by lengthening the ship 5ft by its stem. A 10ft extension would have improved hydrodynamic performance, but it would also have worsened damaged stability. A stern flap (similar to those fitted to RN ships) replaced the earlier stern wedge.

The two aft SPY-1D arrays are raised 8ft (in earlier ships they are at the same level as the two forward arrays) in the Flight IIA ships. This reduces reflections off the raised deck of the helicopter hangar. As partial weight compensation, the towed array and the Harpoon launchers were eliminated, although they can be restored if necessary (space for their canisters is reserved between the two uptakes). A Kollmorgen Mk 46 electro-optical director was added atop the bridge. Earlier ships had underway replenishment cranes in their nests of vertical launch cells, but transfer at sea was never very practical, and it was least practical for heavy missiles such as Tomahawk. In a post-Cold War world, it was clear that Tomahawk was the most important weapon US Navy surface combatants had on board. In Flight IIA the cranes were eliminated so that more missiles could be carried, a total of ninety-six. There was also a survivability upgrade: five blast-hardened bulkheads were added amidships, four of them around the machinery spaces, fore and aft of each engine room. The LM2500 gas turbines were further upgraded. It has also been reported that Flight IIA ships have a redesigned and more survivable electrical system.

The first Flight IIA ship was *Oscar Austin* (DDG-79, FY94). By FY2005, thirty-four such ships had been ordered, supplementing twenty-eight of the original type: FY95: DDG-80 to DDG-82,

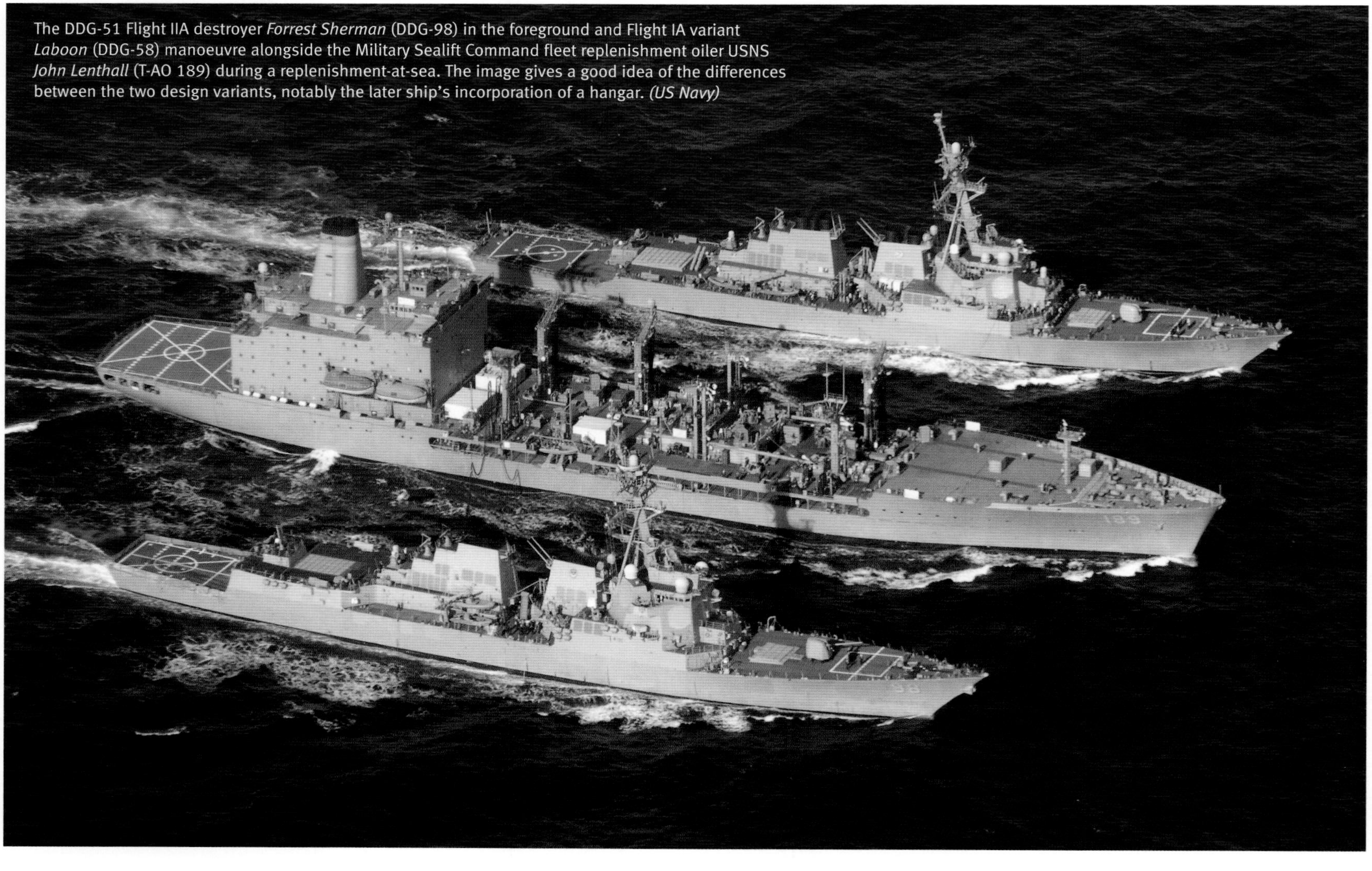

The DDG-51 Flight IIA destroyer *Forrest Sherman* (DDG-98) in the foreground and Flight IA variant *Laboon* (DDG-58) manoeuvre alongside the Military Sealift Command fleet replenishment oiler USNS *John Lenthall* (T-AO 189) during a replenishment-at-sea. The image gives a good idea of the differences between the two design variants, notably the later ship's incorporation of a hangar. *(US Navy)*

FY96: DDG-83 to DDG-84, FY97: DDG-85 to DDG-88, FY98: DDG-89 to DDG-92, FY99: DDG-93 to DDG-95, FY00: DDG-96 to DDG-98, FY01: DDG-99 to DDG-101, FY02: DDG-102 to DDG-104, FY03: DDG-105 to DDG-106, FY04: DDG-107to DDG-109, FY05: DDG-110 to DDG-112.[16] By this time, it was assumed that there would be no further development of the *Arleigh Burkes* as, in 1992, work began on a new-generation destroyer which ultimately became the *Zumwalt* class. Indeed, no orders were placed for further *Arleigh Burkes* under the FY06-09 programmes, pending orders for the *Zumwalts*. However, when the *Zumwalt* class proved disappointing, the *Arleigh Burkes* were placed back into production. The first of these, *John Finn* (DDG-113) was ordered under the FY10 programme for construction at the Ingalls yard; she was delivered in December 2016 and will commission in July 2017. Along with DDG-114 to DDG-115 ordered in FY11, she is designated the Restart version, repeats of earlier ships. From DDG-116 onwards ships are designated as the Flight IIA Technology Insertion version. Authorisations were: FY12: DDG-116, FY13: DDG-117 to DDG-119, FY14: DDG-120. FY15: DDG-121 to DDG-122; FY16: DDG-123 to DDG-124. The US Navy intends to transition to a revised Flight III variant with a more advanced Air and Missile Defence Radar (AMDR). Although it was initially thought this change would first take effect from a third FY16 ship allocated to BIW, on 27 June 2017 it was announced that HII would build the lead Flight III. This will be *Jack H. Lucas* (DDG-125), one of two ships authorised in FY17.[17]

DESIGN CHANGES AND MODERNISATION

In addition to the various Flights, various changes have been introduced to the *Arleigh Burke* design without special designation. Some of the significant changes are listed below:

CIWS

The early Flight IIA ships were completed without the Phalanx CIWS system found in earlier Flights. This reflected a decision to devote six of the vertical launch cells to four 'quad-packed' vertically-launched ESSM (Evolved Sea Sparrow) short-range defensive missiles. ESSM is controlled by the same Aegis system which controls the usual SM-2 area defence missiles. However, delays to the ESSM

The Flight IIA design process included consideration of an *Arleigh Burke* variant equipped with 128 VLS cells. Whilst this option was not adopted, South Korea has done something very similar with its KDX-III design, which owes much to the *Arleigh Burkes*. This picture shows the ROKN *Seoae Ryu Seong-Ryong* (DDG-993), right, and the US Navy's Flight IIA *Preble* (DDG-88) moving into formation during an exercise in October 2013. *(US Navy)*

The DDG-51 Flight IIA destroyer *Halsey* (DDG-97) participates in a simulated strait transit during a group sail training exercise with the *Theodore Roosevelt* Carrier Strike Group in in May 2017. Authorised under the FY2000 appropriations, *Halsey* was launched by Ingalls in January 2004 and commissioned in July 2005 *(US Navy)*

programme meant that the first six ships (through to DDG-84) were subsequently retrofitted with the Phalanx system. In 2008, the navy announced that by 2013 all would have at least one Phalanx; this is the aft Phalanx on later ships.

In 2015 it was revealed four *Arleigh Burke* class destroyers – DDG-64, DDG-71, DDG-75 and DDG-78 – forward deployed to Rota in Spain for European ballistic missile defence duties would be fitted with the SeaRAM anti-ship missile defence system (an eleven-round RAM launcher combined with a Phalanx mount and its autonomous sensors). A single SeaRAM launcher is currently fitted in place of their aft Phalanx system. The installation reflects vulnerabilities to the ships self-defence capabilities when operating in the specialised missile-defence role.[18]

MAIN GUN

From DDG-81 on, ships have the long-barrel Mk 45 Mod 4 5in/62 gun in place of the earlier 54-calibre Mod 1. This gun was to have fired a rocket-boosted guided shell (ERGM) to a maximum range of 40 nautical miles, but that munition was cancelled in 2008.

PROPULSION

In 2016 the navy began refitting the thirty-four original Flight IIA ships with hybrid electric drive to reduce fuel cost. It employs a motor attached to the main reduction gear to drive ships at 13 knots and below.

STRUCTURE

Mustin (DDG-89) had her visible funnel caps eliminated and this change has been implanted in all subsequent ships. DDG-91 through to DDG-96 were modified to operate the WLD-1 mine-countermeasures drone in a starboard hangar, but this system was landed in 2010.

SYSTEMS

The most important changes have been the least visible, involving ongoing upgrades to the Aegis system's capability in line with various Baselines. For example, Aegis Baseline 6, installed in the early Flight IIA *Arleigh Burkes*, introduced Cooperative Engagement Capability (CEC). Baseline 7, first fitted to *Pinckney* (DDG-91), was hosted entirely on commercial-off-the-shelf (COTS) hardware and also saw integration of the improved SPY-1D(V) radar and SQS-53D sonar. However, perhaps most signif-

Table 3.1.2: ARLEIGH BURKE (DDG-51) CLASS DESTROYERS: CLASS LIST

FLIGHTS 1,1A AND II (28 SHIPS)

Number	Name	FY Ordered	Builder	Laid Down	Launched	Commissioned	Notes
DDG-51	Arleigh Burke	1985	BIW	6 Dec 1988	16 Sep 1989	4 Jul 1991	Flight I
DDG-52	Barry	1987	Ingalls	26 Feb 1990	10 May 1991	12 Dec 1992	First Flight IA ship
DDG-53	John Paul Jones	1987	BIW	8 Aug 1990	26 Oct 1991	18 Dec 1993	
DDG-54	Curtis Wilbur	1989	BIW	12 Mar 1991	16 May 1992	19 Mar 1994	
DDG-55	Stout	1989	Ingalls	8 Aug 1991	16 Oct 1992	13 Aug 1994	
DDG-56	John S. McCain	1989	BIW	3 Sep 1991	26 Sep 1992	2 Jul 1994	
DDG-57	Mitscher	1989	Ingalls	12 Feb 1992	7 May 1993	10 Dec 1994	
DDG-58	Laboon	1989	BIW	23 Mar 1992	20 Feb 1993	18 Mar 1995	
DDG-59	Russell	1990	Ingalls	24 Jul 1992	20 Oct 1993	20 May 1995	
DDG-60	Paul Hamilton	1990	BIW	24 Aug 1992	24 Jul 1993	27 May 1995	
DDG-61	Ramage	1990	Ingalls	4 Jan 1993	11 Feb 1994	22 Jul 1995	
DDG-62	Fitzgerald	1990	BIW	9 Feb 1993	29 Jan 1994	14 Oct 1995	
DDG-63	Stethem	1990	Ingalls	11 May 1993	17 Jun 1994	21 Oct 1995	
DDG-64	Carney	1991	BIW	3 Aug 1993	23 Jul 1994	13 Apr 1996	SeaRAM modification 2016
DDG-65	Benfold	1991	Ingalls	27 Sep 1993	9 Nov 1994	30 Mar 1996	
DDG-66	Gonzalez	1991	BIW	3 Feb 1994	18 Feb 1995	12 Oct 1996	
DDG-67	Cole	1991	Ingalls	28 Feb 1994	10 Feb 1995	8 Jun 1996	
DDG-68	The Sullivans	1992	BIW	27 Jul 1994	12 Aug 1995	19 Apr 1997	Some Flight II upgrades applied
DDG-69	Milius	1992	Ingalls	8 Aug 1994	1 Aug 1995	23 Nov 1996	
DDG-70	Hopper	1992	BIW	23 Feb 1995	6 Jan 1996	6 Sep 1997	
DDG-71	Ross	1992	Ingalls	10 Apr 1995	22 Mar 1996	28 Jun 1997	SeaRAM modification 2016
DDG-72	Mahan	1992	BIW	17 Aug 1995	29 Jun 1996	14 Feb 1998	First Flight II ship
DDG-73	Decatur	1993	BIW	11 Jan 1996	10 Nov 1996	29 Aug 1998	
DDG-74	McFaul	1993	Ingalls	26 Jan 1996	18 Jan 1997	25 Apr 1998	
DDG-75	Donald Cook	1993	BIW	9 Jul 1996	3 May 1997	4 Dec 1998	SeaRAM modification 2016
DDG-76	Higgins	1993	BIW	14 Nov 1996	4 Oct 1997	24 Apr 1999	
DDG-77	O'Kane	1994	BIW	8 May 1997	28 Mar 1998	23 Oct 1999	
DDG-78	Porter	1994	Ingalls	2 Dec 1996	12 Nov 1997	20 Mar 1999	SeaRAM modification 2016

icant of all is the adaptation of Aegis and supporting radar and other equipment to support a ballistic-missile defence role.

The same long vertical launcher cell which can accommodate Tomahawk can also accommodate a Standard Missile with a large finless Mk 104 booster. That can provide greater range, but perhaps more importantly the same envelope can accommodate the SM-3 anti-ballistic missile weapon (which requires the booster). The software-controlled combat system can be modified to control the weapon, and the SPY-1 radar can also be modified. Originally anti-missile operation required a different radar power tube (high power at longer pulse intervals), so ships were capable of either the anti-missile or the air-defence mission. Changeover was a shipyard operation. Later it became possible to switch the radar back and forth between the two modes, although only the most recent Baseline allows both to be undertaken simultaneously.

Restart ships (DDG 113–115) introduce an Open Architecture Computing Environment (OACE) which makes it possible to continuously upgrade a ship's computers using evolving commercial chips. The previous special military computers, such as UYK-43, have long been overtaken by commercial equipment, but the problem in substituting commercial computers has always been to ensure that the same capabilities are maintained. That was particularly difficult in the tightly-coupled Aegis system. OACE forms part of Aegis Baseline 9, which is also being retrofitted in many earlier ships. It includes a multi-mission processor that allows

Table 3.1.2 (contd): ARLEIGH BURKE (DDG-51) CLASS DESTROYERS: CLASS LIST

FLIGHT IIA (ORIGINAL CONSTRUCTION – 34 SHIPS)

Number	Name	FY Ordered	Builder	Laid Down	Launched	Commissioned	Notes
DDG-79	Oscar Austin	1994	BIW	9 Oct 1997	7 Nov 1998	19 Aug 2000	First Flight IIA Ship, 2 x CIWS
DDG-80	Roosevelt	1995	Ingalls	15 Dec 1997	10 Jan 1999	14 Oct 2000	2 x CIWS
DDG-81	Winston S. Churchill	1995	BIW	7 May 1998	17 Apr 1999	10 Mar 2001	First ship with Mk 45 Mod 4 5in/62 gun, 2 x CIWS
DDG-82	Lassen	1995	Ingalls	24 Aug 1998	16 Oct 1999	21 Apr 2001	2 x CIWS
DDG-83	Howard	1996	BIW	9 Dec 1998	20 Nov 1999	20 Oct 2001	2 x CIWS
DDG-84	Bulkeley	1996	Ingalls	10 May 1999	21 Jun 2000	8 Dec 2001	2 x CIWS
DDG-85	McCampbell	1997	BIW	15 Jul 1999	2 Jul 2000	17 Aug 2002	1 x CIWS (aft) onwards
DDG-86	Shoup	1997	Ingalls	13 Dec 1999	22 Nov 2000	22 Jun 2002	
DDG-87	Mason	1997	BIW	20 Jan 2000	23 Jun 2001	12 Apr 2003	
DDG-88	Preble	1997	Ingalls	22 Jun 2000	1 Jun 2001	9 Nov 2002	
DDG-89	Mustin	1998	Ingalls	15 Jan 2001	12 Dec 2001	26 Jul 2003	First ship with funnel caps eliminated
DDG-90	Chafee	1998	BIW	12 Apr 2001	2 Nov 2002	18 Oct 2003	
DDG-91	Pinckney	1998	Ingalls	16 Jul 2001	26 Jun 2002	29 May 2004	Modified to house WLD-1 MCM drone
DDG-92	Momsen	1998	BIW	16 Nov 2001	19 Jul 2003	28 Aug 2004	Modified to house WLD-1 MCM drone
DDG-93	Chung-Hoon	1999	Ingalls	14 Jan 2002	15 Dec 2002	18 Sep 2004	Modified to house WLD-1 MCM drone
DDG-94	Nitze	1999	BIW	17 Sep 2002	3 Apr 2004	5 Mar 2005	Modified to house WLD-1 MCM drone
DDG-95	James E. Williams	1999	Ingalls	15 Jul 2002	25 Jun 2003	11 Dec 2004	Modified to house WLD-1 MCM drone
DDG-96	Bainbridge	2000	BIW	7 May 2003	30 Oct 2004	12 Nov 2005	Modified to house WLD-1 MCM drone
DDG-97	Halsey	2000	Ingalls	5 Feb 2003	9 Jan 2004	30 Jul 2005	
DDG-98	Forrest Sherman	2000	Ingalls	12 Aug 2003	30 Jun 2004	28 Jan 2006	
DDG-99	Farragut	2001	BIW	7 Jan 2004	9 Jul 2005	10 Jun 2006	
DDG-100	Kidd	2001	Ingalls	1 Mar 2004	15 Dec 2004	9 Jun 2007	
DDG-101	Gridley	2001	BIW	30 Jul 2004	28 Dec 2005	10 Feb 2007	
DDG-102	Sampson	2002	BIW	14 Mar 2005	17 Sep 2006	3 Nov 2007	
DDG-103	Truxtun	2002	Ingalls	11 Apr 2005	17 Apr 2007	25 Apr 2009	
DDG-104	Sterett	2002	BIW	17 Nov 2005	20 May 2007	9 Aug 2008	
DDG-105	Dewey	2003	Ingalls	4 Oct 2006	18 Jan 2008	6 Mar 2010	
DDG-106	Stockdale	2003	BIW	10 Aug 2006	24 Feb 2008	18 Apr 2009	
DDG-107	Gravely	2004	Ingalls	26 Nov 2007	30 Mar 2009	20 Nov 2010	
DDG-108	Wayne E. Meyer	2004	BIW	17 May 2007	19 Oct 2008	10 Oct 2009	
DDG-109	Jason Dunham	2004	BIW	11 Apr 2008	2 Aug 2009	13 Nov 2010	
DDG-110	William P. Lawrence	2005	Ingalls	16 Sep 2008	15 Dec 2009	19 May 2011	
DDG-111	Spruance	2005	BIW	14 May 2009	6 Jun 2010	1 Sep 2011	
DDG-112	Michael Murphy	2005	BIW	12 Jun 2010	8 May 2011	5 Sep 2012	Last of original Flight IIA

This image of the *Arleigh Burke* class destroyer *Wayne E. Meyer* (DDG-108), pictured in April 2017, provides a good indication of the current configuration of many of the Flight IIA variant. The Flight IIA ships were originally completed without a Phalanx CIWS but all now have at least one installed, normally in the aft position, and 25mm cannon – increasingly the latest Mk 38 Mod 2 version – are fitted on either side between the bridge structure and forward funnel. From DDG-81 onwards, the 5in gun is the longer-barrelled Mk 45 Mod 4 variant. *(US Navy)*

Six of the early Flight IIA series of DDG-51 class destroyers were fitted with a hangar or 'kennel' for the autonomous WLD-1 minehunting system, involving a modification to the starboard superstructure aft the second funnel. Although WLD-1 is no longer shipped, the hangar remains, as evidenced in this March 2016 view of *Chung-Hoon* (DDG-93). *(US Navy)*

The four Flight IA and Flight II variant *Arleigh Burke*s forward-deployed to Europe as part of a missile defence shield were all equipped with the SeaRAM missile system in place of their aft Phalanx CIWS in 2016. The images show the location of SeaRAM in *Carney* (DDG-64) in May 2017 (top left and below)and a test firing of the system from *Porter* (DDG-78) on 4 March 2016 (top right). *(Leo van Ginderen, US Navy)*

Although not immediately apparent from external observation, the *Arleigh Burke* class have benefitted from ongoing hardware and software improvements to the Aegis combat system and associated weapons that have significantly improved the breadth of their capabilities. A particularly significant development has been the adaptation of systems to allow the ships to be used to counter the increasing threat from ballistic missiles. This picture shows *Fitzgerald* (DDG-62) firing a Standard SM-3 missile during missile defence tests in October 2012. *(Missile Defense Agency)*

Spruance (DDG-111) pictured operating with *Momsen* (DDG-92) and a MH-60R Seahawk helicopter in the Indian Ocean in July 2016. The *Arleigh Burke* class have proved tough and flexible ships that remain valuable assets today in spite of their Cold War origins. *(US Navy)*

ships to target ballistic missile threats and other threats simultaneously.

FLIGHT I/II UPGRADE PROGRAMME

A DDG Upgrade programme for Flight I/II ships was announced in 2010. This included modifications of vertical launching cells to fire quad-packed ESSM, a new electro-optical surveillance system, an upgraded sonar system with the new SQR-20 multi-

Notes

1. The story is inevitably considerably more complex. For example, the *Seawolf* (SSN-21) class submarine was conceived as an ideal means of operating inside the bastions, forcing the Soviets to concentrate on dealing with the threat it represented rather than operating freely outside the bastions against NATO shipping and against the carriers. In the case of the carriers, a key element was to demonstrate very long strike ranges (using the A-6 Intruder) to force the Soviets to fight at maximum range, without fighter support. A forthcoming book by Dr John Lehman Jr, who as Secretary of the Navy was largely responsible for articulating the navy's Maritime Strategy and for making sure that it could be executed, will describe the strategy and its execution in much greater detail. The US Navy promoted the Maritime Strategy as a national strategy, in part an alternative to the army's concentration on the Central Front in Germany. At the end of the Cold War it was still contentious, and it had not formally been adopted by either the US Government or by NATO. However, it certainly guided US naval development during the 1980s.

2. The bastion idea had been raised by studies conducted by the Center for Naval Analyses in the 1970s based on published Soviet writings, but apparently many in the fleet found the idea unconvincing. The reference to the role of US naval intelligence was to a study released during a US-Russian conference on the Cold War at Sea held in 1996.

3. Dr Lehman participated in the 'Seaplan 2000' study; he later said that it had had an important impact on his thinking. It was also significant that as a naval reservist he was a bombardier-navigator on board an A-6 and hence thought about what his kind of aircraft could contribute.

4. The earlier escort of shipping operations during the Iran-Iraq War were more like the convoy escort operations NATO envisaged during the Cold War.

5. Interest in a new destroyer can be traced much further back. For details, see the author's revised edition of his *U.S. Destroyers: An Illustrated Design History* (Annapolis: Naval Institute Press, 2004). It was widely held at the time that the earlier design studies had embodied far too much

function towed array, a SPY-1 radar transmitter upgrade, a multi-mission signal processor, CEC and upgraded command/control. The enhancements to the Aegis system were in line with Aegis Baseline 9. Contracts for the first upgrades were announced in July 2010 but later announcements make it likely that only seven ships (DDG-51, DDG-52, DDG-53, DDG-57, DDG-61, DDG-65 and DDG-69) will receive the full mid-life upgrade. The others will be limited to ship system overhauls and an anti-submarine upgrade, presumably including the SQR-20 towed array.[19] As of 2014, this partial upgrade was estimated to cost c. US$170m compared with US$270m for the full scheme.

CONCLUSION

The *Arleigh Burkes* are ships which have enjoyed an extensive production run that – with the latest Flight III modifications – is likely to extend well into the future. Moreover, they have been proven as an extremely successful design in service. Notably, their high survivability was tested when *Cole* (DDG-67) was attacked in Aden on 12 October 2000. The charge of roughly 2,000lbs was considerably larger than any which might have been expected in normal combat, yet she was not sunk. She returned to service on 19 April 2002.

---

risky technology; they were seen as vehicles to justify technology developments of intense interest at the time.

6. In principle the SPQ-9A radar used by the *Kidd* class's gun system could have been used to engage a third target.

7. The emphasis on energy conservation is probably best understood in political terms. For some years Anthony Battista, the powerful staff member of the House Armed Services Committee, had pressed the US Navy to improve gas turbine reliability by adopting an energy recovery system called RACER (Rankine Cycle Energy Recovery). It used some turbine exhaust heat to generate steam; in the early 1980s it was claimed that installing one RACER would add thirty percent to a destroyer's endurance. Proponents claimed that RACER was the most effective means of fuel conservation which would be available before 2000. To opponents it was an unnecessary complication which would drastically reduce reliability. The navy was already eliminating steam plants, which were far more efficient than gas turbines but also more complex and less reliable. When the new destroyer was being designed, RACER was still in the concept stage. Secretary Lehman opted to make the ship compatible with RACER but without having to wait for it for the initial ships. That avoided a fight. As it happened, Mr Battista retired before the ships were built, and when that happened the provision for RACER was dropped.

8. The Naval Sea Systems Command (NAVSEA) has overall responsibility for the design, build, delivery and maintenance of both ships and systems for the US Navy.

9. The contemporary project for a fast replenishment ship (AOE) was a good illustration. The initial design, which incorporated full military features, was judged too expensive, so many features were eliminated. When shipyards bid on the project, it turned out that all of their bids were well below the estimated price. The military features could have been incorporated without breaking the budget. By the time that was apparent, it was too late to do so. Note that the *Burke*s have turned out heavier than planned: currently approaching 9,000 tons full load for the Flight IA/Fight II series.

10. Arleigh Albert Burke (1901–96) was a distinguished US Navy admiral who ultimately reached the position of the Department of the Navy's most senior officer as CNO between 1955 and 1961. The '31-knot' title was a reference to his aggressive handling of Destroyer Squadron 23 during the Second World War.

11. New Threat Upgrade was often called a 'poor man's Aegis.' It used the midcourse update feature of Aegis, but it lacked the precision SPY-1 radar, so its target trackers had to guide the missile for a much greater fraction of its trajectory. New Threat Upgrade certainly did increase the fleet's ability to overcome saturation attacks, but it did not match the quick-reaction capacity of Aegis, whose tactical picture was merged with its fire-control system. When the Cold War ended, saturation attacks became very unlikely, but pop-up shots by concealed shore missiles and from missile boats became a very important Third World threat. Aegis or something very much like it was the only way to deal with that problem.

12. It was important politically that the new destroyer be very easily distinguished from what was considered a very expensive *Ticonderoga* class cruiser. Building what looked like a slightly degraded *Ticonderoga* would have been difficult, as Congress would have seen it as a dodge rather than a sincere attempt to buy a less-expensive ship in large numbers. The fall-back to a modified *Ticonderoga* was consistent with Secretary Lehman's practice of controlling weapon system costs by always having two alternatives for any function. The Secretary was very concerned that the money which became available for the Reagan administration build-up would disappear into inflated defence costs, and he applied vigorous cost control. That included requiring that all changes to the approved design be subject to high-level examination, as a way of limiting the usual overruns due to changes during construction. Such overruns had plagued the previous administration, particularly in the *Knox* (FF-1052) class frigates (among Secretary Lehman's first tasks was to clear up the overhanging cost claims).

13. The Mk 26 twin-arm launcher could handle the SM-1 or -2 anti-aircraft missile, Harpoon and ASROC. It could not launch anything as large as a Tomahawk. A vertical launcher in the same space had greater capacity and could launch much larger missiles, with a maximum diameter of about 21in (SM-1 and -2 diameter was 13.5in). Whilst the early members of the *Ticonderoga* class were fitted with the Mk 26, ships from the sixth ship, *Bunker Hill* (CG-52), onwards were fitted with the Mk 41 VLS.

14. No vertically-launched version of the Harpoon anti-ship missile was developed, despite substantial interest at various times. The earlier *Burke*s therefore had a group of eight Harpoon canisters aft.

15. This – and earlier studies – also looked at significant increases in VLS capacity, perhaps to as many 128 cells. The South Korean KDX-III destroyer design, based on that of the Flight IIA *Arleigh Burke*s, demonstrates the feasibility of this idea. By contrast, Japan's *Kongou* and *Atago* classes, respectively derived from the Flight I/II and Flight IIA designs, have retained the same VLS capacity as their American counterparts.

16. From March 1998 onwards, ships were purchased under extended multi-year contracts with BIW and Ingalls. These contracts were subject to necessary Congressional authorisation. This has resulted in some discrepancies in published sources as to under which FY programme certain ships were authorised.

17. For more detail on the truncated Zumwalt programme, see Scott Truver's 'Zumwalt (DDG-1000): Past and Future Tense', *Seaforth World Naval Review 2017* (Barnsley: Seaforth Publishing, 2016), pp.136–55. It is hoped to provide further detail on the new Flight III ships in a future edition of *Seaforth World Naval Review.*

18. This vulnerability relates to ships without the latest Baseline 9 upgrade to engage ballistic missile targets and other threats simultaneously. Unmodernised Flight I and Flight II ships also lack the additional defence layer provided by ESSM.

19. SQR-20 has also been installed on some of the Flight IIA ships, giving them a towed array capability for the first time.

F222

3.2 **SIGNIFICANT SHIPS**

BADEN-WÜRTTEMBERG CLASS

Germany's F125 type stabilisation frigates

Author:
Conrad Waters

The end of the Cold War had a marked impact on the structure of many European navies. This was particularly the case for Germany's *Deutsche Marine*. Previously largely configured to support forward coastal defence in the Baltic, protect the Baltic approaches and conduct anti-submarine operations in the North Sea and beyond, it found the *raison d'être* for both missions greatly diminished following the Soviet Union's collapse. Like many of its European counterparts, the following years saw a gradual evolution to more expeditionary roles in the face of a more unstable global environment. However, the Second World War's continuing legacy inevitably resulted in this movement being accompanied by a degree of political hesitation. This hesitancy impacted the type of expeditionary tasks assumed. Much emphasis was placed on activities towards the lower end of the operational spectrum, such as ensuring maritime security and peacekeeping. Conversely, there was much less appetite for more robust, power-projection activities.[1]

Inevitably, it became apparent that a generation of warships designed for 'hot-war' fighting in a largely European regional context were not best-suited for the changed concept of operations. The emphasis on undertaking distant global operations aimed at ensuring peace and stability in the face of a range of largely low-intensity threats required new designs of warship with dramatically different operating characteristics than their predecessors. At the same time, Germany's track record of success in developing and building a series of successful frigate classes based on MEKO technology provided a strong legacy from which to meet the new requirements. The result is the new *Baden-Württemberg* class of F125 type stabilisation frigates. These innovative new ships essentially see the wholesale updating of previous design concepts to meet the demands of a new era.

The lead F125 class frigate *Baden-Württemberg* pictured on sea trials in July 2016. Designed for lengthy deployment in support of low-intensity stabilisation operations around the globe, she reflects the new mission profile of the *Deutsche Marine* in the post-Cold War era. *(ThyssenKrupp Marine Systems)*

CLASS ORIGINS

The most important influence on modern Germany Navy frigate design has undoubtedly been the MEKO (*Mehrzweck Kombination*) modularised warship concept first developed by Blohm & Voss in the 1970s. This basically involves the installation of weapon, sensor, electrical and other key ship service components as standardised modules with similarly standardised connections to power, ventilation and data networks. As well as facilitating selection of a wide range of equipment and speeding building and refit times, MEKO design practices are also associated with a high level of survivability. For example, vertical distribution arrangements for electrical systems and ventilation significantly improve the integrity of compartments below the waterline. These are typically arranged as independent units

Recent German frigate designs have been heavily influenced on the MEKO modularised warship concept first developed for export markets during the 1970s. The first German Navy frigates using MEKO concepts were the four F123 *Brandenburg* class ships, commissioned between 1994 and 1996. *Schleswig-Holstein* is pictured here docking at Djibouti in May 2010 whilst participating in the European Union Operation 'Atalanta' counter-piracy mission. It was stabilisation operations such as these that drove the key requirements of the F125 programme. *(German Armed Forces)*

The F123 class design was subsequently used as the basis for the F124 *Sachsen* class, which were equipped with the specialised radars and missiles required for the air-defence role. They also saw a further evolution of the stealth technology incorporated in the early ships, including use of X form technology. This picture shows *Sachsen* test-firing a SM-2 surface-to-air missile in August 2004. *(German Armed Forces)*

with their own energy, air-conditioning and fire-fighting arrangements.

The MEKO concept was initially applied to export designs. The earliest was the Nigerian MEKO 360 frigate *Aradu*, which was ordered in 1977 and commissioned in February 1982. The first application in a German Navy frigate was in the F123 *Brandenburg* class, which entered service between 1994 and 1996. Developed by the ARGE F123 consortium of Blohm & Voss, Bremer Vulkan, HDW and Thyssen Nordseewerke, the anti-submarine orientated F123 design was an evolution of the MEKO 360 type combined with an overlay of equipment used in the previous F122 *Bremen* class. A further evolution saw the commissioning of three air-defence frigates of the F124 *Sachsen* class between 2004 and 2006. These introduced a reshaped, X-form upper hull and superstructure to reduce radar signature. Both designs reflected the very high priority the German Navy placed on survivability, including the use of three continuous armoured box girders at the strength deck and the provision of blast-resistant double bulkheads by way of lateral sub-division.[2]

At the time deliveries of the *Sachsen* class were underway, work was already well advanced on the follow-on F125 programme. The new ships were intended to replace the aging *Bremen* class, with initial studies in the late 1990s apparently focusing on a multi-role combatant. However, the German Navy's steady shift towards international stabilisation operations became an increasingly important influence as time progressed. Although the *Brandenburg* and *Sachsen* classes were successful designs, it was clear that the requirements of this new mission profile could not be met by a further simple evolution of these existing types. The detail of these requirements changed over time but can essentially be summarised as:

- The ability to undertake sustained deployments at long distance from Germany.
- Cost-effective operation, including reduced manning.
- The ability to undertake a wide-range of missions, but particularly in stabilisation scenarios.
- Particularly good abilities to support task group command and Special Forces requirements.
- High levels of survivability.

By mid-2005, German industry had developed an

The F125 design went through several iterations between the submission of initial industry proposals in 2005 and the start of construction in 2011. These graphics show the design (i) as initially envisaged around 2005 (when a navalised 155mm MONARC gun and GMLRS rocket launcher formed part of the armament), (ii) shortly before orders were placed in 2007 and (iii) around the time construction commenced. By 2011 the planned design had also increased in size from original plans. *(ARGE F125, ThyssenKrupp Marine Systems)*

initial design to meet the navy's specification. With overall length of 139.4m and breadth of 18.1m, the new ship was broadly similar in size to the *Sachsen* class (overall length 143m and breadth 17.4m) and retained the core survivability features seen in earlier frigates. However, other elements of the configuration were dramatically different. Visually, the most apparent change was the adoption of a twin 'island' structure that split key sensors and communications into two pyramid-like deckhouses fore and aft. This provided a degree of redundancy in key equipment. The principle of dispersion and redundancy was also seen in the new combined diesel-electric and gas propulsion (CODLAG) system, with twin diesel generator compartments widely separated. Considerable automation allowed a significant reduction in manning; core crew of 120 was less than half that of previous ships. However, provision of accommodation for a total of up to 190 personnel made it easier to embark additional specialists, for example command staff or Special Forces operatives. Also integral to the new design was the ability to undertake sustained deployments of up to two years in duration and to spend longer operating at sea; up to 5,000 hours p.a. compared with 2,500 hours p.a. for previous ships. A two-crew operating arrangement was envisaged to support this planned level of availability.

Also evident in the new design was the focus on stabilisation activities in the littoral. The layered air-defence capabilities seen in earlier frigates were reduced to a pair of Mk 49 launchers for the RAM close-in weapons system (CIWS). A housing for the proposed navalised GMLRS guided rocket system took the place of the Mk 41 vertical launch system (VLS) found in previous ships. Also new was a turret for the MONARC (modular naval artillery concept) 155mm gun, based on the PzH 2000 howitzer used by the German Army. Lighter-calibre guns – and non-lethal systems such as water cannon – were also provided to allow a graduated response to asymmetric threats.

Some two years were to elapse between the submission of this initial design and the award of construction contracts and it was only in May 2011 that actual construction commenced. Significant revisions to the original design proposal took place

The two members of the ARGE F125 consortium split construction of the F125 class, with Lürssen being responsible for constructing fully-outfitted bow sections and TKMS undertaking construction of the remainder of each ship, as well as integrating the two halves and conducting final outfitting and testing. This work was assigned to the Blohm & Voss shipyard in Hamburg, where the two separate sections of *Baden-Württemberg* are pictured being moved into the assembly hall in November 2012. *(ThyssenKrupp Marine Systems)*

A picture of *Baden-Württemberg* in a covered assembly hall at Blohm & Voss in Hamburg in December 2012, with her two constituent parts now joined together. She would subsequently be christened in the assembly hall in December of the following year before being floated out for final outfitting to be completed. *(ThyssenKrupp Marine Systems)*

over this period. The proposed MONARC gun was replaced with Oto Melara's new 127mm/64 lightweight mounting and the navalised GMLRS was abandoned.[3] The ship also grew in size, ultimately to a length approaching 150m. This probably reflected the limited remaining growth margins in the hull dimensions of the previous F123/F124 types, particularly given the emphasis on duplicating systems and equipment in the new ships.

CONSTRUCTION AND DELIVERY

Formal construction contracts for four F125 class frigates were awarded to the ARGE-125 consortium comprising ThyssenKrupp Marine Systems (eighty percent) and Lürssen (twenty percent) by Germany's Federal Office for Defence Technology & Procurement on 26 June 2007. Reports at the time suggested that the value of the programme was some €2.6bn or €650m per ship (US$900m at then current exchange rates). In line with its dominant role in the consortium, ThyssenKrupp Marine Systems (TKMS) was allocated the majority of construction work. Its Blohm & Voss yard in Hamburg was responsible for fabricating each ship's aft section and also undertook integration and final outfitting of all four ships. For its part, Bremen-based Lürssen fabricated and fitted out the remainder of each ship up to and including the forward deckhouse. Some assemblies of the first two ships were undertaken at Peenewerft at Wolgast. Following acquisition by Lürssen, this shipyard also completed fabrication of the outfitted bow sections of the third and fourth ships.[4]

At the time of contract signature it was envisaged that deliveries of the new class would take place from the end of 2014 onwards. This timescale encompassed a considerable period for detailed design work and the keel of the lead ship – *Baden-Württemberg* – was not formally laid until November 2011. By then, it was already anticipated that delivery of the first ship would be delayed by around fifteen months. Subsequently, difficulties arose during the construction phase. These reportedly included problems with the application of fire-resistant coatings and issues with the integrated platform management system. It also appears that the hull sections produced at different yards did not always precisely match the contours of their counterparts, with differences of up to 25mm on each side of the relevant sections requiring a degree of re-working. As such, it was not until 6 April 2016 that *Baden-Württemberg* departed

Hamburg for the first time to conduct preliminary sea trials in the North Sea.

Current plans envisage *Baden-Württemberg* being formally commissioned before the end of 2017. The second ship, *Nordrhein-Westfalen*, has also now commenced sea trials and is expected to become operational before the end of 2018. The other two ships have both been launched and will follow at approximately annual intervals. Table 3.2.1 provides further details on key construction dates. When finally commissioned, all four ships will be based at Wilhelmshaven, forming the 4th Frigate Squadron.

OVERALL DESIGN

The final F125 class design has an overall length of 149m and a beam of 18.8m. Hull depth to the main deck is 9.8m and designed draught is c. 6.6m when the propeller protrusions are taken into account. Full load displacement is in the order of 7,250 tons. This makes the class by far the largest surface combatants operated by Germany since the end of the Second World War. The main deck is at the level of the helicopter flight deck. Higher, lettered decks start with B deck at forecastle level and encompass C deck (upper deck level), D deck (bridge level), E deck (top of bridge) and so on upwards. Working downwards from the main deck, there are a lower deck, a platform deck and a hold. Hull and superstructure are of steel construction and the usual attention to minimising radar cross section is evident. This includes further refinement of the X form concept introduced with the F124 class and also present in the K130 *Braunschweig* class corvettes.[5]

The ship is split into three damage-control sections – DC I aft, DC II amidships and DC III forwards – each further sub-divided into two autonomous damage-control zones. Each zone is separated from its neighbours by blast resistant double bulkheads (five in total) and has fully independent fire-fighting, ventilation, power distribution and platform management capabilities. The system of utilising triple armoured box girders at strength-deck level seen in the previous F123 and F124 classes is continued, with vital spaces such as command and control functions and magazines also provided with armoured protection against splinter damage. The design meets German Shock Standard BV 0230, understood to be amongst the most demanding in Europe in terms of its specifications.

Access to the ship is provided by means of tilting and telescoping ship's brows located to both port

Two images of the second member of the F125 class, *Nordrhein-Westfalen,* seen in the final stages of construction in September 2016. She commenced sea trials in January 2017 and will be commissioned into the 4th Frigate Squadron at Wilhelmshaven in the course of 2018. *(Bruno Huriet)*

Table 3.2.1: BADEN-WÜRTTEMBERG CLASS LIST

NAME	PENNANT	ORDERED	LAID DOWN	CHRISTENED[1]	COMMISSIONED
Baden-Württemberg	F222	26 June 2007	2 November 2011	12 December 2013	[2017][2]
Nordrhein-Westfalen	F223	26 June 2007	24 October 2012	16 April 2015	[2018][3]
Sachsen-Anhalt	F224	26 June 2007	4 June 2014	4 March 2016	[2019]
Rheinland-Pfalz	F225	26 June 2007	29 January 2015	24 May 2017	[2020]

Notes:

1. Relates to the formal christening ceremony. The date of the actual float-out of the assembled ship was different: *Baden-Wurttemberg* and *Sachsen-Anhalt* were christened before float-out; *Nordrhein-Westfalen* and *Rheinland-Pfalz* afterwards.
2. Sea trials commenced 6 April 2016.
3. Sea trials commenced 27 January 2017.

Table 3.2.2.
BADEN-WÜRTTEMBERG PRINCIPAL PARTICULARS

Building Information:[1]	
Laid Down:	2 November 2011
Christened:	12 December 2013
Commenced Sea Trials:	6 April 2016
Builders:	ARGE F125 Consortium of ThyssenKrupp Marine Systems & Lürssen. Ship assembled at Blohm & Voss, Hamburg.
Dimensions:	
Displacement:	c. 7.250 tons full load displacement.
Overall Hull Dimensions:	149.6m x 18.8m x 5.4m. Length at waterline is 141.9m and beam 18.0m.
Equipment:	
Missiles:	2 x 21-cell Mk 49 RAM launchers for RIM-116 surface-to-air missiles. 2 x quadruple launchers for Harpoon surface-to-surface missiles.
Main Guns:	1 x 127mm OTO 127/64 gun, 2 x 27mm MLG 27 cannon, 5 x 12.7mm OTO Hitrole machine guns.
Other:	Various non-lethal weapons systems such as water cannon.
Countermeasures:	KORA-18 RECMS. 4 x Rheinmetall MASS decoy launchers.
Aircraft:	Flight deck and hangars for two medium-sized helicopters.
Principal Sensors:	TRS-4D multi-function radar (4 panels). Integrated IFF. Navigation radar. Portable Cerberus Mod 2 diver detection sonar.
Combat System:	Atlas Naval Combat System (ANCS). Atlas Data Link System (ADLis) with NATO Link 11, 16 and 22 functionality.
Propulsion Systems:	
Machinery:	CODLAG. 4 x MTU 20V 4000 M53B diesel generators each rated at 3MW provide power for 2 x 4.7MW Siemens electric motors and hotel services. 1 x LM2500 GE gas turbine rated at 20MW. Twin shafts. 1 x 1MW bow thruster.
Speed:	26 knots. Endurance is 4,000 nautical miles at 18 knots.
Other Details:	
Complement:	Normal crew is 120 plus 20 aviation detachment. Accommodation is provided for 190 personnel.
Class:	*Baden-Wurttemberg* (F222); *Nordrhein-Westfalen* (F223); *Sachsen-Anhalt* (F224) and *Rheinland-Pfalz* (F225).

Notes:

1. Fabrication commenced in May 2011 and float-out occurred on 31 March 2014. Commissioning is planned for mid-2017.

and starboard. These are housed within covered bays located at main deck level between the two boat bays and within the DC II section. The relevant bays or deck stations are each separated from the internal citadel by a nuclear biological and chemical defence (NBCD) lock and also contain a watch stand. Similar NBCD locks control access, for example, to the forward handling deck underneath the forecastle, to the hangar at main deck level and to the after handling position on the lower deck.

The CODLAG propulsion arrangement in industry's original 2005 proposal has been retained. Two pairs of 3MW MTU 20V 4000 M53B diesel generators – widely dispersed in damage-control sections DCI and DCIII – provide up to 12MW of electrical power for both propulsion and hotel functions. This is distributed by means of a medium voltage network. For propulsion purposes, electricity is supplied to two 4.7MW Siemens electric motors that drive the ship's twin shaft lines to a comparatively high speed of 20 knots. For higher speeds, a single 20MW GE LM2500 gas turbine located in dame control section DC3 provides additional power to enable a maximum 26 knots to be achieved.[6] Linkage of the propulsion system with the shaft lines is by means of a cross-connected gearbox arrangement supplied by RENK AG. The shaft lines are provided with controllable-pitch propellers. There is also a 1MW bow tunnel thruster supplied by Norway's Brunvoll to assist with low-speed manoeuvrability. The integrated platform management system, which also performs damage control monitoring and training functions, is provided by Germany's Siemens.

COMBAT MANAGEMENT SYSTEM AND SENSORS

The heart of the F125's class combat management capabilities is formed by the Atlas Naval Combat

Baden-Württemberg (2016)
1:700 scale

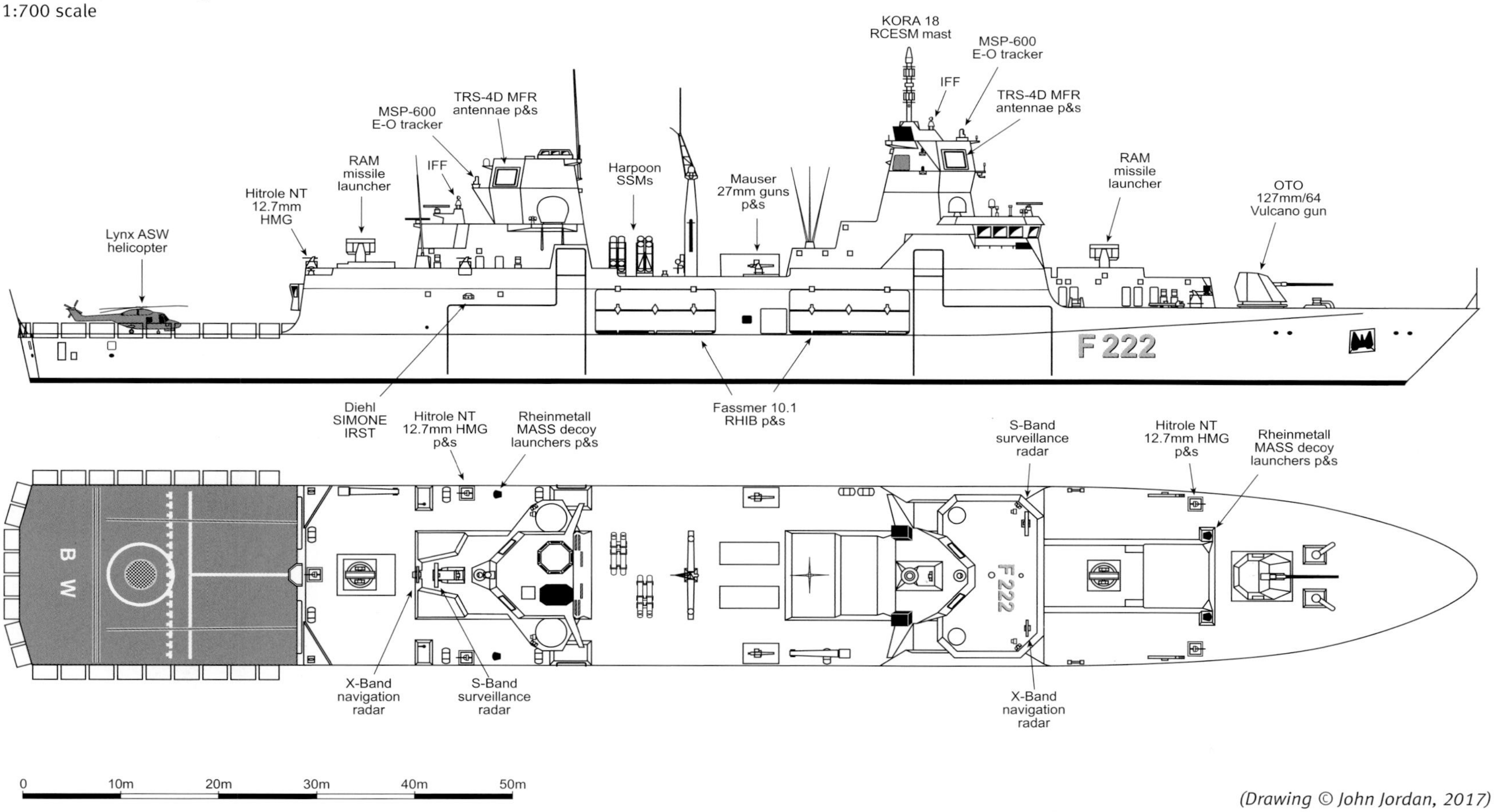

(Drawing © John Jordan, 2017)

System (ANCS). This was developed for the ships under a contract with Atlas Elektronik announced in September 2009. In common with other modern combat-management systems, this links weapons, sensors and communications functions to various multi-function operator consoles by means of a high-speed data network. Precise details of the hardware configuration on the F125 class have not been published. However, Atlas Elektronik-produced schematics suggest that two large display screens and twelve consoles can be found in the main combat information centre – nine along the compartment's walls and three in a central command position – with a further two consoles in a separate annexe. Other consoles and/or display screens can be found on the bridge and in other command spaces, whilst it would appear that access to the system from a portable workstation is also possible. The OMADA operator consoles are supplied in different variants and their display can be configured to specific user requirements. All combat systems are able to be controlled from any one of the consoles.[7]

The software supporting the combat management system is described as having a common core to which various expanded capabilities can be added in a modular fashion. The common core encompasses functions such as tactical picture compilation, track management, threat evaluation and weapons assignment. Options have been tailored to the specific stabilisation mission of the F125 class and include functions supporting activities such as asymmetric warfare, Special Forces operations, gunfire support and search and rescue. It is possible to reconfigure the software to support a changed operating profile. ANCS functions are aligned with NATO standards, facilitating interoperability with other alliance units in joint expeditionary operations.

ANCS is integrated with Atlas Elektronik's ADLis (Atlas Data Link System), which allows the ship to use a variety of datalinks to share information with other German and allied units through radio and satellite communication systems. ADLis supports NATO Links 11, Link 16 and the new Link 22, which will replace Link 11 in due course. The F125 class will be the first German warships equipped with Link 22 capability.

A comprehensive list of sensors and countermeasures equipment is headed by the TRS-4D multi-function radar produced by Hensoldt (formerly part of Airbus Defence and Space). An active phased array, it operates in the 4,000–5,000 MHz NATO G

Baden-Württemberg pictured departing Hamburg by means of the River Elbe on 7 April 2016, the first day of her initial sea trials. The image provides a good view of the two forward faces of the TRS-4D multi-function radar, which is split between the fore and aft deckhouses as a survivability measure. *(ThyssenKrupp Marine Systems)*

This aft image of *Baden-Württemberg* departing on sea trials clearly shows the rear faces of the TRS-4D radar. Other points of interest are the RCESM mast on top of the forward structure and the twin hangars. *(ThyssenKrupp Marine Systems)*

band (US Navy C band). It is able to monitor over 1,000 air and surface targets out to a maximum range of 250km (air surveillance), provide target designation to the combat-management system for anti-air and anti-surface warfare purposes and undertake surface gun fire control. The variant installed in the *Baden-Württemberg* class comprises four fixed arrays split between the forward and aft deckhouses to provide a degree of redundancy. TRS-4D has good capabilities in the littoral waters envisaged to be the F125 class's key area of operation, including a high-resolution surface channel optimised for detecting small surface targets.[8] It is supplemented by surface surveillance and navigation radars which feed into an integrated bridge and navigation system supplied by Raytheon Anschütz.

The ships' radar systems are reinforced by two Rheinmetall MSP (Modular Sensor Platform) 600 electro-optical trackers that can be used for surveillance, tracking and fire control. Located on the forward and aft deckhouse masts, they operate without the emission of any signals and are therefore an excellent complement to the multi-function radar. Another important passive monitoring system is the Diehl Defence SIMONE ship infrared monitoring, observation and navigation equipment. This provides 360-degree coverage of a ship's immediate vicinity and can detect very small objects such as inflatable boats or frogmen. Two multi-headed sensor modules are located at B deck level to the port and starboard hangar sides, with additional sensors located on top of the bridge and on the lower extensions of the aft mast. Interestingly, there is no provision for fixed submarine or mine-detection sonar, although Atlas Elektronik's portable Cerberus Mod2 diver detection sonar is shipped.[9]

Countermeasures include a comprehensive radar and communications electronic support measures (RCESM) suite installed on top of the forward mast to detect emissions from potential enemies. Physical countermeasures are in the hands of four Rheinmetall MASS (Multi-Ammunition Soft-kill System) launchers integrated with the combat management system. Located port and starboard to the aft of the forward gun and on top of the hangar roof, they are equipped with a wide range of decoys that are designed to lure away incoming missiles.

WEAPONS SYSTEMS

The F125 type's stabilisation function is most apparent from its weapons outfit, which is clearly based on providing a proportionate response to the level of threat posed. This commences with the use of non-lethal systems such as water cannon and then progresses through a range of light weaponry through to the much more coercive potential inherent in weapons such as the main 127mm gun. The integrated gunnery outfit includes no fewer than eight weapons, viz.

- **1 x 127mm Leonardo Defence Systems OTO 127/64 lightweight gun:** An evolution of the previous 127/54 mount, this weapon was first installed onboard the Italian FREMM type frigate *Carlo Bergamini* in 2011. It has also been used to equip the Algerian Navy's new MEKOA-200 AN class frigates and is located in the typical 'A' position forward of the superstructure in the F125 class. The new gun is capable of fully-automated loading through installation of an optional automated ammunition-handling system and can engage sea, land and – as a secondary function – air targets. It can be used with the newly-developed Vulcano series of extended-range munitions as well as standard ammunition, with the guided long range (GLR) variant of Vulcano using GPS technology to allow engagement of land targets in excess of 100km (54 nautical miles). It has not been confirmed whether or not the German Navy will use the new munitions but their acquisition would go a long way to compensating for the land attack capability lost with the deletion of the proposed navalised GMLRS.
- **2 x 27mm Rheinmetall *Marine-Leicht-Geschütz* MLG 27 cannon:** The MLG 27 has replaced older manually-operated 40mm L/70 and 20mm guns in German Navy warships. It utilises the Mauser BK 27 gun fitted to German Luftwaffe (and previously Navy) aircraft. The mount is equipped with an on-mount, multi-sensor package for fully autonomous operation but – as is the case for the F125 – can also be integrated into the overall combat-management system. It is normally remotely controlled from an operator console. The system is designed to provide point-defence protection against air and small sea-targets at ranges from 100m out to around 2.5km; it can also be used against larger surface vessels and targets to a maximum range of c. 4km. The gun can fire a wide range of ammunition at up to 1,700 rounds per minute and has a quoted maximum eight seconds reaction time between target acquisition and engagement. The two mounts on the F125 class are installed port and starboard between the two deckhouses.
- **5 x 12.7mm Leonardo Defence Systems OTO Hitrole remotely-controlled naval turrets:** Like the MLG 27, these weapons interface with the F125's overall combat-management system but also have onboard sensors for autonomous operation. It is optimised for close-range defence against

These two views of *Baden-Württemberg* on sea trials in July 2016 provide a good indication of overall armament layout. A 127mm gun is located in 'A' position whilst RAM launchers are positioned forward of the bridge and on top of the hangar aft. Lighter armament is located just aft of the main gun and around the superstructure to provide 360 degree coverage. Mounts for Harpoon surface-to-surface missiles are located just forward of the aft deckhouse. *(ThyssenKrupp Marine Systems)*

asymmetric threats. Five mounts are positioned on top of the hangar (three) and just aft of the main gun (two) to provide 360-degree coverage.

This comprehensive complement of weapons can also be supplemented by additional, stand-alone machine guns of up to 12.7mm calibre.

By way of contrast to the gunnery outfit, missiles are much less in evidence. In line with the initial 2005 design, surface-to-air missiles are limited to the close-range RAM Mk 31 Guided Missile Weapon System. This originated as a collaborative programme between the United States and Germany and has currently been developed to Block II configuration. Originally derived from the AIM-9 Sidewinder air-to-air missile, the RIM-116 RAM Rolling Airframe Missile used by the system combines data gathered by the ship's combat-management system with onboard infrared and radar homing for defence against anti-ship missiles out to a range of around five nautical miles (9km). It can also be used to counter other threats, such as helicopter, aircraft and surface targets. The two distinctive 21-cell Mk 49 launchers for this system are located forward of the bridge and on top of the aftermost part of the hangar.

Anti-surface missile capabilities are in the hands of the venerable Boeing RGM-84 Harpoon. There is provision for two quad-packed launchers just forward of the after deckhouse. The sub-sonic and increasingly dated Harpoon is probably an interim solution given increased German industrial collaboration in the area of surface-to-surface missiles with other European countries. For example, Saab's RBS15 Mk 3 weapon is already installed in the recent K130 class corvettes. However, recently announced German collaboration with Norway's Kongsberg in the form of a joint programme to enhance the Norwegian Naval Strike Missile (NSM) suggests that this weapon is most likely to be installed in the F125 class in due course.

Hanger and flight deck facilities are provided for the operation and support of up to two medium-sized helicopters, significantly increasing the class's flexibility. The hangar arrangement follows previous German practice in providing for two separate hangars. This reduces the risk of both machines being destroyed by a single incident. A helicopter handling system is installed for automatic transfer of helicopters to the hangars after landing. The German Navy sea control helicopter currently deployed on frigate-sized vessels is the Mk 88A Sea Lynx. This is likely to be replaced around the 2025 timeframe. The German Navy is already acquiring the tactical transport variant of the NH-90 helicopter for search and rescue and transport tasks and this is probably the leading contender for the replacement contract. However, the AW159 Lynx Wildcat is another possible alternative. At present, the F125 class's helicopters provide the only inherent anti-submarine capability in the new ships.

A picture of *Baden-Württemberg* turning at speed. The enclosed bays for the ship's 10.1m RHIBs are prominent. Four of these large speedboats are shipped to undertake a range of missions in littoral waters. *(German Armed Forces)*

OTHER KEY DESIGN FEATURES

The *Baden-Württemberg* class's rather unusual armament configuration is possibly the most obvious indicator of the new ships' focus on low and medium-intensity stabilisation operations. However, this mission-focus runs through the four vessels' overall design characteristics, being evident in features ranging from the considerable amount of surplus accommodation available for specialist personnel to the extensive level of redundancy provided across key equipment to maximise time on station. The logistical requirements of long-distance deployment are also evidenced by provision of internal transport routes for palletised cargo. These facilitate replenishment and speed offloading, for example when supporting disaster relief. Even the large fitness room installed in the space originally allocated for the GMLRS system evidences attention to the needs of crew members deployed on long-distance expeditionary operations.

A notable design feature is the provision of extensive command and other support facilities for joint assignments undertaken with other branches of the German armed forces or in conjunction with allies. The extensive combat-management and communications capabilities inherent in the F125 class have already been described. These are supplemented by features such as the availability of dedicated command spaces for planning and control functions and the furnishing of an extensive medical facility. The latter is valuable both in ensuring the health of crew members deployed at distance and assisting with humanitarian relief missions.

Another indicator of the ship's specialist role is the shipping of four, large 10.1m rigid-hulled inflatable boats (RHIBs). For stealth purposes, these are housed within covered bays port and starboard amidships. Weighing up to 8 tons loaded, they are capable of deploying fifteen personnel at speeds of 35 knots out to a range of 130 nautical miles. Specially designed to support Special Forces operations, they can also be used, for example, for convoy protection, the transport of injured personnel or evacuation operations. The *Baden-Württemberg* and her sisters are also capable of carrying two TEU (Twenty-foot Equivalent Unit) containers between the two deckhouses and above the boats. These can be used to house specialist equipment and perform something of the functions of the specialist mission bay found in many of the most recent warship designs.

Perhaps of as much importance to the F125's actual performance as the design itself is the dual crew operating concept. This will see eight separate crews – designated Alpha through to Hotel – assigned to the class and rotated through various stages of training, work-up and deployment. The personnel of ships undertaking actual operations will be rotated at roughly four-monthly intervals at the location of the deployment. This is a key element in

Nordrhein-Westfalen pictured on the River Elbe in May 2017. The F125 class will utilise a dual-crew operating concept to maximise the amount of time they can spend deployed away from Germany. *(Hummel under Creative Commons Attribution Share Alike 4.0 Licence)*

the plan to keep ships on station in their assigned operational area for periods of up to two years, as well as doubling actual operating sea-time to around 5,000 hours p.a. The concept is also intended to improve overall service morale by limiting time away from home and provide clear time slots for crew training and development. Similar arrangements have been used with success when such swaps have taken place at a vessel's home base, for example with strategic missile submarine crewing. However, the German Navy's plan to swap crews in an operational zone is more ambitious and gives rise to a number of questions, not least the need to ensure effective transfer of situational awareness between the relevant personnel.

CONCLUSION

The F125 programme is essentially a combination of a continued evolution of the German Navy's successful MEKO-based designs with a radically new mission profile and operating concept. The resulting design has not been without its critics. In addition, in common with many major defence projects, actual

Baden-Württemberg pictured at sea with a F123 class frigate in mid-2016. The F125 class represents the latest iteration of MEKO technology first introduced to the German Navy with the F123 design. *(German Armed Forces)*

implementation of the programme has been patchy.

On the plus side, the overall idea of a dedicated stabilisation asset fits well with the type of international mission most frequently conducted by the *Deutsche Marine* in the post-Cold War era. In a similar fashion to the Royal Netherlands Navy's equally controversial *Holland* class offshore patrol vessels, *Baden-Württemberg* and her sisters will be superbly well-equipped to carry out their intended function. Given limited budgets and stretched resources, it also makes sense to design the warships assigned to this mission so as to perform their duties in an as efficient and cost-effective manner as possible. Whilst the initial capital outlay involved in acquiring warships often gains most public attention, it is ongoing crewing, maintenance and associated charges that account for the bulk of a programme's through-life cost. The German Navy's solution in terms of achieving increased asset utilisation, extended maintenance intervals and reduced crewing through the use of enhanced automation and system redundancy represents a logical response to this challenge. Although this solution has yet to be tested in practice, the thinking behind it appears sound.

The F125 design is capable of supporting two helicopters, currently the Sea Lynx pictured here. Nevertheless, the lack of any other organic anti-submarine capability in the class is a notable omission. *(German Armed Forces)*

A view of the OTO 127/64 lightweight gun that arguably forms the F125 class's main armament. The lack of a flexible vertical launch system for a range of missiles is the main criticism levied at the design. *(German Armed Forces)*

The main criticism of the *Baden-Württemberg* class appears to be the limitations inherent in the actual specification handed to the ships' designers. The omission of a vertical launch system, with its ability to deploy a wide variety of missiles, is difficult to understand in a 7,000-ton ship and may well have been driven by political considerations. Whilst it has little impact on the class's intended deployment on low/medium-intensity operations, the lack of a layered air-defence capability would be a major disadvantage in more intensive scenarios. Indeed, it is likely that a F125 class vessel would need to be escorted under such circumstances to ensure its survival. The lack of anti-submarine warfare capabilities in a littoral combatant, a feature shared with the K130 class corvettes, is also a notable omission. The changed political environment of recent years – with all-out warfighting now a more likely possibility – only serves to highlight the questionable nature of these choices.

The programme's implementation gives rise to further questions. The overall delivery schedule has been delayed several times and a number of problems have emerged during construction. There have been recent reports that the ships have turned out overweight and have only limited margins available for future enhancement.[10] It is difficult not to conclude that the fragmentation of German warship building that took place following the reversal of TKMS' consolidation of the industry may have played a part in these problems.

Looking forward, the outlook seems more positive. The basic flexibility of the MEKO concept will allow the enlarged F125 hull to be readily adapted to future programmes; it formed the basis of an unsuccessful design proposal for the new Australian surface combatant and the new German MKS-180/F126 programme will likely be a derivative. Meanwhile, the reconsolidation of much of German surface warship shipbuilding under Lürssen's leadership – coupled with TKMS ongoing design skills – offers the prospect of a more cohesive industrial base. In any event, *Baden-Württemberg* represents a robust and technologically-advanced ship that will undoubtedly give the *Deutsche Marine* good service in the years ahead.

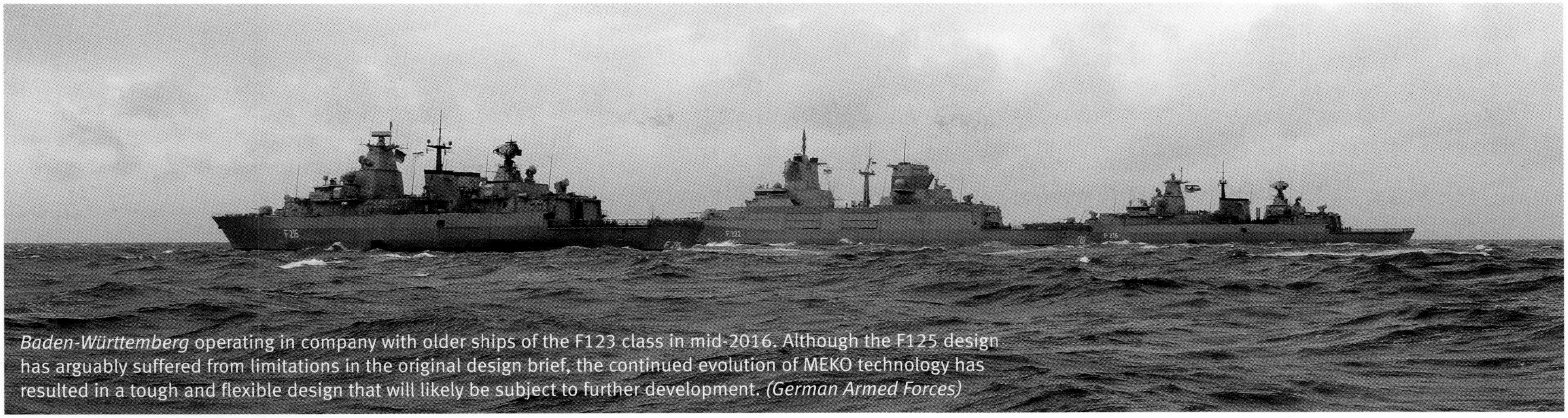

Baden-Württemberg operating in company with older ships of the F123 class in mid-2016. Although the F125 design has arguably suffered from limitations in the original design brief, the continued evolution of MEKO technology has resulted in a tough and flexible design that will likely be subject to further development. *(German Armed Forces)*

Notes

1. The end of the Cold War saw a similar steady reorientation of the mission profile of the overall German Armed Forces or *Bundeswehr*, the German Navy being known as the *Bundesmarine* or Federal Navy until 1995. By the time the F125 programme was approved, it was recognised that out-of-area deployments were the armed forces' most likely task and a new structure of (i) intervention, (ii) stabilisation and (iii) support forces was being adopted. Of these, the high intensity intervention force of 35,000 personnel comprised under fifteen percent of total numbers.

2. A good source of information on the development programmes and technical characteristics of the German frigate designs preceding the F125 class can be found in the Newtown Connecticut-based *Forecast International* group's series of archived reports on specific warship classes. These can be accessed by examining the Warships Forecast archive at: https://www.forecastinternational.com/archive/naval.cfm. Interestingly, the MEKO concept might never have been adopted by the German Navy as a result of the submission of a competing, cheaper proposal from Bremer Vulkan (in association with AEG-Schiffbau) at the time the F123 was being designed. The proposal, based on a modification of the existing non-modular F122 class was drawn up without the knowledge of Bremer Vulkan's partners in the ARGE F123 consortium and gained considerable traction within the German Ministry of Defence. Considerable last-minute work by the ARGE partnership produced the F123 design that ultimately found favour.

3. A prototype MONARC gun – using a flexible mount with special damping elements to help deal with the greater recoil of such a large weapon – was temporarily installed on the F124 class frigate *Hamburg* in 2002. Fire-control arrangements were also later trialled on her sister *Hessen*, using a PzH 2000 howitzer fitted to her flight deck. Although it seems that the tests were reasonably successful, the process of navalisation was greater than expected and the gun's proposed use in the F125 programme was dropped by early 2007. The navalised GMLRS was dropped around the same time. Another factor in the decision may have been a need to reciprocate heavy Italian investment in Type 212A submarine technology by utilising Oto Melara's competing 127mm gun.

4. At the time the contract with the ARGE-125 consortium was awarded, German warship building was largely split between ThyssenKrupp Marine Systems (TKMS) and Lürssen following the former group's acquisition of Howaldtswerke-Deutsche Werft (HDW) in 2005. The consortium had also agreed that the Peene-Weft shipyard in the former East Germany would also be allocated some project work. The subsequent break-up of much of TKMS' shipbuilding activities included the disposal of the Blohm & Voss yard, where TKMS' physical part of the contract was being executed. In the following complicated and often confusing restructuring of German shipbuilding that has taken place in recent years, Lürssen has subsequently acquired both the Peene-Werft (2012) and Blohm & Voss (2016) facilities. However, TKMS still retains its eighty percent share of the ARGE-125 consortium, as well as considerable project management and design capabilities. Sources differ on programme cost; some refer to the contract with the ARGE F125 consortium amounting to 'about €2bn'.

5. For further detail on the K-130 programme, see Guy Toremans' '*Braunschweig* Class Corvettes: Eagerly awaited by the German Navy', *Seaforth World Naval Review 2013* (Barnsley: Seaforth Publishing, 2012), pp.128–47.

6. It appears that the propulsion plant is capable of higher speeds but had been purposely limited to extend service intervals in line with the desire to allow sustained deployments of up to two years. See Tim Becker's 'Class F125 Frigate – First Interim Result', *European Security & Defence 1/2011* (Bonn: Mittler Report Verlag GmbH, 2011), pp.61–4.

7. This information is derived from a December 2010 paper by Heinz Marsau-Rierlof Atlas Elektronik entitled *New Frigate Class F125 for German Navy, Sensors and Effectors, Network Enabling Capability*. A copy can currently be found by searching the web.

8. A rotating, single-faced variant of TRS-4D has also been developed. It will equip later class members of the LCS-1 variant of the US Navy's Littoral Combat Ship, ultimately replacing the TRS-3D radar found in earlier ships in the class. It has also been selected for the modernisation of the three Type 23 frigates the Chilean Navy acquired from the United Kingdom.

9. The lack of a fixed sonar system seems a major omission for a warship intended to operate in littoral waters where the submarine threat is potentially significant.

10. See Sabine Siebold's *Late and overweight – Germany's new frigates found wanting* (Berlin, Reuters, 12 May 2017). The article cited a confidential report stating that a persistent list of 1.3 degrees to starboard, as well as the weight problem, emerged during sea trials. German defence ministry sources stated that appropriate measures had been taken to remedy the slight list – a not-unusual occurrence – and that design and performance parameters would still be met.

11. The author acknowledges with gratitude the assistance of Dipl-Ing Hartmut Ehlers in reviewing an early draft of this chapter. All errors and opinions expressed remain the author's own.

3.3 **SIGNIFICANT SHIPS**

OTAGO CLASS OPVs

The Royal New Zealand Navy's day-to-day 'workhorses'

Author:
Guy Toremans

New Zealand is inextricably linked to the sea.[1] Given its geographical isolation and dependence on maritime trade, it is inevitable that New Zealand is a maritime nation. This inherent reliance on sea lanes and the need to prevent any disruption to free passage is a major influence on Royal New Zealand Navy (RNZN) tasking. Others include maintaining the security of the country's c. 15,000km coastline, of its c. 4,000,000km^2 Exclusive Economic Zone (EEZ) (the fifth largest in the world), and of the Ross Dependency in Antarctica, where New Zealand has a permanent scientific presence at Scott Base. The legacy left by the former British Empire in the Pacific provides further security responsibilities, including for the dependent territory of Tokelau and for the associated states of the Cook Islands and Niue.[2] The challenges associated with managing this immense and intrinsically valuable maritime domain must be met with pragmatic, innovative and cost-effective solutions. This underlines the significance of maintaining a credible maritime patrol capacity. The navy's two *Otago* class offshore patrol vessels (OPVs) form the heart of this capability.

Left: New Zealand's *Otago* class offshore patrol vessel *Wellington* pictured against an impressive coastal backdrop in December 2016. Acquisition of the two ships has significantly improved the RNZN's constabulary abilities; an important and necessary development given responsibilities that extend southwards to Antarctica and northwards towards the equator. *(RNZN)*

CLASS ORIGINS

In June 2000, the then New Zealand government published a *Defence Policy Framework* that formed the basis for a restructuring of the New Zealand Defence Force (NZDF).[3] The Framework acknowledged that a strong priority for future investment was the maintenance of New Zealand's maritime patrol capabilities. To a large extent, this reflected

A picture of *Otago*'s maiden arrival in New Zealand waters in April 2010. The two new *Otago* class OPVs are at the heart of New Zealand's ability to police a vast and intrinsically valuable maritime domain. *(RNZN)*

The *Otago* class were acquired as part of the larger NZ$500m Project Protector programme that was aimed at bolstering the RNZN's maritime patrol and transportation capabilities. These images show the two offshore patrol vessels and four smaller *Rotoiti* class inshore patrol vessels acquired under the same programme operating in consort in 2010. *(RNZN)*

contemporary views that a lack of constabulary vessels capable of operating over long distances in difficult weather conditions was hindering effective policing of New Zealand's maritime interests. Subsequent studies encompassing a wide range of government bodies in addition to the Ministry of Defence and NZDF took place during 2001–2. These looked at, *inter alia*, overall maritime patrol capabilities and the optimum fleet mix. They culminated in the publication of key findings from a *Maritime Forces Review* in January 2002. As a result of these findings, the government announced it would embark on a NZ$500m (c. US$210m at then current exchange rates) acquisition programme designed to bolster both maritime patrol and sealift capabilities. This was to encompass a multi-role transport ship, at least two OPVs and four or five inshore patrol vessels.[4]

The new acquisition programme – quickly named Project Protector – moved forward with relative speed. Initial indications of interest from international shipbuilding groups were sought in mid-2002. Six companies from Australia, Germany, the Netherlands, Singapore and the UK were subsequently invited to submit detailed proposals. From these, Australia's Tenix Defence was ultimately selected as preferred supplier. In July 2004, a NZ$499.7m contract was signed with Tenix for a total of seven vessels, including a multi-role ship and two offshore and four inshore patrol vessels.[5] The two OPVs were to become the *Otago* class, officially known as the 'Protector' class OPVs.

Assembly of the two OPVs was allocated to Tenix's Williamstown yard in Melbourne. However, the project also involved significant opportunities for local industry. Around NZ$110m work across the total programme was contracted with New Zealand-based countries, much channelled to the then Tenix Shipbuilding New Zealand yard in Whangarei. As well as building the four inshore patrol vessels, the Whangarei facility constructed the bridge and helicopter hanger modules for both OPVs. These were then barged across the Tasman Sea to Williamstown. Manufacturing started on the first of class – HMNZS *Otago* – in February 2005 and she was laid down in December of that year. Launch subsequently took place on 18 November 2006. The second unit, named HMNZS *Wellington*, was laid down on 2 June 2007 and floated out on 27 October 2007.[6]

The two ships were originally intended to enter

service from 2007 onwards. However, they were considerably delayed as a result of problems that emerged during construction, most notably a significant excess over their planned displacement. It had originally been intended that the vessels would include a generous margin for future growth but a decision taken relatively late in the design process to add limited Class 1C ice protection so as to allow the ships to sail in new-ice waters appears to have used up much of this flexibility. Without careful management, this increased displacement could actually impact the effectiveness of the steel belt fitted to offer protection from ice damage, as it would not extend high enough up the hull due to the ships' sitting lower in the water. Systems have been put in place to monitor weight-gain throughout the ships' service lives to mitigate this risk. In early 2010, BAE Systems – which had acquired Tenix Defence in 2008 – agreed to pay New Zealand compensation reported as amounting to NZ$85m in settlement of this and a number of other, widely-reported problems relating to Project Protector.

Otago was finally formally accepted into the RNZN at Williamstown on 18 February 2010 but her delivery voyage was subsequently delayed by engine problems. She finally departed from Australia on 3 April 2010 and arrived at Devonport Naval Base, Auckland on the 9th of that month. *Wellington* was delivered on 6 May 2010, arriving in Auckland on 11 June 2010 after also experiencing minor propulsion defects. She was the last of the Project Protector vessels to enter service.

The second *Otago* class vessel, *Wellington*, pictured on builders trials in September 2008. The need to resolve problems relating to excess weight meant that it would be almost two more years before she was finally accepted by the RNZN. *(RNZN)*

PLATFORM DESCRIPTION

The RNZN's *Otago* OPV design is one of a series of patrol vessel designs developed by what is now the Canadian Vard Marine consultancy, part of the wider Italian Fincantieri shipbuilding group.[7] Currently forming design Vard 7 085 in the consultancy's portfolio, it shares a common heritage with patrol vessels such as the Irish Naval Service's *Róisin* class and the Mauritius National Coast Guard's former *Vigilant* (all Vard 7 080 series vessels). Essentially designed and built to Lloyd's Register commercial shipbuilding standards overlaid with military specifications in areas such as damage control and stability, the ships have an overall length of 85m, a beam of 14.2m and a draft of 3.6m. Full load displacement is some 1,900 tons. The class incorporates some stealthy technologies, with the topside and hull sections designed to reduce radar cross-section.

The design features a total of six decks and a hold. Running from top to bottom, these first encompass a Bridge Deck and Deck 02. The latter is level with the main armament and accommodates an operations room, communications facility and secure intelligence compartment below the bridge. Deck 01 equates with the level of the forecastle, hangar and flight deck and also incorporates offices and a sick bay in the forward superstructure. Deck 1 is effectively the main deck, housing workshops, stores and accommodation facilities for senior personnel, as well as the aft working deck. Deck 2 is the principal accommodation deck but is split amidships by the two-storey engine room and the diesel outlets. Propulsion machinery and generators are concentrated on Deck 3.

Each ship comprises ninety-six separate compartments. A major design emphasis has been on ensuring survivability in the event of damage, including the provision of nine main watertight sections and full redundancy of propulsion and generation systems. The ships can withstand total vertical flooding to the waterline in one watertight section and remain within Lloyd's classification requirements for stability and survivability. There are two damage-control zones, each equipped with firefighting pumps, a power distribution panel and a fixed foam fire-extinguishing system. The basic safety concept for detecting fire and flooding is based on an 'Intruder Detection System' composed of six CCTV cameras (four internal/two external)

Table 3.3.1: OTAGO CLASS LIST

NAME	PENNANT	ORDERED	LAID DOWN	LAUNCHED	COMMISSIONED
Otago	P148	July 2004	16 December 2005	18 November 2006	18 February 2010
Wellington	P55	July 2004	2 June 2007	27 October 2007	6 May 2010

A close-up view of the 25mm Rafael Typhoon gun on *Otago*. This fully-stabilised gun, which forms the ship's principal permanent armament, replaced the MSI Austig mount previously fitted. *(Guy Toremans)*

monitored on displays throughout the ship and a fire-detection system. Fires can be combatted by a salt water fire extinguishing system, a CO_2 extinguishing system in the machinery spaces and hazardous compartments, an aqueous film-forming foam system encompassing the flight deck, hangar and fuelling compartment, a deep fat fryer extinguishing system in the galley, and portable fire-fighting equipment. Fighting fires, fumes and flooding is based on isolating the relevant rooms and providing redundant controls for the relevant installations. This minimises the risk of any potential spread of smoke and heat.

The aft section in each ship is dominated by a 16.6m x 14m (232.7m^2) helicopter deck. Helicopter operations are supported by refuelling and traversing systems, as well as a night landing capability comprising a Helicopter Visual Landing Aids (HVLA) suite with a glide path indicator, a horizon bar, deck lighting and status lights. Operations can be sustained in weather conditions up to Sea State 5. The 15.8 x 6.5m (102.7m^2) hangar allows the OPVs to accommodate the Kaman SH-2G (I) Super Seasprite SH-2G (I) and NH Industries NH-90 helicopters operated by the Royal New Zealand Air Force (RNZAF), providing a basic maintenance facility. Alternatively, unmanned aerial vehicles (UAV) can be embarked. The incorporation of the flight deck and hangar significantly increase the ships' operational flexibility.

The RNZN's role of policing of NZ's extensive waters obviously requires the *Otago* class's crews to conduct frequent boarding operations. The OPVs are well fitted-out for this task, with their main assets arguably comprising two 7.4m rigid-hull inflatable boats (RHIBs). They are launched and recovered by wave-compensated single point davits port and starboard, which allow the RHIBs to be launched in up to Sea State 4 conditions whilst the ships are travelling at 10 knots.

To maximise versatility, the ships have an extended working deck aft that provides space for up to three Twenty-foot Equivalent Unit (TEU)

A view of *Otago*'s foremast shows navigation and communications equipment but few of the sensors typically found on front-line warships. An exception is the ball-like Rafael Toplite EO sensor located on the forward face of the mast, which controls the main gun. *(Guy Toremans)*

Otago's spacious bridge is fitted with an integrated bridge system that allows operation by just three crew during transit cruising. The IPMS enables full propulsion plant monitoring from the bridge. *(Guy Toremans)*

containers. These can incorporate workshops with equipment to support specialist tasks – such as anti-pollution control, diving or MCM support activities – or be outfitted as a medical facility. The space can also be used for racks for larger, 11m RHIBs. If required, the ships can also be modified into hydrographic research vessels. The working deck is serviced by a crane with a safety working load (SWL) of 10 tons at a 10.5m outreach. The OPVs also have a standardised NATO 1 Probe Receiver replenishment station on their starboard side.

A noteworthy feature of the *Otago* class design is its focus on seaworthiness and stability so as to support safe deployment in all but the most extreme sea states. The ships are very stable in high seas and strong winds, whilst turning well and accelerating quickly. A slender, bulbous bow allows for higher speeds in appropriate conditions. The ships are fitted with non-retractable Rolls Royce Gemini 30 electro-hydraulic fin stabilisers, assisting maintenance of full operational capabilities up to Sea State 6 and supporting the helicopter and RHIB recovery characteristics already described. The class's excellent seaworthiness has already been demonstrated in deployments to the Ross Sea in Antarctica, where very adverse sea conditions have been encountered. A degree of 'winterisation' is installed to assist with Antarctic conditions.

ENGINEERING ASPECTS

The *Otago* class are powered by two MAN B&W 12V RK280 diesels engines. Each develops a maximum power output of 5,400kW and drives one of two shafts fitted with five-bladed Wärtsilä controllable-pitch propellers by means of twin reduction gearboxes. The engine room is fully automated and the overall propulsion arrangement is managed by an engine control system accessed from the machinery control room or bridge.

The propulsion configuration provides 14,680shp – nearly twice that of the RNZN's *Anzac* frigates' diesels – and allows speeds up to 22 knots in Sea State 4 with minimal discomfort to the crew. The ships can operate on one shaft at 19 knots. Considering New Zealand's long coasts and extended area of maritime interest, an important design criterion was overall endurance. This amounts to an impressive 6,000 nautical miles at a cruising speed of 15 knots or 8,000 nautical miles at 12 knots. The class is also designed to sustain autonomous operations for up to twenty-eight days.

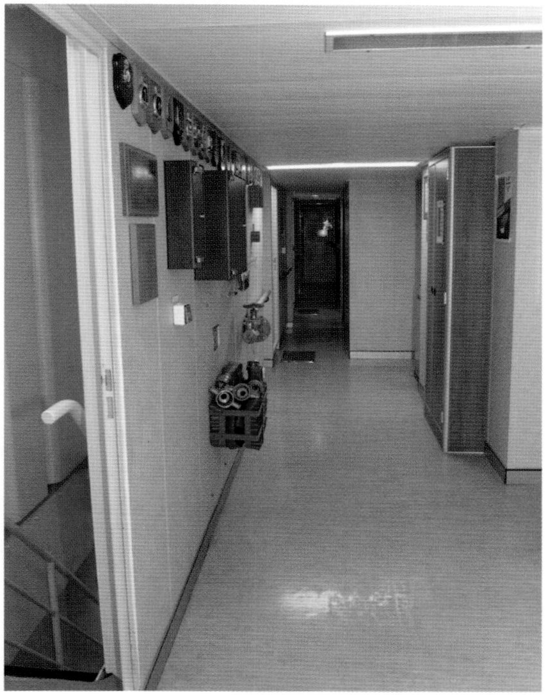

A combination of lean manning and relatively large size makes the *Otago* class spacious ships, a feeling reinforced by the use of wide gangways and corridors. Accommodation standards are also high, as evidenced by this view of a pretty officer's cabin. *(Guy Toremans)*

The main propulsion system is supplemented by three diesel generators which are located forward of the engine room and which produce electricity for hotel services. There is also a supplementary emergency generator located at Deck 01 level forward of the hangar. The class uses twin rudders with independent movement and a Wärtsilä CT150H transverse bow thruster with an output of 450kW to improve low speed manoeuvrability.

The platform management function is carried out by the usual Integrated Platform Management System (IPMS). This monitors and controls the ships' propulsion, generating, electrical and air-conditioning equipment and performs a central role in the damage-control function previously described. As previously mentioned, the IPMS enables full propulsion plant monitoring from the bridge. The steering system has nine separate positions, six of which can be controlled from the bridge by either the officer of the watch (OOW) or the helmsman. The command centre, also integrated into the IPMS, synergises the management of navigation and security functions in order to achieve greater operational efficiency, especially during maritime constabulary operations.

An integrated bridge system (IBS) provides for all navigation, meteorological and associated requirements. It includes an electronic chart display and information system (ECDIS), heading and speed sensors, an autopilot, a magnetic compass, an automatic identification system (AIS), an echo sounder, a speed log, a GPS navigator, and an integrated radar/navigation/AIS display system. The level of functionality provided is a key factor in reducing the number of personnel required for ship operation.

Indeed, the achievement of lean manning arrangements was a critical factor underlying the overall Project Protector. As such, the *Otago* class design aims to minimise crew numbers by maximising automation and reducing maintenance wherever possible. The platform can be operated by just thirty-five personnel, although the core crew is normally forty-two. During transit cruising the ships can be controlled by a crew of just three on the bridge. Additionally, when leaving harbour, the commanding officer has the ability to conduct all propulsion-control functions from a bridge pitch control lever console. One consequence of this 'lean manning' concept is a requirement for the crew to be sufficiently well-trained to carry out multiple taskings.

WEAPONS SYSTEMS AND SENSOR SUITE

The *Otago* class is not intended to be used for warfighting operations. This is reflected, for example, in the lack of a combat-management system found in front-line surface combatants. The class's essentially constabulary role is also evidenced by their minimal weapons fit, focused on a single Rafael Typhoon 25mm stabilised naval gun mounted forward of the bridge. This weapon system has a rate of fire of 200rpm at an effective range of 1,800m. Line-of-fire stabilisation enables the crew to engage targets with high levels of precision from a safe stand-off distance and in rough sea conditions. The Typhoon replaced the MSI DS 25M Autsig 25mm gun originally fitted to the ships under an upgrade programme carried out in 2014. The ships also mount two General Dynamics/US Ordnance M2HB .50 calibre (12.7mm) Browning machine guns located port and starboard. They have a rate of fire of 840rpm. This weapon suite is supplemented by light machine guns as required, whilst boarding crews can be equipped with personal weapons.

When embarked, the SH-2G (I) Super Seasprite helicopter can provide the ships a modest anti-surface (ASuW) and anti-submarine (ASW) capability. The SH-2G (I) variant replaced older SH-2G (NZ) helicopters in the course of 2016, also bringing the ability to deploy Kongsberg Defence & Aerospace AGM-119 Penguin Mk 2 Mod 7 anti-ship missiles in lieu of the previous Raytheon AGM-65 Maverick. The helicopters can also deploy up to two Mk 46 Mod 5 lightweight homing torpedoes, which can be used in water as shallow as 40m, as

Table 3.3.2.

OTAGO (PROTECTOR CLASS OPV) PRINCIPAL PARTICULARS

Building Information:	
Laid Down:	16 December 2005
Launched:	18 November 2006
Delivered:	18 February 2010
Builders:	Tenix Defence (now BAE Systems Australia) at its Williamstown yard in Melbourne, Australia[1]
Dimensions:	
Displacement:	1,900 tons full load displacement.
Overall Hull Dimensions:	85.0m x 14.2m x 3.6m. Length at waterline is 77.6m.
Equipment:	
Armament:	1 x 25mm Rafael Typhoon, 2 x Browning machine guns.[2]
Aircraft:	1 x SH-2G (I) Super Seasprite helicopter (platform and hangar).[3]
Principal Sensors:	Surface-search and navigation radars. Rafael Toplite EO director for main gun.
Combat System:	No combat-management system. Communications package includes HF/VHF and UHF radio links plus satellite communications.
Other:	2 x 7.4m RHIBs for boarding operations. Working deck can transport up to three containers with various equipment fits or larger RHIBs.
Propulsion Systems:	
Machinery:	2 x MAN B&W 12RK 280 diesels each rated at 5.4MW. Two shafts. Bow thruster.
Speed:	Maximum speed on main diesels is 22 knots. Range is 6,000 nautical miles at 15 knots.
Other Details:	
Complement:	Normal crew is 42 (9 officers) plus a 12-strong flight team. Accommodation is provided for 80 personnel.
Class:	Two ships: *Otago* (P148), *Wellington* (P55). A planned third OPV is likely to be of different design.

Notes:

1. Superstructure blocks were constructed at Tenix Shipbuilding New Zealand, Whangarei, where fabrication commenced in February 2005.
2. Originally 1 x 25mm MSI DS 25M Autsig, replaced by the Typhoon in 2014.
3. Other helicopters in the NZDF inventory can be embarked. The SH-2G (I) Super Seasprite variant replaced the older SH-2G (NZ) type during 2016.
4. Classification is LRS 100 A1 SSC PATROL MONO G5 - ICE CLASS 1C, LMC UMS

well as depth charges and/or a M60 machine gun.

The ships are endowed with two surface-search/navigation and search radars, a Warrlock direction finding system, an obstacle avoidance sonar (OAS) and Rafael's Toplite multi-sensor optronic payload, the last-mentioned replacing the ships' original Vistar 350 EO system. The Toplite EO system incorporates a laser designator, an advanced correlation tracker and a forward-looking infrared sensor, thus providing the OPVs the capability to detect and track targets in all weather conditions,

An indispensable requirement for effective performance of constabulary duties is the provision of a comprehensive suite to enable the ships to communicate with all of the appropriate authorities. As such, the *Otago* class are provided with a full range of military and civilian communication capabilities, including radio links, Inmarsat and Satcom.

ACCOMMODATION

The combination of the *Otago* class's relatively large size with a lean-manned crew makes for spacious ships with high standards of accommodation. As mentioned previously, the base line crew is currently forty-two (of which nine are officers), to which must be added a twelve-strong flight crew if a helicopter is embarked. However, there are sufficient berths for as many as eighty personnel, with up to thirty-six spare berths for crew under training, specialised personnel or staff from other government agencies.[8]

The sense of spaciousness is reinforced by wide gangways and corridors, as well as excellent accommodation facilities. Interiors are very comfortable and suited for a mixed-gender crew. The commanding officer has his own living and office space, a bedroom and a private bathroom. The heads of department have single cabins, while the majority of the crew are housed in two-berth accommodation with private facilities. Junior ratings are allocated four-berth cabins with private lavatories and

HMNZS *Wellington* (2010)
1:500 scale

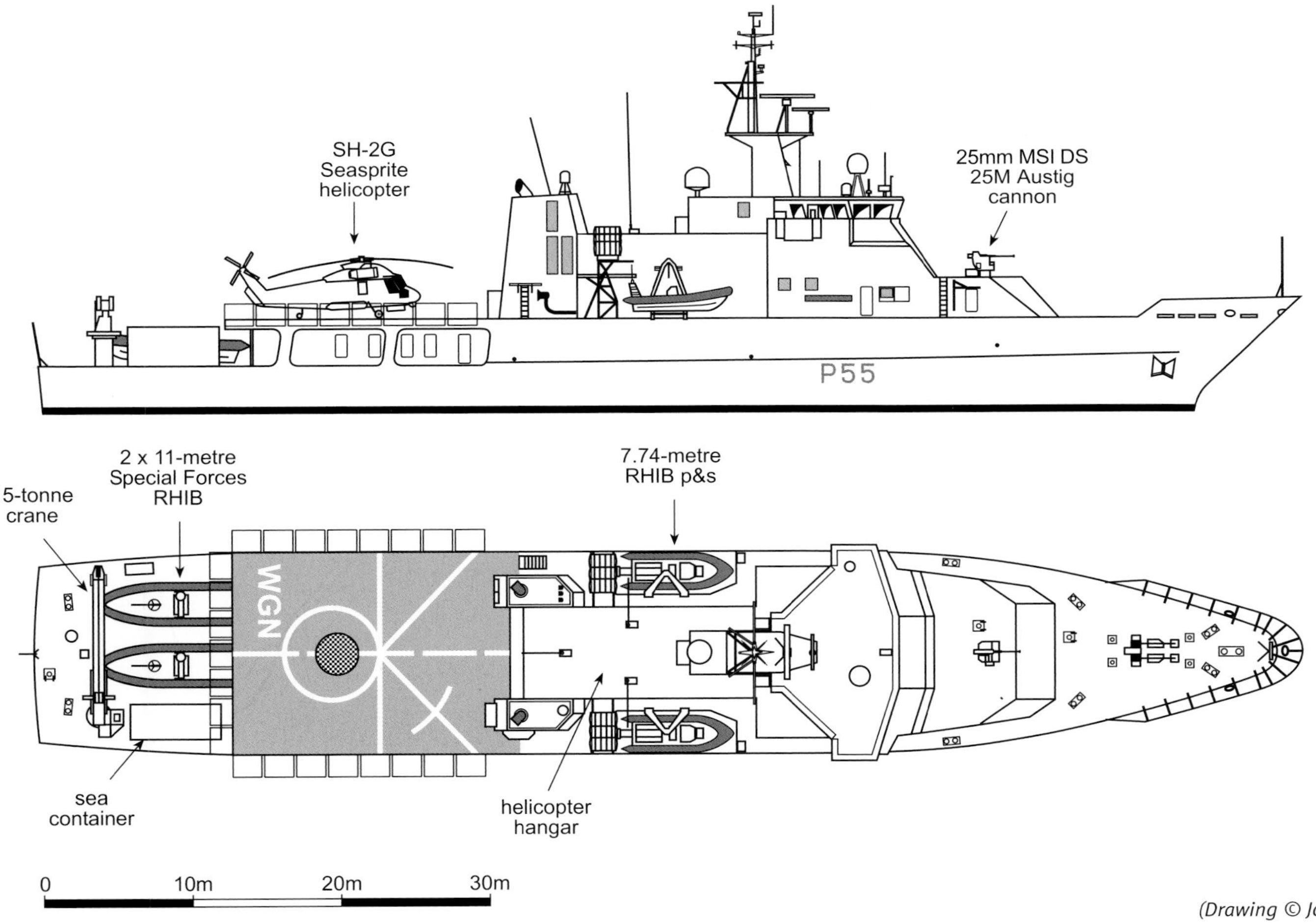

(Drawing © John Jordan, 2017)

The *Otago* class are primarily intended for peacetime policing operations, making the ability to launch seaboats in a wide range of weather conditions an important attribute. This image shows one of *Otago*'s 7.9m RHIBs, which are launched and recovered by wave-compensated single point davits. *(Guy Toremans)*

showers. Embarked trainees and supplementary personnel have six-berth cabins, all with en suite toilets and washrooms. The ships feature a well-equipped galley, a wardroom for the officers, a mess for the petty officers and a cafeteria for the junior ratings. There is also a laundry, library, gym and a sick bay.

Considerable attention has been paid to ensuring crew safety and comfort in the often hostile sea conditions the OPVs can encounter. A number of revisions have already been made to the initial design, benefitting from experience that has been gained. The need to operate in potentially appalling weather conditions across a demanding maritime environment has seen additional handrails installed around most ladders and walkways. The fin stabilisers have been supplemented by an anti-heeling tank that helps to reduce roll at speeds below those where the stabilisers are effective.

This stern view of *Otago* turning at speed in May 2011 shows the effectiveness of the ship's stabilisers in maintaining stability during high-speed manoeuvres. Post completion improvements have included the addition of an anti-heeling tank to reduce roll at speeds below those where the stabilisers are effective. *(Royal Australian Navy)*

MISSION PROFILE

Primarily intended to carry out peacetime policing tasks, the *Otago* class performs a wide range of missions. These include surveillance operations, fishery protection, border and customs patrol, counter-drug and counter-terrorism operations, the prevention of illegal immigration, search and rescue, the provision of humanitarian aid/disaster relief, defence diplomacy and the simple maintenance of a maritime presence. The ships are also able to collect environmental data and provide limited logistical support for smaller vessels. Many operations are carried out in coordination with a wide range of other New Zealand government agencies such as the police, customs service or Ministry of Fisheries or in conjunction with similar organisations maintained by partners amongst the Pacific island nations. For example, fisheries inspection officers may be embarked to monitor and carry out compliance checks on fishing vessels to deter illegal activities.

The two ships routinely deploy throughout New Zealand's own extensive EEZ in support of these duties, for example under Operation 'Kauwae' (inshore fishery protection) and Operation 'Zodiac' (offshore fishery protection). However, they also regularly travel further afield – as far south as the Southern Ocean and the Ross Sea in Antarctica as well as northwards into the south-western Pacific. The deployments to the Southern Ocean and Antarctica are frequently undertaken under the Operation 'Castle' and Operation 'Endurance' banners; those to the south-western Pacific and beyond often form part of Operation 'Calypso'. The 'Castle' series of deployments are essentially focused on regulating the Antarctic fishery in the Commission for the Conservation of Antarctic Marine Living Resources (CCAMLR) regions and involve surveillance and boarding operations. The 'Endurance' series support the Department of Conservation (DoC) and the MetService in resupplying and conducting maintenance on their outposts and assets on the remote sub-Antarctic islands. For this mission the ships also embark inspectors, a maritime survey team and hydrographers. The 'Calypso' deployments include a similar

A picture of *Wellington* operating in Antarctic waters in October 2015. The arduous nature of such deployments is reflected in a four-week-long work-up and one-week period of cold-weather training before setting sail. *(RNZN)*

fishery protection focus as Operation 'Castle' but also involve a considerable element of regional defence diplomacy.

The arduous nature of some of these missions is reflected in the degree of forward preparation involved. For example, prior to deployments to the Southern Ocean and Antarctica, ships undergo a four-week-long work-up and a one-week period of cold-weather training. These preparations encompass survival, first aid and medical proficiency; damage control, firefighting and flood containment exercises; boarding-team training; and force protection serials. All are carried out under the watchful eyes of the Maritime Operational Evaluation Team (MOET) 'sea riders'.

OPERATIONAL EXPERIENCE

Upon completion of their contractor trials both OPVs went through extensive acceptance trials. Subsequently, a period of work-up was concluded by evaluation by the MOET. In December 2010 both ships headed south for a week-long operation around the Campbell and Auckland Islands; the first time the RNZN's new OPVs had sailed together. Since then the two ships have been working hard earning their keep, as demonstrated by the following highlights of their operational activity:

HMNZS *OTAGO* (P148)

Otago sailed for her very first operational mission on 22 February 2011, providing humanitarian aid and disaster relief support to the people of the badly damaged city of Christchurch on South Island after the city was hit by a severe earthquake. After providing supplies, assisting with food distribution, clearing debris and using the ship's osmosis plant to produce fresh water, the ship was released from Operation 'Christchurch Quake' on 6 March.

Two months later, *Otago* departed for her maiden long-range deployment, a fishery protection mission to the south west Pacific under the Operation 'Calypso' banner. Ports of call included Raoul Island, Tonga, Samoa, Tokelau, Rarotonga and Niue The mission demonstrated the new OPVs' capability to assist the Pacific islands with the protection of their EEZ and resources.

In 2012 – between 7 May and 18 May – *Otago* conducted comprehensive first-of-class flight trials with a Kaman SH-2G (NZ) Super Seasprite helicopter of the RNZAF's No 6 Squadron in the Hauraki Gulf and Bay of Plenty. These tests involved sixty-one day and night landings and take-offs in varying wind and sea conditions, winching operations, inflight refuelling trials, and simulations of various operational emergencies. The trials demonstrated that the Super Seasprite was well suited to conduct airborne surveillance tasking from the OPVs, with flights of up to two and a half hours duration enabling them to cover about 4,500 square nautical miles in a single sortie.

The summer of 2012 saw *Otago* undertake a further deployment under Operation 'Calypso' to the south west Pacific. The ten-week deployment included fishery protection patrols and a three-week long 'forward basing' trial in Nuku'alofa in Tonga. The main purpose was to put the ship into a maintenance period away from Devonport Naval Base to prove the ships' capability of maintaining patrols over a prolonged period far from their homeport. The ship also provided a diplomatic presence at celebrations to mark the fiftieth anniversary of Samoan independence and provided passage for the Governor-General of New Zealand to the Tokelau Islands for his inaugural visit.

In September 2012, *Otago* conducted a further series of flight trials in the Hauraki Gulf, this time with the Iroquois helicopters of RNZAF's No 3 Squadron.[9] Thereafter, another long-range mission to the Southern Ocean and Antarctica under Operation 'Castle' took the ship away from home until the year's end.

2013 was another busy year. After refresher force protection training and a MOET evaluation, *Otago* set sail for Operation 'Endurance' to the Southern Ocean. This was followed by a deployment to the south-west Pacific for Operation 'Calypso' to conduct fisheries patrols and undertake visits to Samoa, Tokelau and the Cook Islands. During her port call in Samoa, the OPV took part in the UN-led Small Island Developing States' (SIDS) symposiums in Tokelau.

The period between October 2014 and June 2015 saw *Otago* undergo a major maintenance period, including armament and sensor modifications. This was followed by work-up and MOET safety and readiness checks. Subsequently, she departed Devonport Naval Base at the end of June 2015 for a ten-week mission to the Pacific Islands. During this period she took part in celebrations marking the fiftieth anniversary of independent governance in Rarotonga. Refresher training in October then prepared her for deployment southwards for the 2015/16 summer season in Antarctica.

Much of 2016 was spent in home waters. In August 2016 she successfully completed an aviation safety and readiness check in the Hauraki Gulf. Thereafter, she sailed to the Kermadec Islands under Operation 'Harve'. This deployment marked a quantum leap in capability for the *Otago* class as, for the first time since their commissioning, a Super Seasprite embarked for the duration of an operation. Upon her return, the crew prepared for a key role in the RNZN's seventy-fifth anniversary celebrations in Auckland. During the International Naval Review on 19 November 2016 she assumed the role of reviewing ship, embarking New Zealand's Governor-General, the Right Honourable Dame Patsy Reddy, and numerous other dignitaries for the occasion.

The start of 2017 saw *Otago* depart Devonport for another re-supply mission to the DoC's outposts in the Auckland and Campbell Islands. After a period of rest and recreation, the OPV then departed for yet another ten-week deployment patrolling the Pacific islands' fisheries.

Otago pictured in company with one of the *Rotoiti* class inshore patrol vessels at the time of her maiden arrival in New Zealand waters in 2010. *Otago* and her sister *Wellington* have been used far more extensively than the Project Protector inshore patrol vessels, reflecting the greater usefulness of larger vessels with better seakeeping qualities and endurance in policing New Zealand's extended maritime domain. This is reflected in the recent decision to acquire a third OPV, which will effectively replace the *Rotoiti* class ships. *(RNZN)*

HMNZS *WELLINGTON* (P55)

Wellington's first major operation took place between late December 2010 and February 2011 when she undertook the new class of OPVs' initial deployment to Antarctic waters as part of cold-weather trials. During the summer, she embarked six specialist fishery officers for the Operation 'Zodiac' deep water fishery protection mission, a patrol which reportedly saw the first ever boarding and inspection of foreign-flagged vessels by New Zealand on the high seas. Subsequently, in October 2011, she departed Devonport to participate in the Australian-led Operation 'Render Safe' to remove remnants of Second World War ordnance around Rabaul in Papua New Guinea. Operating in conjunction with the now-retired HMNZS *Resolution*, the Australian MCMVs *Diamantina* and *Gascoyne,* as well as Papua New Guinea service personnel, *Wellington*'s involvement was a good example of collaborative inter-agency work promoting New Zealand's wider regional interests.

Crewing constraints impacted *Wellington*'s availability during the latter half of 2012 and into 2013. However, she undertook further re-supply and fishery protection missions, as well as additional ordnance disposal activities in the Solomon Islands area (Operation 'Pukaurua'). The last-mentioned were carried out in conjunction with the diving support vessel HMNZS *Manawanui*, personnel from the navy's Littoral Warfare Unit (LWU) and specialists from other nations.[10]

The ship's operational tempo picked up during 2013, partly as a result of a decision to prioritise

A picture of *Wellington* shortly after completion. She has been extremely active in her career to date, with deployments reflecting the varied nature of New Zealand's constabulary requirements. As well as undertaking the class's initial deployment to Antarctic waters in 2011, she also became the first of the class to travel as far north as the equator, which she crossed as part of the Operation 'Calypso' fishery protection and defence diplomacy mission in 2016. *(RNZN)*

crewing of the OPVs over the smaller inshore vessels. Between March and September, she spent only thirty-nine days in Auckland compared with over 145 days at sea and twenty days in foreign ports. Deployments included further fishery protection missions under Operations 'Zodiac' and 'Kauwae', support for further Second World War ordnance disposal and the international exercises Pacific Partnership 13 and Southern Katipo 13. These were followed by further resupply activities at the start of 2014 during which the ship hosted secondary school students and subject experts as part of a 'Young Blake Expedition'. A maintenance period at Devonport saw *Wellington* come out of refit in June 2014 with a new Typhoon gun and other improvements.

Subsequent years have seen a continuation of this established pattern of deployments supporting constabulary activities, resupply and scientific activities and diplomatic engagement. For example, highlights in 2015 included a mission to the Kermadec Islands with scientists of the Institute of Geological and Nuclear Sciences Limited and Woods Hole Oceanographic Institute embarked. The scientists conducted survey operations on the Macauley Caldera, an underwater volcano about 100km south of Raoul. Amongst the ten tons of scientific research equipment taken on board was a state-of-the-art Sentry autonomous underwater vehicle (AUV), a fully autonomous submersible capable of exploring the ocean down to 6,000m. The Sentry produced bathymetric, side scan, sub-bottom, and magnetic maps of the seafloor, along with digital photographs of a variety of deep-sea terrain.

Another aspect of the class's capability was demonstrated following the devastating Category 5 cyclone 'Winston' that struck Fiji in February 2016. *Wellington,* accompanied by the multi-role vessel *Canterbury,* embarked 60 tons of aid supplies – including canned food, bottled water and vaccines – and the LWU as part of a major humanitarian response. Upon arrival on station, the LWU team conducted seabed surveillance, and surveyed the coastline to determine mooring spots for larger aid-carrying vessels.

Later in 2016, *Wellington* became the first RNZN OPV to cross the equator as part of an Operation 'Calypso' deployment. Upon her return from this mission she conducted trials with the Scan Eagle remote piloted aircraft system, supported by a four-person team from Insitu Pacific. The use of unmanned aerial vehicles offers the potential of augmenting the OPVs' intelligence, surveillance and reconnaissance (ISR) capabilities. A further technological development has been the installation of a new weather station developed by the Defence Technology Agency and which is designed to provide enhanced forecasting of weather conditions for both helicopter and boat operations. The system proved its worth during an emergency deployment to Kaikoura on the country's South Island on another disaster relief after the region was struck by a magnitude 7.8 earthquake in November 2016.

Over the coming years *Otago* and *Wellington* will both undergo sequential upgrades to their sensor, communication and propulsion systems in order to extend their life well into the 2030s.

A THIRD OFFSHORE PATROL VESSEL

Nations are increasingly looking to the sea for additional food and energy resources. However, as such resources become depleted elsewhere, the pressure on under-exploited sea areas such as those typified by New Zealand's EEZ and the Southern Ocean are increasing in attraction. This is resulting in an increase in illegal activity, notably unregulated and unreported fishing. Another source of pressure is the increase in human activity across the Antarctic

Left: Although the *Otago* class's mission profile is largely focused on constabulary duties, it is important to note that they can be required to operate in a wide range of environmental conditions. This is evidenced by these two views of *Wellington*, both taken during the course of 2016. Although the view of Antarctica was taken in sunny weather, the adverse weather that can be encountered in the region can make southern deployments quite challenging. This is a factor in the planned acquisition of a third OPV, which is likely to be larger and incorporate additional ice protection. *(RNZN)*

continent, including both scientific activity and tourism. An enhanced framework of international controls is already being put in place to counter these issues; the Ross Sea Region Marine Protected Area will come into force in December 2017 and the requirements of the International Maritime Organization's new Polar Code are steadily coming into effect. Given this backdrop, it is not surprising that the need to protect the Southern Ocean and Antarctica was given increased emphasis in the recently released *Defence White Paper 2016.*[11]

This change in emphasis will be accompanied by a number of alterations to the RNZN's future force structure, notably the ice-strengthening of the long-planned replacement for the replenishment tanker *Endeavour* and the acquisition of a third, ice-strengthened OPV. These plans were confirmed in a revised *Defence Capability Plan 2016* published in November 2016.[12] The new patrol vessel is likely to be larger than the current ships and has been referred to as an 'ocean patrol vessel' by Rear Admiral John Martin, the RNZN's head. The aim is to give the navy the ability to conduct patrols across a wider range of ice conditions than are possible with the *Otago* class. A consequence of the decision to expand the OPV fleet from two to three vessels will be the steady withdrawal of the existing fleet of inshore patrol vessels, which will ultimately be entirely removed from service.

The NZDF is currently in the early stages of drawing up the requirements for the new ship, including the specific design features required for operating in the Polar region. This project planning phase is expected to be completed by early 2018, with the construction phase likely to be initiated in 2019 for an in-service date early in the 2020s. Clearly, the navy's experience with the *Otago* class design will provide a solid foundation for developing design requirements, although a larger hull is likely. It is also anticipated that the new platform will

Two pictures of *Otago* carrying out first-of-class trials with an RNZAF Super Seasprite helicopter in May 2012. The ability to support helicopter operations is likely to remain an important feature of the planned, third OPV but greater provision for the new generation of unmanned systems now coming into service is also likely to be a priority. *(RNZN)*

Views of *Otago* operating with a Samoan patrol boat off Apia during her first 'Calypso' deployment in 2011 and of *Wellington* in the Antarctic. The acquisition of the two *Otago* class vessels has provided a major boost to New Zealand's constabulary tasking. *(Royal Australian Navy/RNZN)*

leverage off new technologies, such as unmanned systems, to deliver a more efficient and effective patrol capability. The RNZN is anticipating significant interest from internal shipbuilders and design consultancies once specific requirements are finalised. Vard certainly has a wide range of designs – including for the new Canadian Arctic Offshore Patrol Ship (AOPS) – that could potentially meet the navy's needs, although a derivative of the Damen Sigma type or of Fassmer's OPV80 could be contenders.

CONCLUSION

The induction of the two *Otago* class OPVs into the fleet has provided a significant uplift to New Zealand's maritime patrol capability. These platforms provide the RNZN with greater capacity to carry out sustained operations and to maintain a maritime presence under almost all climatic condi-

Although the *Otago* class have proved capable and useful in service, Antarctic weather conditions can prove a challenge, as evidenced by this view of *Wellington*. The improved capabilities for the planned third OPV should help resolve this problem. *(RNZN)*

tions across the vast maritime domain that falls under its responsibility. This has included an ability to extend patrols to areas that were not previously subject to regular oversight.

Another major benefit has been the OPVs' ability to fulfil many of the missions which were formerly carried out by the frigates *Te Mana* and *Te Kaha*. The expense of operating the frigates in secondary roles had been putting a strain on the RNZN's budget. The OPVs can perform many of these missions at far less cost, whilst freeing up the frigates for more appropriate deployments.

Whilst the *Otago* class are undoubtedly very capable platforms for the constabulary role, the RNZN does acknowledge that it can still be quite challenging to operate them in the extreme weather that can be encountered in the Antarctic regions, particularly if they are required to remain at sea in these areas for longer periods. The planned acquisition of a third, possibly larger OPV will, however, go a long way to resolving this problem.

Notes

1. New Zealand is made up of two large islands, the 114,000km^2 North Island and the 151,000km^2 South Island. There are also a large number of smaller islands, of which the 1,700km^2 Stewart Island is the most extensive. Others include the Antipodes, Auckland, Bounty, Campbell, Chatham, Kermadec, Snares and Solander islands, some located at considerable distance from the main island group.

2. New Zealand proper, together with the Antarctic claim of the Ross Dependency, the dependent territory of Tokelau and the associated states of the Cook Islands and Niue officially form the Realm of New Zealand.

3. See *The Government's Defence Policy Framework* (Wellington: Her Majesty's Government in New Zealand, June 2000). A copy can be found at: http://www.nzdf.mil.nz/downloads/pdf/public-docs/defencepolicyframework june2000.pdf

4. The *Defence Policy Framework* set provision of effective air and naval transport capabilities as another key priority, explaining the acquisition of the multi-role transport ship. Patrol capabilities were also informed by a *Maritime Patrol Review* (Wellington: Department of the Prime Minister and Cabinet, 2001), which reported in February 2001.

5. Contemporary press reports suggested the two OPVs cost around NZ$90m (US$40m) each.

6. *Otago* is named after the Royal Navy *Rothesay* class (Type 12) frigate *Otago*. Launched on 11 December 1958 the frigate was commissioned into the RNZN on 22 June 1960. She paid off into inactive reserve in November 1983, subsequently being sold for scrap in 1987 and broken up at Auckland. *Wellington* is named after the former Royal Navy *Leander* class frigate HMS *Bacchante*, acquired by the RNZN in 1982 and renamed HMNZS *Wellington*. Decommissioned in 1999 the frigate was deliberately sunk in Houghton Bay, off the south coast of Wellington on 13 Nov 2005. A previous HMS *Wellington*, a *Grimsby* class sloop that served in New Zealand waters in the 1930s, remains afloat on the River Thames as headquarters ship of the Honourable Company of Master Mariners.

7. The Vard Marine OPV portfolio has gone through several changes of ownership, most recently being part of the STX Canada Marine business before the latter's acquisition by Vard in 2014. It was previously part of the Aker design business and was part of Kvaerner Masa Marine when the *Otago* class were ordered.

8. These could include personnel from the New Zealand Customs Service, the Ministry of Foreign Affairs and Trade, the Ministry of Fisheries, Maritime New Zealand and the New Zealand Police.

9. The Bell UH-1H Iroquois helicopters formerly operated by the RNZAF have now been withdrawn from service. Their replacement is the TTH variant of the Eurocopter NH-90 helicopter.

10. The LWU is the collective name for a group of units operating primarily in coastal and shallow waters, including divers, hydrographers and mine countermeasures specialists. It is intended to ensure safe access to and use of harbours and coastal zones.

11. See *Defence White Paper 2016* (Wellington, Ministry of Defence, 2016). A copy is currently available at: http://www.nzdf.mil.nz/downloads/pdf/public-docs/2016/defence-white-paper-2016.pdf

12. See *New Zealand Government Defence Capability Plan 2016* (Wellington: Ministry of Defence, 2016). A copy can currently be found at: http://www.defence.govt.nz/assets/Uploads/2016-Defence-Capability-Plan.pdf

4.1 TECHNOLOGICAL REVIEW

Author:
David Hobbs

WORLD NAVAL AVIATION

An Overview of Recent Developments

2017 saw many navies making progress with their ship and aircraft re-equipment programmes. However, the technical complexity of new systems has led to time slippage with delivery of the aircraft carriers *Gerald R. Ford* (CVN-78) and *Queen Elizabeth*, together with the three variants of the F-35 Lightning II. China has continued its ambitious programme of naval aviation expansion and India, too, plans a larger carrier force but has yet to translate all its ideas into hard realities. Among the medium-sized navies, Australia and France seem to be achieving long-term goals with pragmatic approaches to platform procurement and deployment.

The US Navy took delivery of the lead *Ford* class carrier *Gerald R. Ford* (CVN-78) on 31 May 2017, rebuilding the fleet to the Congressionally-mandated level of eleven aircraft carriers. She had previously undertaken a series of builders' sea trials and acceptance trials; this image shows her heading for sea for the first time at the start of builders' trials on 8 April 2017. *(Huntington Ingalls Industries)*

US NAVY AIRCRAFT CARRIERS AND THEIR AIRCRAFT

Gerald R. Ford arrived at Naval Station Norfolk, Virginia, on 14 April 2017 to prepare for acceptance trials originally scheduled by the US Navy for 2016.[1] Project staff described 'complexities introducing several first of class technologies into service' which have led to slippage. The necessary trials were conducted from 24 to 26 May 2017 and were followed by formal delivery on 31 May. Once commissioned, she is to undergo extensive flying trials that will eventually clear every type of US naval aircraft to use her catapults and arrester wires or – in the case of vertical landing aircraft – to operate from her deck. Construction of the remainder of the class has undergone an objective and technically deep review aimed at cost reduction. This will lead to the replacement of the Dual Band Radar (DBR) by the more affordable Enterprise Air Surveillance Radar (EASR) in subsequent ships. This is based on the Raytheon S band (NATO E/F band) SPY-6 Air and Missile Defence Radar (AMDR) that will equip the Flight III *Arleigh Burke* (DDG-51) class destroyers. The first EASR was originally to be installed in

Although most media attention has inevitably focused on the delivery of *Gerald R. Ford* (CVN-78), the ongoing series of midlife refuelling and complex overhauls (RCOHs) of existing ships is equally important. 12 May 2017 saw Huntington Ingalls Industries redeliver *Abraham Lincoln* (CVN-72), the latest carrier to have gone through this process. These images taken the previous day show the ship undertaking high-speed turning tests during end of refit trials. *(US Navy)*

Enterprise (CVN-80) – the third ship in the class – but it has now been found possible to fit it in the second, *John F. Kennedy* (CVN-79). EASR is to be produced in two versions: a fixed-face array for the *Ford* class nuclear-powered aircraft carriers (CVNs) and a rotating variant for amphibious assault ships (LHAs/LHDs). Raytheon has been awarded a US$92m contract to develop the new system, with options of US$723m covering sixteen ship-sets of equipment: six for carriers and ten for assault ships.

The Pentagon now seemingly accepts that much of the root cause of the eye-watering costs of projects such as the F-35 Joint Strike Fighter and *Ford* class carriers stems from Donald Rumsfeld's policy as Secretary of Defence in the early 2000s when he directed that future weapons systems were to be transformational, seeking for the limits of the technically possible. His successors now have to attempt to manage those projects into affordability. In this regard, some progress appears to be being achieved. For example, US Navy has elected to retain the Electro-Magnetic Aircraft Launch System (EMALS) and Advanced Arrester Gear (AAG) rather than revert to earlier technologies in CVN-79 and CVN-80 because development problems are now being successfully overcome and they have greater potential for future growth. In addition, the costs of reversion in the next two ships would also have been high.

Another transformational system making progress is the Naval Integrated Fire Control-Counter Air (NIFC-CA) architecture. This is a form of co-operative engagement capability (CEC) that seeks to combine aircraft and ship-based weapons and sensors into an integrated network to improve air-defence capabilities. The US Navy describes the E-2D Hawkeye as the central node of NIFC-CA and its ability to defeat hostile air and missile threats. Under this construct, the E-2D's AN/APY-9 radar acts as a primary sensor to cue, for example, Raytheon AIM-120 AMRAAM air-to-air missiles launched by an F/A-18E/F Super Hornet onto

Table 4.1.1: US NAVY PLANNED AIRCRAFT PROCUREMENT: FY2016–FY2018

TYPE	MISSION	FY2016[1] Authorised	FY2017[2] Requested	F2017[2] Authorised	F2018[3] Requested
Fixed Wing (Carrier Based)					
F-35B Lightning II JSF	Strike Fighter (STOVL)	15	16	18	20 (20)
F-35C Lightning II JSF	Strike Fighter (CV)	6	4	8	4 (6)
FA-18E/F Super Hornet	Strike Fighter (CV)	5	2	14	14 (14)
EA-18G Growler	Electronic Warfare	10[4]	0	0	0 (0)
E-2D Advanced Hawkeye	Surveillance/Control	5	6	6	5 (5)
Fixed Wing (Land Based)					
P-8A Poseidon	Maritime Patrol	17	11	11	7 (6)
C-40A Clipper	Transport	1	0	2	0 (0)
KC-130J Hercules	Tanker	2	2	2	2 (2)
Rotary Wing					
AH-1Z/UH-1Y Viper/Venom	Attack/Utility	29	24	26	22 (27)
VH-92A	Presidential Transport	0	0	0	0 (0)
CH-53K Super Stallion	Heavy-Lift	0	2	2	4 (4)
C/MV-22B 22 Osprey	Transport	19	16	19	6 (6)
MH-60R Seahawk	Sea Control	29	0	0	0
Unmanned Aerial Vehicles					
MQ-8C Fire Scout	Reconnaissance	5	1	5	0 (2)
MQ-4 Triton	Maritime Patrol	4	2	3	3 (3)
RQ-21A Blackjack	Tactical reconnaissance	6	8	8	4 (4)
Totals:		153	94	124	91 (99)

Notes:

1. FY2016 numbers relate to the authorised procurement programme. This varied quite significantly from the initial Presidential budget request for just 124 aircraft, with the difference largely comprising additional orders for F-35 and, particularly, FA-18E/F and EA-18G fast jets.
2. Numbers for 2017 relate to the initial FY2017 budget plans requested by the Obama administration and the numbers actually authorised in May 2017. Both columns include a handful of aircraft requested from overseas contingency operations funds.
3. Numbers in brackets reflect purchases for FY2018 previously envisaged in the Future Years Defense Programme numbers provided in the FY 2017 budget.
4. Includes three aircraft funded by Boeing as part of a 2014 settlement of a long-running dispute related to the cancellation of the A-12 Avenger programme in the early 1990s.

targets detected by the E-2D using Link 16 to exchange data between the platforms. APY-9 is also seen as the primary sensor to guide Raytheon Standard SM-6 surface-to-air missiles launched from Aegis cruisers and destroyers against targets beyond their own radar horizons. However, other aircraft can also fulfil the E-2D role. In a September 2016 demonstration a US Marine Corps F-35B detected a target and data-linked track information to the *Desert Ship* (LLS-1) test facility in the White Sands missile range. This integrated the information with its Baseline 9 Aegis combat system and launched a Standard SM- 6 missile to destroy the target.

The future direction of American naval aviation will be heavily influenced by the new Trump administration that took power on 20 January 2017. The administration plans to enhance and enlarge the US armed forces over a twenty-five to thirty-year period. Promises include an enlarged fleet – including twelve aircraft carriers – but it is not clear to what extent or how these are to be realised. The US omnibus spending bill for FY2017 agreed in May 2017 approved more naval aircraft than requested by the previous Obama administration but the subsequent FY2018 budget request actually reduces aircraft purchases previously planned for that year. Greater priority has been given to funding maintenance and other readiness measures, which have suffered badly under the financial constraints of sequestration. Table 4.1.1 provides further detail. Particularly noteworthy features were the re-affirmed plans to continue purchase of F/A-18E/F Super Hornets to try to alleviate a growing strike-fighter shortage but a reduction in acquisition of F-35C carrier variant Lightning IIs. In contrast with the usual situation, no details were publicised on longer-term plans pending conclusion of the new administration's internal deliberations on spending priorities.

Immediately after taking office, the administration's new Defence Secretary James Mattis ordered a review to compare capabilities and cost between the much-delayed F-35C Lightning II and an advanced Block 3 version of the F/A-18E/F Super Hornet. Overseen by Deputy Defence Secretary Bob Work, who has a reputation for careful scrutiny of complicated defence projects, the review was also tasked to examine ways to reduce the cost of the whole F-35 programme. Admiral John Richardson, Chief of Naval Operations, has said publicly that the US Navy 'needed the capability the F-35C brought to the service, supplemented by a healthy cadre of advanced Super Hornets and unmanned aircraft in a blended air wing of existing and future aircraft'. Pentagon spokesman Captain Jeff Davis described Work's review as a 'prudent step to incorporate additional information into the budget preparation process and to inform the Secretary's recommendations to the President regarding critical military capabilities'.[2] Interestingly, NIFC-CA is leading the Navy to think of the F-35C in a way that differs significantly from the Air Force F-35A concept. The F-35C is now seen as a NIFC-CA node that will fuse and distribute information from ships and aircraft, passing targeting information to other platforms that will launch weapons. It is therefore more likely to be part of a strike package largely composed of advanced Super Hornets than to form a strike force comprising only F-35Cs. The Navy's Director of Air Warfare, Rear Admiral DeWolfe Miller, told the Senate Armed Services Committee that the service did not view the study as a competition between the two aircraft. Rather, he said, 'the Navy views our

Although the eye-watering cost of leading-edge 'transformational technologies' is leading to some equipment being cancelled or scaled-back in the interests of economy, some of these systems are starting to show their war-winning potential. This graphic shows data on an incoming over-the-horizon missile attack gathered by a F-35B Lightning II being used to direct a SM-6 surface-to-air missile as part of the Naval Integrated Fire Control-Counter Air (NIFC-CA) architecture; a successful trial of the system on these lines was carried out in September 2016. *(Lockheed Martin)*

F/A-18 Super Hornet and the capabilities that come with the F-35C as complementary'.

Boeing's proposed Block 3 Super Hornet absorbs lessons learned from legacy Hornet and Super Hornet operations and is built for an airframe life of 9,000 hours. This is a big improvement over the Super Hornet's original 6,500 hours and could potentially remove the requirement down the line for yet another service life extension programme (SLEP). It features a balanced approach to survivability including electronic warfare and self-protection systems but is not as stealthy as the F-35. It would incorporate a large-area cockpit display for improved user interface and a more powerful computer, known as the distributed targeting processor network (DTPN). An advanced infrared search and track sensor would also be an option and an enhanced 'data pipe' for passing information known as tactical targeting network technology (TTNT). The latter is already a programme of record for the E/A-18G Growler and the E-2D Advanced Hawkeye. DTPN is also in production for the Growler and the adoption of both systems in the F/A-18E/F would pose minimal risk and relatively early deployment. F-35s cannot transmit on any other data link without compromising their stealth characteristics.

Two Boeing F/A-18E Super Hornets being readied for launch from the deck of the carrier *Dwight D. Eisenhower* (CVN-69) during operations in the Mediterranean in December 2016. Super Hornet production is continuing due to a growing strike fighter shortage and Boeing has proposed an upgraded Block 3 variant to encourage further orders. *(US Navy)*

At present up to thirty percent of the Super Hornet sorties flown within an embarked carrier air wing are in the air-to-air tanker role. This uses up a valuable portion of the aircraft's valuable fatigue lives, further exacerbating the fighter shortage. In order to release these aircraft for their primary mission the US Navy wants the MQ-25A Stingray unmanned aerial vehicle (UAV) to enter service as soon as possible. It has evolved from a conceptual long-range intelligence, surveillance and reconnaissance (ISR) aircraft capable of penetrating contested airspace with an air-to-air tanker capability. Commander Naval Air Forces has conceded that industry has struggled to blend the requirements of ISR and air-to-air refuelling because the two missions require very different types of airframe. Early attempts to identify a 'sweet spot' that satisfied both requirements have obviously proved unsuccessful and air-to-air tanking has been emphasised as the most important mission. This will push the competing design teams towards a wing-body-tail design capable of carrying the maximum amount of fuel internally rather than the projected stealthy flying wing design some had considered the best for penetrating contested airspace. General Atomics, Boeing, Lockheed Martin and Northrop Grumman are all expected to respond to a final US Navy

US Navy sailors prepare to launch an E-2D Hawkeye assigned to the 'Tigertails' of Airborne Early Warning Squadron (VAW) 125 from the flight deck of the forward-deployed aircraft carrier *Ronald Reagan* (CVN-76). VAW-125 was the first to convert to the new type in 2014 and has replaced the 'Liberty Bells' of VAW-115 as part of Carrier Air Wing 5 in Japan. *(US Navy)*

An MV-22B Osprey, from Marine Operational Test and Evaluation Squadron (VMX) 1, lifts off from the flight deck of the aircraft carrier *Carl Vinson* (CVN-70) in June 2016. The V-22 Osprey was in the middle of a successful trial as the planned replacement for the C-2A Greyhound as the logistics platform for future carrier on-board delivery (COD) operations. *(US Navy)*

request for proposals during the latter part of 2017 and a contract for the winning design is expected to follow in 2018. Whichever design is chosen, its control systems and data link packages are expected to remain broadly unchanged from those used by the X-47B unmanned aircraft that demonstrated carrier launch and take-off capabilities in 2013

During 2017 the E-2D Advanced Hawkeyes of VAW-125, the 'Tigertails', forward deployed to Marine Corps Air Station (MCAS) Iwakuni in Japan to form part of Carrier Air Wing 5 (CVW-5). The carrier airborne early warning squadron was the first to convert onto the new type in 2014. It has replaced VAW-115, the 'Liberty Bells', which has returned to the United States for conversion onto the E-2D in due course. Also during 2017, CVW-5's fixed-wing squadrons carried out a phased relocation from the naval air facility at Atsugi to MCAS Iwakuni as part of an agreement between the US and Japanese governments on basing and co-operative defence policy.

The E-2D's UHF-band AN/APY-9 radar operates at frequencies between 300 MHz and one GHz. Fighter-sized stealth aircraft are typically optimised to defeat higher-frequency radar transmissions, meaning UHF-band radars should be able to detect aircraft such as the F-22 Raptor, F-35 Lightning II and foreign equivalents. Indeed, many experts have admitted this may be possible in broad terms but have suggested that such systems cannot generate weapons-quality tracks. However, electronic scanning and new signal processing techniques may be overcoming these problems. As previously mentioned, successful live firings of SM-6 missiles against targets undetected by their own platform's radars using APY-9 data have already taken place as part of NIFC-CA testing. This proves, at least, that the E-2D is actually capable of weapons-quality track generation.

The CMV-22B variant of the CV-22 Osprey is due to enter service in the carrier-on-board delivery (COD) role from 2020 onwards. Its potential was evaluated in 2016 with Ospreys of VMX-1 embarked in the *Carl Vinson* (CVN-70) to allow US Marine Corps (USMC) pilots to familiarise themselves with CVN deck-operating procedures and examine how best to incorporate it into carrier operations. Vice Admiral Mike Shoemaker, Commander Naval Air Forces, subsequently told the US Naval Institute that the evaluations had been very successful. 'By the end of the experiment', he said, 'the crew of *Carl Vinson* had figured out how to land and unload the Osprey in about 20 minutes for passenger delivery missions and about 30 minutes for cargo delivery missions'.[3] That fitted in well with the flight deck operating cycle and he added that the CMV-22 improves upon the C-2 Greyhound it will replace by greatly reducing the manpower burden on the ship because it lands and takes off like a helicopter instead of requiring catapults and arrester wires like a fixed-wing aircraft. 'It takes about 6 folks to launch and recover an Osprey; it would take about 40 or so to man up the ship to recover a C-2, so that is a big benefit.' Also, the Osprey can land on a CVN at night whereas the C-2 cannot; when it enters service the CMV-22B will, therefore, be able to land in day or night conditions and even on days when the rest of the air wing is not flying and the catapult and arresting gear systems are not manned and running. Further manpower reductions are possible because the CMV-22B can move people and goods around the entire strike group like a helicopter. At present a C-2 has to deliver to the carrier for unloading; stores which then have to be reloaded into helicopters for distribution around the fleet, using up both time and other scarce assets. The evaluations clearly suggest that the CMV-22B was the right choice for the replacement COD role.

OTHER AIRCRAFT CARRIER OPERATORS

FRANCE

From February 2017, *Charles de Gaulle* has been undergoing the second of four planned refuelling and major refit periods. This will leave France without an operational aircraft carrier until around the end of 2018. Previously, on 23 June 2016, the US Navy's Chief of Naval Operations, Admiral John Richardson, had awarded the French naval strike group based on *Charles de Gaulle,* with the prestigious US Meritorious Unit Commendation for service in support of Operation 'Inherent Resolve' against the so-called Islamic State (IS) during the 'Arromanches II' deployment. The ceremony took place in Paris where Captain Eric Malbrunot FN, the ship's commanding officer, accepted the award on behalf of the group and listened as Richardson described it as 'a symbol of how far and how developed our relationship has come'. The French carrier served as the command element for US Central Command's Task Force 50 from December 2015 to March 2016, the first time that a non-US Navy ship

France's sole aircraft carrier, *Charles de Gaulle*, undertook a final Mediterranean deployment in the autumn of 2016 before being taken into dockyard hands for a major refit and refuelling. For some of this time she operated alongside the US Navy's *Dwight D. Eisenhower* (CVN-69). The work will leave France without an aircraft carrier until around the end of 2018. *(US Navy)*

had done so. A subsequent 'Arromanches III' deployment to the Eastern Mediterranean for further strikes against IS in the autumn of 2016 marked the first operational deployment of an all-Rafael based strike group following the Super-Étendard's official withdrawal from French service on 12 July 2016.

INDIA

The last twelve months also saw the end of another era when the world's oldest operational aircraft carrier, *Viraat* – the former HMS *Hermes* first commissioned in 1959 – was formally paid off for the last time on 6 March 2017. She is likely to be sold for scrap or sunk as an artificial reef after plans to convert her into a museum and conference centre have seemingly come to nothing. Much useful equipment has now been removed and the Indian Navy's (IN's) few remaining Sea Harrier fighters were previously retired in 2016. Most of her 1,500 ship's company will eventually transfer to the 40,000-ton *Vikrant II*, which is being fitted-out in Cochin Shipyard.

Viraat's withdrawal leaves the IN with a single aircraft carrier, the former Soviet *Vikramaditya*, which – like *Vikrant II* – uses short take off but arrested recovery (STOBAR) technology and has an air wing of MiG-29K fighters and Kamov helicopters. India continues to study options for a second indigenously-built aircraft carrier, however, which has been tentatively named *Vishal*. She is expected to displace about 65,000 tons and in addition to arrester wires will possibly be fitted with catapults taking advantage of a collaborative agreement with the United States on carrier design. The IN has asked BAE Systems, DCNS, Lockheed Martin and Russia's Rosoboronexport for proposals for a partnership to build *Vishal*, although by mid-2017 no decision on the way ahead had been reached.

In December 2016 the IN withdrew from the Tejas light combat aircraft project, which had failed to offer a viable carrier-borne fighter, despite its protracted development. In its place, the following month it issued a formal requirement for fifty-seven multi-role carrier-borne fighters as an off-the-shelf purchase. The document detailing the requirement includes over fifty-nine pages of specifications for the air-defence, surface-strike, reconnaissance, electronic warfare and air-to-air refuelling roles and expresses interest in licence production and technology transfer. Logical contenders include the Boeing F/A-18E/F Super Hornet, Dassault Rafale M and RAC MiG-29K. Only the latter – already in IN service – has demonstrated compatibility with the STOBAR technology the navy currently uses but this advantage might be outweighed by reported unreliability and design defects in those already acquired.

BRAZIL

In 2015 it was announced that Brazil's only aircraft carrier, *Sao Paulo* – the former French *Foch* – would be taken into dockyard hands in Rio de Janeiro for a comprehensive hull inspection that was to be followed by installation of a new diesel-electric propulsion system with modernised catapults and arrester gear. However, in early 2017, the Brazilian Navy decided that the work was not viable in such an old hull. France's DCNS, acting in a consultancy role, had apparently advised that the work could take up to ten years at an unknown cost, leaving disposal as the only viable option. For the short-term, her air wing of modernised A-4 Skyhawks and S-2 Trackers will operate ashore. In the longer term – although a longstanding requirement for a replacement aircraft carrier remains – it seems questionable whether a new ship will be affordable given Brazil's parlous economic state and the need to fund submarine and surface escort construction.

CHINA

In April 2017 Chinese media reported that the People's Liberation Army Navy (PLAN) had floated its second aircraft carrier out of dry dock at the Dalian Shipbuilding Industry Company yard and that it had begun fitting out. Referred to as Type 001A, this ship is externally similar to China's first carrier, *Liaoning*, the former Soviet *Varyag*. Chinese analysts believe, however, that the Type 001A's designers have made detailed improvements, including a reduction in 'ski-jump' size from fourteen degrees to twelve to give similar effect from a smaller structure. A larger hangar may allow up to eight more J-15 fighters to be embarked and the flight deck parking space has been improved in the light of operating experience with the earlier ship. The island is reportedly ten percent smaller to give more deck space and is to be fitted with a new S-band (NATO E/F band) radar.

China's second domestically-designed carrier, Type 002, is reported to be a much larger vessel estimated at about 85,000 tons. Some sources state that she was laid down as early as 2015. It is expected to be launched in 2021, by which time analysts believe that a second Type 002 will be under construction. The *South China Morning Post* has reported that the PLAN has decided not to install 'highly advanced electromagnetic take-off technology' in these ships but has instead opted for three steam catapults. It also quoted a naval source as saying that there are still technical problems with nuclear propulsion and so the ships are to be conventionally powered.[4] In September 2016 a J-15 fighter was observed flying with what appeared to be a catapult launch bar on its nose oleo. There are also reports that the training air station in Liaoning Province is being fitted with two catapult tracks measuring 140m (460ft) long and buildings consistent with shore-based steam catapult facilities.

Meanwhile China's first aircraft carrier, *Liaoning*, achieved operational status in 2017 with at least thirteen J-15 Flying Shark fighters and three Z-18 helicopters embarked. She sailed with a task force of three destroyers, two frigates and a tanker for exercises in the South China Sea, passing between the Japanese islands of Okinawa and Miyako with Admiral Wu Shengli and his staff embarked. The Chinese newspaper *Global Times* subsequently commented that 'aircraft carriers are strategic tools which should be used to show China's strength to the world and shape the outside world's attitude toward China' and this deployment certainly shows that the PLAN now has the capability to deploy a significant carrier battle group although it has some way to go to reach the full potential of such a strike force.[5] Similar deployment exercises will follow in 2018 to develop procedures and operational tactics.

RUSSIA

In October 2016 the sole Russian aircraft carrier *Admiral Kuznetsov* deployed from the Northern Fleet at Murmansk to the eastern Mediterranean where she took part in strike operations against targets in Syria from early November. Her voyage gave Russian propaganda organisations an opportunity to publicise Russia's post-Cold War era surface fleet. However, even from that perspective, the deployment was not a complete success as *Kuznetsov* lost two aircraft in preventable accidents. Problems with her arrester gear due to poor maintenance were blamed as the root cause but these were exacerbated by poor command decisions. The first accident happened on 14 November 2016 when a

MiG-29KUBR ran out of fuel whilst orbiting the ship waiting as the flight deck crew attempted to fix a broken arrester wire that had become entangled with others. The second aircraft was lost on 5 December when a Sukhoi Su-33 went over the side after another arrester wire broke. There was no explanation of why the first aircraft was not diverted ashore as, since no Russian naval aircraft are fitted with air-to-air refuelling systems, a tanker could not be launched to extend the endurance of the waiting MiG-29 even though it was itself capable of in-flight refuelling. During the deployment Russian sources said that naval aircraft had flown over 400 combat sorties in support of the Syrian Government in two months but failed to specify how many of these were flown from an airfield ashore to which the majority of Su-33s are thought to have disembarked. The Russian Navy has only a small cadre of fixed-wing pilots and is finding it difficult to recruit more. These accidents will not have helped to solve the problem.

Admiral Kuznetsov arrived back in Russia in February 2017 and there have been reports that a long-awaited modernisation programme will soon commence. The Russian Navy continues to be interested in a replacement carrier programme and the Krylov State Research Centre has offered its Project 23000E 'Storm' class design to both the Russian and Indian governments. However, it appears that financial constraints make this a longer-term ambition and construction is apparently not envisaged to start during the recently revised State Armaments Programme 2018-25.[6] This corresponds with remarks by Deputy Defence Minister Yuri Borisov, who told reporters at the HeliRussia 2016 Helicopter Exhibition in Moscow that Russia's Ministry of Defence is considering starting construction of a new aircraft carrier in about 2025.

ITALY

The Italian Navy is studying the potential cost of modifying *Cavour* to operate F-35B Lightning IIs; it is procuring fifteen and hopes to be able to embark up to twelve of these at any one time.[6] At present the ship operates eight AV-8B Harriers on a regular basis; to deploy a larger air wing of the heavier and more powerful F-35B she will have to be extensively and expensively modified with digital maintenance, logistic support and flight planning facilities. The F-35B produces much more heat than the Harrier it

The Russian aircraft carrier *Admiral Kuznetsov* pictured transiting the English Channel in January 2017 after the conclusion of a Mediterranean deployment. Her air group – which included both MiG-29K and Su-33 strike fighters – undertook operations in support of the Syrian government against rebel targets, the first time a Russian carrier has been used in a warfighting role. Two aircraft were lost to accidents during the deployment. *(Crown Copyright 2017)*

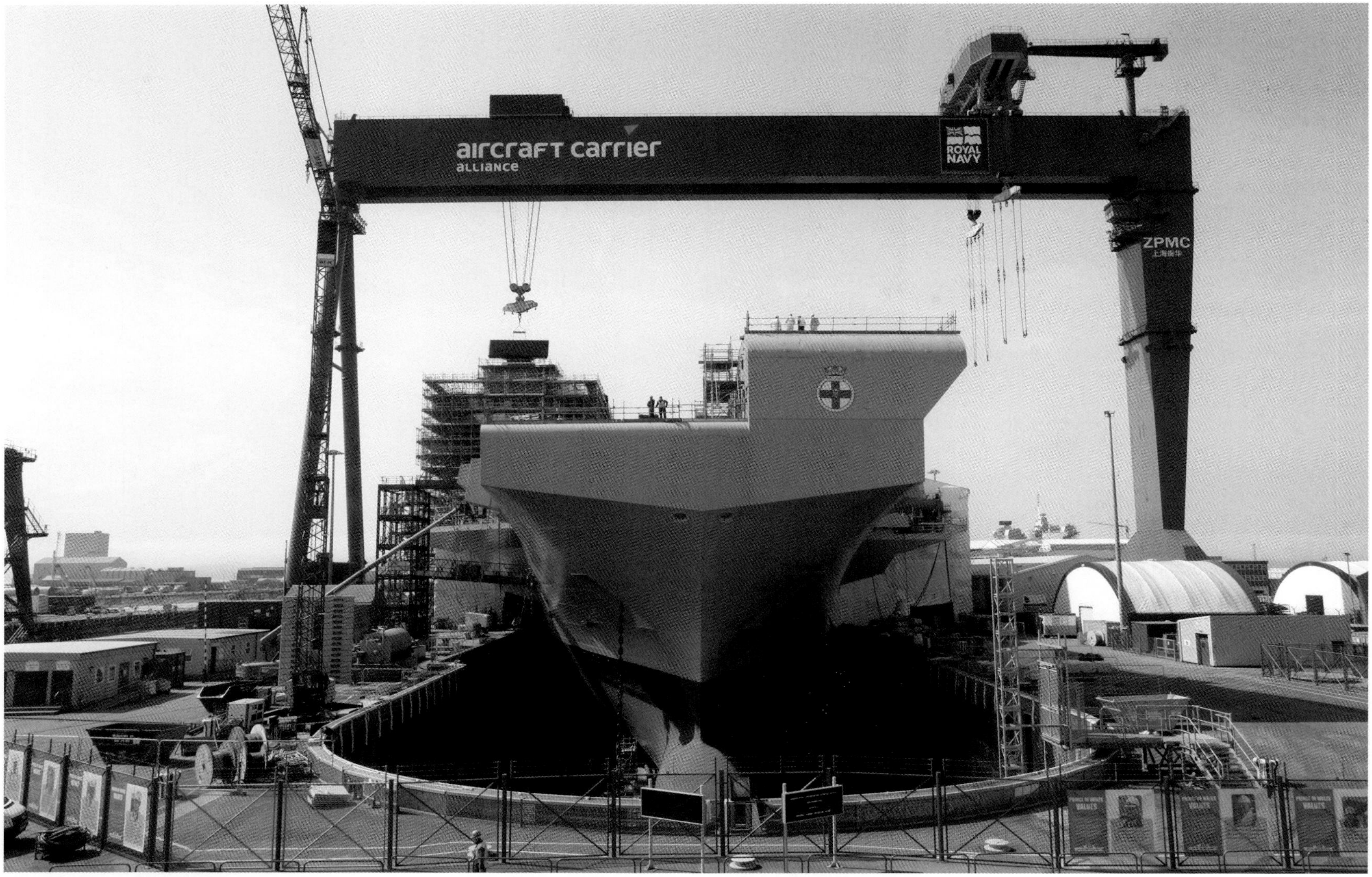

The Royal Navy's *Queen Elizabeth* class carrier programme is making solid progress but continued challenges need to be overcome to deliver a strike carrier capability in a timely and cost-effective manner. This picture shows the second ship in the class, *Prince of Wales*, in the course of construction at Rosyth in mid-2016. *(Aircraft Carrier Alliance)*

replaces and during take-off and landing its thrust peaks at 186 kilonewtons, compared with the Harrier's 106. This means that the flight deck has to be covered with a heat-resistant non-slip coating and that structures under and alongside the deck have to be re-located or protected from heat, blast and noise. Re-siting of cable runs, lighting, ventilation and pipework is required, as well as the complete re-design of some compartments. Engineering and logistic support for the F-35 will also require enlarged secure spaces as will flight planning and mission rehearsal in simulators. A clean bay for maintenance of the aircraft's skin coating will have to be created in the hangar. All of this will have to be built into the 28,000-ton ship by 2023 when the F-35B is due to achieve initial operating capability (IOC) with the Italian Navy

UNITED KINGDOM

The Royal Navy (RN) has already applied these solutions to the new aircraft carriers *Queen Elizabeth* and *Prince of Wales* during their construction at Rosyth. *Queen Elizabeth* had been expected to complete harbour trials at 'about the end of 2016' but that slipped to 'Spring 2017' and more recently 'Summer 2017'. The 'need to prove and set to work the myriad of complicated and technologically-advanced features of her design' was said by the Aircraft Carrier Alliance (ACA) to be the reason for the slippage. One defect became public knowledge after divers found that paint on the underwater part of the hull had failed to adhere properly and has started peeling off. *Queen Elizabeth* finally sailed from Rosyth to commence Part 1 sea trials in the

North Sea on 26 June 2017. She was escorted, initially, by the frigates *Sutherland* and *Iron Duke*. Two Merlin HM 2 helicopters of 820 NAS also operated with the ship and one carried out the first deck landing. After some days at sea, debris was caught around one of her propellers. This was cleared but, as a precautionary measure, it was intended that divers would carry out a survey during a pre-planned refuelling stopover in Invergordon during July. Problems were also reported with some watertight doors. These may need to be rectified prior to her delivery voyage to Portsmouth, where she will ultimately be commissioned.

In March 2017 the UK National Audit Office produced a report called 'Delivering Carrier Strike' in which it spoke well of the progress made thus far in constructing the ships as part of plans to achieve an IOC by December 2020. However, it raised concerns about what it called 'the ambitious master schedule which brings together the interdependent schedules of the three core programmes to achieve the full [carrier strike] capability by 2026'.[7] These include a shortage of military personnel causing gaps in the engineering, operations and aircrew categories. The RN is attempting to increase the number of pilots going through the military flying training system by using 727 Naval Air Squadron (NAS) at Royal Navy Air Station (RNAS) Yeovilton, which normally grades potential aircrew prior to their formal training, to provide elementary flying training for up to twelve pilots per year. On successful completion of their course they will move on to either fast-jet or helicopter training squadrons.

The three core programmes referred to in the report are the carriers themselves, the F-35B Lightning II and the Crowsnest airborne surveillance and control system. All three involve not just new hardware but the training of their people, supporting logistic and operational infrastructure and communications. The ongoing cost of the various elements has been hard to calculate, especially in the case of the F-35B because of fluctuations in the value of the pound against the US$. However, total approved spend on the F-35B programme amounted to £9.1bn (c. US$12bn) as of early 2017, of which the forecast spend up to March 2021 was £5.8bn (c. US$7.5bn). The latter figure that could still rise significantly as development testing will continue until 2019 at least and aircraft procured before then will need to be modified to the operational standard. These modifications could limit the number of aircraft available for the first two squadrons in the early 2020s.

A F-35B Lightning II in British markings overflying RAF Marham during the course of the type's maiden deployment to the United Kingdom. The so-called Joint Force Base Marham is being significantly upgraded in advance of the first planned British operational squadron, 617 Squadron RAF, in early 2018. *(Crown Copyright 2016)*

The RN Fleet Air Arm concentrated during 2017 on keeping the art of embarked naval aviation alive using whatever assets it can. 849 NAS has retained seven Sea King ASaC 7 helicopters at RNAS Culdrose to provide continuity in the airborne surveillance and control role until sufficient Merlin HM 2s have been modified to accept the Cerberus mission system under Project Crowsnest. The squadron has three Flights named Palembang, Okinawa and Normandy after a selection of its battle honours and their personnel are rotated through detachments to the Gulf on surveillance missions with the Combined Maritime Force (CMF). Since this deployment began, Sea King ASaC 7 crews have flown for over 2,000 hours tracking up to 240 contacts at any one time ranging from oil rigs to small boats or even small floating objects and deciding what is part of the normal pattern of Gulf life and what is not. Meanwhile, the Merlin HM 2 force comprises four front-line squadrons, one of which has a training commitment. Thirty aircraft have been upgraded and delivered to HM 2 standard but they need a further upgrade by Leonardo Helicopters, formerly AgustaWestland, to enable them to change role equipment in a reasonable number of hours between the anti-submarine warfare package and the Cerberus airborne surveillance and control package. The Crowsnest Merlins have yet to receive a designation but either HM 2* or HMASaC 5 would seem to be appropriate, with the latter being more descriptive, if lengthier. It is not yet clear whether 849 NAS will remain in commission to draw airframes from the operational pool of Merlins together with 814, 820, 824 and 829 NAS and, given the scale of commitments facing the RN, the NAO report was undoubtedly correct to observe that thirty Merlins will not be sufficient for the level of tasking likely to follow the introduction of the new carriers into service. Twelve Merlin airframes were not converted to HM 2 standard but these have been stripped to provide components for the remainder and their restoration to a usable standard may not now be affordable or practical.

The British Joint F-35B Lightning II Force has been set up as a unique entity that is neither part of the Royal Navy nor the Royal Air Force (RAF). It remains to be seen how successful the concept will be in operation against a sophisticated enemy. Squadron manpower is split roughly equally between the two services and, in theory at least, squadrons will be able to operate on board or ashore. The weakness in the concept is that pilots have to re-qualify in day and night deck landings every time they go to sea and non-naval maintenance personnel have to be trained to live as sailors in a warship. Previously RN squadrons qualified when they embarked and then stayed with their parent carrier using that qualification for significantly long periods and gaining in experience and capability all the time. As Joint Force Harrier demonstrated prior to its disbandment in 2010, temporary detachments never achieved the same level of skill and professional advancement that RN air wings used to achieve and the US Navy still does. Joint Force Harrier relied heavily on the skill and experience built up by the RN Sea Harrier squadrons during its short existence but it is now eleven years since the last Sea Harrier squadron was prematurely disbanded. Captain Jerry Kyd, the commanding officer designate of *Queen Elizabeth*, told the press in late 2016 that the limited F-35 buy-rate thus far would lead to embarked numbers in 2020 being 'very modest indeed'. The carrier air wing will be 'fleshed out with helicopters' but 'a lot will depend on how many USMC F-35Bs are embarked for the ship's first deployment in 2021'. He added that, by 2023, there will be twenty-four UK F-35Bs capable of embarkation. This assumes that both 617 Squadron RAF and 809 NAS will be operational with twelve aircraft each and none in repair. Until then, the UK has confirmed plans for the deployment of USMC F-35Bs on *Queen Elizabeth* on a regular basis.[8]

The purchase of forty-eight F-35Bs has now been formally authorised following approval for the remaining thirty aircraft in the so-called tranche 1 fleet in January 2017. These are being acquired in small annual batches alongside American and other foreign orders. By 2023, the UK is expected to have taken delivery of forty-two, including twenty-four in the first two front-line squadrons and at least five in in 207 Squadron, recently identified as the UK training unit at the Joint Force Base Marham. Three or four aircraft are to remain permanently in the United Sates with 17 Squadron RAF in a test and evaluation role. In March 2017 Sir Michael Fallon, the UK Secretary of Defence, announced during a press conference with his US counterpart General Jim Mattis that Lockheed Martin had been awarded a US$102 million contract to deliver initial training, engineering, maintenance and logistics support for the UK F-35 fleet. Much of the work has been subcontracted to BAE Systems and the two companies will provide over 100 technicians at Marham. Earlier contracts funded a joint operations centre, integrated training facilities and both maintenance and finish buildings which form part of a wider construction project known as Project Anvil. Prior to the planned arrival of 617 Squadron in early 2018, work will have been completed to equip Marham with vertical landing pads capable of absorbing the heat generated by the F-35, new hardened aircraft shelters, and refurbished servicing areas, runways and taxiways. In 2016 Air Commodore Harvey Smyth RAF, Commander of the UK Lightning II Force at the time, stated that the eventual plan is to build up to a force of four operational squadrons although there has been no confirmation of timing or unit identities. Soberingly, some analysts believe that full British operational capability with a fully-trained air wing capable of all that the new carriers were designed to achieve might not be reached until 2026, nearly three decades after the project started.

A US Marine Corps F-35B Lightning II Joint Strike Fighter operating from the deck of *America* (LHA-6) in November 2016. USMC F-35B operations from the deck of the new British aircraft carrier *Queen Elizabeth* will also be a common sight by the early 2020s. *(US Navy)*

THE F-35 PROGRAMME

The Pentagon announced in March 2017 that Rear Admiral Mat Winter is to replace Lieutenant General Christopher Bogden as director of the F-35 Joint Programme Office (JPO). After confirmation of his appointment he is to be promoted to Vice Admiral. Winter brings considerable experience to this task having previously served as head of the Office of Naval Research, Navy unmanned aviation development and, since December 2016, as Bogden's deputy. By 2017 two USMC and a single USAF unit had achieved IOC with others about to

The F35 programme continues to face developmental challenges but is moving towards operational maturity. The autumn of 2016 saw seven F-35B aircraft embark in the new amphibious assault ship *America* (LHA-6) for the third shipboard phase of developmental testing, as well as for operational testing to help prepare for the first deployment on *Wasp* (LHD-1) around the end of 2017. *(US Navy)*

do so. However, the type's development and demonstration phase is still not over and the US Government Accountability Office (GAO) believes that it will not be completed until May 2018, a year after the present target date and US$1.7 billion over budget. The JPO agrees that the development phase is running late but believes that completion by October 2017 is still possible with an overspend of US$530 million. Whenever the development phase is officially completed, the type will then enter a phase of initial operational test and evaluation which is due to culminate in full operational capability.

The biggest problems are software related. Development test pilots flying with different software iterations have experienced system shut-downs and had to reboot computers on the ground and in flight. Lockheed Martin engineers have resolved the problem for block 3i software, the version used by the USAF to declare IOC in 2016. However, the problem has recurred with the later block 3F which is to give greater operational capability allowing use of external weapons and real-time data download. Lockheed Martin engineers have reduced the average time between these failures to about one in every fifteen hours but the problem is being compounded over time because delays completing one iteration of software lead to a domino effect of delay on the next. As serious, the operational version of the autonomic logistics information system (ALIS) 3.0, which is used for flight and weapons planning, maintenance support and personnel qualification is not expected to be ready until mid-2018 and, even then, several capabilities have had to be deferred until later in the year. Mission data files, effectively a classified library of every friendly and potentially hostile radar signature that F-35s are likely to encounter in the theatre in which they are operating, are also delayed and the GAO does not believe that they can be verified until 2019. All of these software problems could delay the date by which full operational capability for all versions can be achieved.

In addition to software shortcomings, rectifica-

A US Marine Corps F-35B Lighting II assigned to Marine Fighter Attack (VMFA) Squadron 121, the 'Green Knights' undergoing air-to-air refuelling practice in June 2016. In January 2017 the squadron deployed from MCAS Yuma in Arizona to MCAS Iwakuni in Japan and will embark periodically on *Wasp* (LHD-1). *(US Navy)*

tion is needed for structural problems that have emerged during testing. These include the outer, folding, portion of the F-35C wing, which was found to have inadequate structural strength to support the loads induced by pylon-mounted AIM-9X Sidewinder missiles during manoeuvres that cause buffet. Lockheed Martin engineers have designed an improved outer wing which is undergoing flight test during 2017 and which, if found satisfactory, will have to be retrofitted to existing F-35Cs and those on the production line. The problem has delayed the introduction of AIM-9X on the F-35Cs with block 3F software and without it the type is incapable of high off-boresight shots at close range. F/A-18E/F Super Hornets are already flying with AIM-9X block II and AIM-120D advanced medium-range air-to-air missiles. Another significant F-35C problem has been unacceptable oscillation during catapult launch. Test pilots have reported that aircraft 'bobbed up and down excessively' on their nose oleos, making it impossible to read instruments, difficult to reach emergency controls and even causing pain. Fleet pilots reported oscillations during 105 test launches from the USS *George Washington* (CVN-73), which were so severe that seventy-four percent suffered moderate pain and eighteen percent severe pain. A number of factors are thought to have contributed to the problem including the pilot's seat restraint and hand-hold (grab) bar location, the mass and centre of gravity of the pilot's helmet and display unit, the physical characteristics of the nose landing gear strut and the length and release load of the repeatable-release hold-back bar. Early efforts to overcome the problem have centred on adjusting the pullback force during launch. If this does not work, more extensive modifications to the nose gear and helmet display are being considered. The last resort might have to be a complete nose-gear redesign.

The US Marine Corps increasingly see their amphibious assault ships operating as surrogate light aircraft carriers or 'Lightning carriers' as more F-35Bs are delivered. Meanwhile, the second *America* class LHA-type amphibious assault ship *Tripoli* (LHA-7) was floated out ahead of schedule by Huntington Ingalls Industries at the start of May 2017. *(Huntington Ingalls Industries)*

Meanwhile Martin Baker has proposed a three-part solution to the problem with its US16E ejection seat that has prevented pilots under 136lbs (61.7kg) flying the F-35; these comprise a lighter helmet to ease strain on the neck during the first phase of ejection, a lightweight switch on the seat that delays deployment of the main parachute and a fabric panel sewn between the parachute risers, known as a head support panel (HSP) that will protect the pilot's head from moving backwards during parachute deployment. Tests have already shown that the latter significantly reduced opening shock for lighter pilots. However, the weight restriction is unlikely to be lifted until all F-35s have the modified seat fitted and all pilots have the generation III 'light' helmet, expected to be available from November 2017.

Despite its problems, the Lightning II has moved further towards operational maturity in the past year. In January 2017 Marine Fighter Attack Squadron VMFA-121, the 'Green Knights', deployed from MCAS Yuma to MCAS Iwakuni in Japan, where it joined the 1st Marine Air Wing. By July 2017 it will have sixteen aircraft and is planned to embark regular detachments in the USS *Wasp* (LHD-1), itself base-ported in Sasebo, Japan. A second USMC F-35B squadron, VMFA-211 the 'Avengers', achieved IOC in 2016. However, in February 2017 the Marine Corps' Deputy Commandant for Aviation, Lieutenant General Jon Davis, told the press that the USMC has changed the order in which units are to replace their legacy aircraft with F-35s. Under earlier plans VMA-311, the 'Tomcats', at present equipped with AV-8B Harriers at MCAS Yuma, was to be the third unit to transition onto the F-35B, starting in 2018. Now VMFA-122, the 'Werewolves', at present based at MCAS Beaufort and equipped with F/A-18C Hornets is to be converted instead, moving to MCAS Yuma to do so. The change was made because the AV-8B is demonstrating significantly better serviceability than the older Hornets and is better able to retain operational capability for its last projected decade in service. Also, reducing the number of Hornets will reduce the SLEP work load in maintenance yards. The fourth unit to convert is to be VMFA(AW)-225, the 'Vikings', at present operating F/A-18D Hornets at MCAS Miramar. A further Miramar unit, VMFA-314, the 'Black Knights', is to become the first Marine unit to convert onto the F-35C.

The Royal Navy amphibious helicopter carrier *Ocean* will pay off during 2018 and may be sold overseas. This picture shows her during a deployment to the Gulf in late 2016; she was flagship of the US Central Command's Task Force 50 for three months during this time. *(Crown Copyright 2016)*

The identity of the first operational US Navy F-35C squadron was revealed as VFA-147 the 'Argonauts' during the annual Tailhook Association Convention at Reno in October 2016. It currently operates F/A-18Es at NAS Lemoore, California, and is to begin conversion onto the new type in 2018 and enter the carrier deployment cycle after achieving IOC in 2020. The convention was also informed that VFA-101, the 'Grim Reapers', has begun to move from Eglin Air Force Base, where it carried out its own transition, to NAS Lemoore to become the first F-35C fleet replacement squadron. Four classes of newly-qualified pilots are to be trained by the end of 2017 and a second fleet replacement squadron, VFA-125 the 'Raiders', is to be re-formed at NAS Lemoore tasked with converting fighter squadrons from legacy types onto the F-35C. The US Navy hopes to achieve IOC for the F-35C with block 3F software in August 2018.

BIG-DECK AMPHIBIOUS SHIPS AND THEIR AIRCRAFT

The ability of big-deck amphibious ships to act as surrogate aircraft carriers was demonstrated by *Makin Island* (LHD-8) with AV-8B Harriers, MV-22 Ospreys, helicopters and 2,400 marines embarked when her amphibious group filled a gap in the 5th Fleet area of operations between the departure of the *Eisenhower* (CVN-69) battle group in November 2016 and the arrival of the *George H W Bush* (CVN-77) group in January 2017. Although less capable, Britain's *Ocean* became flagship of the US Central Command's Task Force 50 in the Arabian Gulf for three months after November

2016. Her tailored air wing comprised RN Sea King ASaC 7s and Merlin HM 2s, Army Air Corps Apache AH 1s and RAF Chinook HC 4 helicopters.

This surrogacy potential was given due weight in the USMC's annual Marine Aviation Plan, released in March 2017.[9] This described 'high-end operations against a near-peer adversary' using an amphibious assault ship acting as a 'Lightning carrier' or CVL with up to twenty F-35Bs embarked. The document noted that between 2017 and 2027 the USMC will possess the majority of naval F-35s and that by 2025 the service expects to operate 185, enough to equip all the LHDs. It states that 'while the amphibious assault ship will never replace the aircraft carrier, it can be complementary if employed in imaginative ways. The CVL concept has been employed 5 times utilising AV-8B Harriers in a Harrier carrier concept … a Lightning carrier … can provide the naval and joint force with significant access, collection and strike capabilities'. The CVL concept envisages four MV-22s embarked alongside the F-35Bs to provide air-to-air tanking support. In late 2017 *Wasp* (LHD-1) will have F-35Bs embarked to form part of a new expeditionary strike group that will combine an amphibious ready group with a surface action group of three destroyers using the same command structure as a carrier strike group. It will be led by Commander Task Force 76, who will fly his flag in *Wasp*. A Marine Corps general will also be embarked. This strike group will lack the airborne early warning capability provided by a carrier air wing's E-2 Hawkeyes. However, the navy has pointed to the success of the 2016 demonstration that used a USMC F-35B to provide target data for an SM-6 missile engagement to show what will soon be possible. These early deployments will allow the new technologies to get used to working together.

For more conventional amphibious operations, the Marine Aviation Plan recommends fitting new sensors to the MV-22 Osprey fleet to enable landing sites to be examined with greater magnification from longer distances, and to enable the carriage of precision weapons, further enhancing capability. Lieutenant General Jon Davis says that the corps aims to fit Harvest Hawk bolt-on/bolt-off kits to the entire KC-130J Hercules and MV-22 Osprey fleets. Three different nose sensors will be tested before a main gate decision is taken and Davis said the Corps has already procured ten Harvest Hawk packages for evaluation on the Hercules and its early adoption for the Osprey 'would be a good start'. Forward-firing and door-mounted guns are also under consideration. The initial Harvest Hawk kit includes an MX-20 sensor ball, Intrepid Tiger electronic warfare pod and kinetic weapons such as Hellfire, Griffin and Viper missiles capable of precision strikes. Other projected MV-22 upgrades include carriage of a portable battlefield surgical unit to provide a forward surgical suite and shock trauma section at assault company level. However, the most important short-term MV-22 improvement is the V-22 aerial refuelling system (VARS), intended to enable in-flight refuelling of all the corps' embarked air wings from within their own resources. Cobham Mission Systems are due to deliver VARS kits from 2018 with equipment based on the proven FR300 hose-drum unit and IOC is planned for late 2019. Meanwhile, trials with the new CH-53K King Stallion heavy-lift helicopter are going well and the corps hopes to begin replacing the CH-53E with it during 2019. Tests successfully completed to date include carrying a 27,000lb (12,247kg) external load over a distance of 110 nautical miles and low-rate initial production is expected to begin during late 2017. Legacy USMC aircraft continue to carry out front-line operations including participation by Harriers from *Boxer* (LHD-4) and *Wasp* in Middle East strikes against IS.

Australia's new amphibious assault ship *Canberra* carried out first-of-class flying trials off the east coast of Australia with a number of different types of helicopter embarked. This picture shows two Australian Army Tiger ARH helicopters stowed in her hangar with a number of other types in March 2017. *(Royal Australian Navy)*

Other nations are deploying an increasing number of big-deck amphibious ships although the RN intends to withdraw *Ocean* from service in 2018 as she reaches the end of her twenty-year design life. She may be offered for sale rather than scrap.[10] Her amphibious role will be assumed by the new *Queen Elizabeth* class aircraft carriers, which are large enough to operate combined strike, sea control and amphibious air wings but could also deploy a more focused aviation component under the Tailored Air Group (TAG) concept. To keep the art alive, seventy RN and RM personnel with two Merlin HC 3i helicopters will embark in the French helicopter carrier *Mistral* for a five-month deployment to the Far East in late 2017. The first Merlin helicopter upgraded to HC 4 standard also began acceptance flying trials in 2017 and is to be delivered to the Commando Helicopter Force at RNAS Yeovilton by the end of the year. The upgrade includes a glass cockpit instrumentation system similar to that fitted in the HM 2, power-folding main rotor blades and tail pylon, a modernised avionics suite and a new grey paint scheme. Once testing is complete, HC 4s will re-equip 845 and 846 NAS as they come off the modification line, restoring their ability to embark in a range of warship types. The twenty-fifth and last HC 4 is now to be delivered in 2020, somewhat earlier than originally thought.

In early 2017 *Canberra* carried out first-of-class flying trials off the east coast of Australia with Royal Australian Navy (RAN) and Australian Army MRH-90 Taipans, Navy MH-60Rs and Army Tiger ARH and Chinook helicopters embarked. Teams from navy and army flight trials units combined to validate ship/helicopter interfaces and operating limitations for all six helicopter spots on the flight deck. The LHDs *Canberra* and *Adelaide* offer a quantum leap in capability for the Australian Defence Force.

Reports in the *South China Morning Post* during March 2017 indicated that the PLAN is building a class of LHD-type amphibious assault ships, identified as the Type 075. They are expected to be able to 'launch various types of helicopter to attack naval vessels, enemy ground forces or submarines' and are to displace 40,000 tons. With a length of 250m they will closely resemble the US Navy's *Wasp* class. The first is anticipated to be completed by Shanghai's Hudong Zhonghua Shipbuilding Company in 2019 and become operational around 2020.[11]

A US Navy MH-60S Sea Hawk helicopter acting in the cargo transfer role during a Pacific deployment in February 2017. Deliveries of the type were completed in 2015 and the 'sister' MH-60R will complete production for the US Navy in 2018. *(US Navy)*

SEA CONTROL HELICOPTERS

The US Navy operates the MH-60R/S series of helicopters from both carriers and other warship types. It expects to take delivery of its 280th and last MH-60R during 2018; the 275th and last MH-60S was delivered in 2015. Both versions have a projected 10,000 flying-hour life which the first 'S' is expected to reach in 2024 and the first 'R' in 2029. A SLEP is being considered as part of a mid-life upgrade for both variants, starting in the mid-2020s. This would include fleet-wide fitting of the 'generation four' open-architecture computers introduced in the last production batch and a number of other spiral upgrades to mission systems to keep the aircraft operationally effective.

In September 2016 the US Navy ordered ten further MQ-8C Fire Scout unmanned helicopters from Northrop Grumman as the type nears the end of its development testing. With the basic airframe of the Bell 407, it shares a common ground station

A picture of the production line for Northrop Grumman's MQ-8C Fire Scout. The type has a larger airframe than the earlier MQ-8B but shares many internal components with the earlier type. *(Northrop Grumman)*

with the earlier, smaller, MQ-8B and has many of the same internal components. However, no further acquisitions were included in the FY2018 budget request, reflecting a programme restructuring that will probably cap the fleet at sixty (thirty 'B's and thirty 'C's). A single MQ-8B was embarked in *Coronado* (LCS-4) during 2017, whilst the first MQ-8C flight is to deploy in 2018.

Imaginatively, the US Navy announced in mid-2016 that AeroVironment had been contracted to supply 150 Blackwing mini-UAVs to equip the navy's entire fleet of submarines and unmanned underwater vehicles. According to the USN's Director for Undersea Warfare, Rear Admiral Charles Richard, Blackwing has been developed from the successful Switchblade backpack-transportable UAV. It is 20in (50.8cm) long, 4in (10.2cm) in diameter, weighs only 4lbs (1.8kg) and can carry an electro optical/infrared sensor, GPS and a secure digital data link. Details of endurance have not been revealed but it can be launched from a dived submarine. In service it is expected to be capable of intelligence-gathering, surveillance, reconnaissance, communications relay and third-party targeting

In Australia, the RAN flight trials team has cleared the MH-60R Seahawk and MRH-90 Taipan to operate from *Anzac*-class frigates and is preparing for similar clearance work on the new *Hobart*-class destroyers. In 2016 the team cleared the 'Bay'-class landing ship dock *Choules* to operate ScanEagle unmanned aircraft and 2017 will conclude with clearance trials for the new EC-135 training helicopters to operate from a variety of decks. By 2040 the RAN expects to have thirty helicopter-capable ships, with MH-60R flights deployed by 816 NAS and MRH-90 flights deployed by 808 NAS. In February 2017 it also procured a single Schiebel S-100 unmanned helicopter to evaluate its performance ahead of a potential larger buy. By the end of 2017 two flights of unmanned air vehicles could be at sea, wherever possible partnered with a manned capability.

Conversely, the RN has withdrawn ScanEagle from service as an economy measure. At the end of March 2017 the last Lynx helicopters were also retired, as planned, after forty-one years in service. Their replacement, the Leonardo Wildcat HMA 2, equips 815 and 825 NAS, with the twenty-eighth and last delivered in late 2016. It is both heavier and larger than the Lynx but still more suited than the large Merlin to a small ship deck. It is designed around a digital avionics management system that reduces aircrew workload and has a paperless maintenance system shared with Army Wildcats but not with Merlin or F-35. Incredibly, it does not have a data link so tactical information has either to be passed by secure voice or brought back to the ship by the aircrew, a weakness that must limit the aircraft's usefulness in fast-moving joint or coalition operations. In the short term, air-to-surface missile strike capability was lost with the Lynx as the Martlet and Sea Venom missiles intended for the Wildcat are not expected to achieve IOC until 2020 and will probably not be available throughout the fleet until after that.[12] Deliveries of the eight Wildcats ordered by the Republic of Korea Navy were also completed at the end of 2016, whilst it has been reported the pair ordered by the Philippines will be equipped with Israeli Spike missiles. The Wildcat is also being considered by other countries, including Bangladesh.

All eight of the P8-I variant of the P-8 maritime patrol aircraft so far delivered to India by Boeing seen lined up together at INS Rajali in Tamil Nadu. Options for another four were exercised in July 2016, with the United Kingdom and Norway also placing orders for the aircraft over the past twelve months. *(Boeing)*

Japan's largest warship, *Izumo*, broadly fits into the sea control heading because of the composition of her air group although she is actually larger than the RN's now discarded *Invincible* class aircraft carriers. In May 2017 she sailed for a programme of exercises and port visits in the South China Sea and beyond and will call in Singapore, Indonesia and Sri Lanka before joining the Malabar joint naval exercise with Indian and US warships in the Indian Ocean in July 2017. Her air wing comprises seven Mitsubishi-built SH-60K sea control helicopters, as well as additional Leonardo MCM-101 mine-countermeasures helicopters. The MV-22 Osprey – being acquired by Japan – is one of a number of other rotorcraft that can also be embarked. A sister-ship, *Kaga*, was delivered on 22 March 2017.

MARITIME PATROL AIRCRAFT

The re-equipment of US Navy patrol squadrons with the P-8A Poseidon has continued steadily since 2013 and export deliveries have already been made to the Indian Navy and the Royal Australian Air Force. Moreover, further overseas deals are being secured. The first two of an eventual nine aircraft for the RAF were ordered by the UK in August 2016; in

November 2016 the Norwegian Government ordered five P-8As; and in May 2017 the New Zealand Government confirmed it is considering a US$1.46 billion deal to replace its P-3K Orion fleet with four P-8As. A formal order is expected to follow in 2018, with deliveries in the early 2020s. With Saudi Arabia also signalling its intent to order the aircraft in May 2017, these deals confirm the Poseidon as the most important contemporary maritime aircraft programme.

The Indian Navy operates eight P-8Is and took up an option for a further four in July 2016. The aircraft has exceeded expectations and proved cheaper to operate than the Russian-made Tu-142 aircraft that it has replaced. They are being used in their primary maritime role over the Indian Ocean where they monitor shipping and participate in naval exercises, especially those investigating improved anti-submarine warfare tactics. Meanwhile, the US Navy has signed contracts with Boeing for spiral improvements to the P-8A which include the Minotaur system which integrates sensors, computers and communications to gather and process surveillance data for transmission to shore and surface ship operations rooms and what are known as multi-static active coherent (MAC) enhancements. This system of systems includes new computing and security architecture, automated digital network common data link upgrades, anti-surface-warfare signals intelligence, combat system architecture improvements and communications capability upgrades.

In January 2017 the Pentagon announced an agreement for a joint UK/US P-8A logistic support facility for both British and deployed US Navy P-8s at RAF Lossiemouth in Scotland. The first two RAF P-8s are expected to arrive there in 2019 and it would also make sense for UK crews to be trained in the USA to allow all nine British aircraft to be used operationally. Since the Nimrod MR 2 was withdrawn from service a decade ago, the RAF has kept MPA skills alive through a 'seed-corn' programme in which British aircrew have been embedded within the MPA communities of Australia, Canada, New Zealand and the USA. Air Marshall Gerry Mayhew, who is responsible for RAF fast-jet and ISR assets, has told reporters that this has made it possible to initiate an MPA squadron at Lossiemouth capable of keeping pace with aircraft deliveries.[13]

The P-8A's intended US Navy unmanned partner, the MQ-4C Triton, has also continued to make progress. The first squadron, unmanned patrol squadron VUP-19, the 'Big Red', commissioned at NAS Jacksonville, Florida, in October 2016 as part of Patrol and Reconnaissance Wing 11 alongside P-8As. By mid-2017 the unit had 130 personnel and a mission trainer, with the first aircraft due in the autumn. The Triton's operational assessment was carried out by a team from VX-1, VX-20 and VUP-19 Squadrons and, having passed, was ordered into low-rate initial production in September 2016. During the assessment an airborne Triton successfully exchanged full motion video with an airborne P-8A via a common data link, demonstrating the Triton's ability to use its sensors to track a surface target and build situational awareness for a Poseidon crew flying many miles away. Operational MQ-4Cs are to fly orbits at high altitude for up to twenty-four hours covering thousands of square miles of sea with their AN/ZPY-3 radars, EO/IR sensors and an identification system that monitors commercial shipping. The USN intends to procure sixty-nine Tritons to complement the planned fleet of 117 P-8As and Australia has committed to buying Tritons to complement its own P-8A force. In February 2017, the US Navy announced that East Coast MQ-4C Tritons are to operate from a FOB to be established at Naval Station Mayport, Florida. Under the remote split operational model to be used the mission crew will be established at an operations centre at Jacksonville while the maintenance personnel will be at Mayport. Construction work at Mayport is to start in late 2017 with the first aircraft due in 2020.

Notes

1. Her arrival followed builder's sea trials that commenced on 8 April 2017.

2. See Sam LaGrone's 'Mattis Orders Comparison Review of F53C and Advanced Super Hornet', *USNI News* – 27 January 2017 (Annapolis: US Naval Institute, 2016) and currently available at: https://news.usni.org/2017/01/27/mattis-orders-comparison-review-f-35c-advanced-super-hornet

3. See Megan Eckstein's 'V-22 Experiment on Carrier Shows Increased Flexibility Over C-2 In COD Mission', *USNI News* – 16 August 2016 (Annapolis: US Naval Institute, 2016) and currently available at: https://news.usni.org/2016/08/18/v-22-experiment-carrier-shows-increased-flexibility-c-2-cod-mission

4. The information was reported in an article by Minnie Chan 'No advanced jet-launch system for China's third aircraft carrier, experts say', *South China Morning Post* – 13 February 2017 (Hong Kong: South China Morning Post, 2017).

5. The comment was made in an article entitled 'China's carrier fleet must sail beyond offshore zone', *Global Times* – 25 December 2016 (Beijing: People's Daily, 2016).

6. A further fifteen F-35Bs are being acquired by the Italian Air Force and could also potentially be available for embarkation.

7. See the Comptroller & Auditor General's *Delivering Carrier Strike* (London: National Audit Office, 2017). A copy can be found at: https://www.nao.org.uk/wp-content/uploads/2017/03/Delivering-Carrier-Strike.pdf

8. Further detail on Captain Kyd's comments and USMC F-35B embarkation on *Queen Elizabeth* can be found in two articles by Andrew Chuter, viz. 'British Naval Commander Wants US Marine Aviation on Aircraft Carrier' dated 29 September 2016 and 'US Marine Fighter Jet Deployment Onboard British Warship Made Official' dated 16 December 2016, both in *Defense News* (Vienna VA: Sightline Media Group, 2016). The UK Ministry of Defence originally thought the 2007 withdrawal of the Sea Harrier would lead to a temporary gap in fighters at sea until 'about 2012'. How wide of the mark that proved to be!

9. The USMC no longer hosts the annual plan on its own website. However, a number of independent organisations host it on their own sites and it can be found by searching the internet.

10. A sale to Brazil for up to £80m (c. US$105m) has been rumoured in a number of press reports.

11. See Minnie Chan 'China building navy's biggest amphibious assault vessel, sources say', *South China Morning Post* – 29 March 2017 (Hong Kong: South China Morning Post, 2017).

12. A more detailed review of these missile programmes is contained in Chapter 4.3.

13. In the author's opinion, the UK is, at last, doing the sensible thing and buying into a major international, collaborative MPA programme that should continue with spiral upgrades for decades rather than pursuing a lonely, expensive project with no overseas interest like the failed Nimrod MRA 4 project.

4.2 TECHNOLOGICAL REVIEW

Author:
Norman Friedman

A NEW AGE OF NAVAL WEAPONS?

We may be on the verge of a new era of naval weapons, relying on electric rather than chemical power (though ultimately the electric power will usually come from chemical sources).[1] The decisive argument to adopt an integrated turbo-electric powerplant for the new *Zumwalt* (DDG-1000) class destroyer was apparently that it was needed to power lasers and rail guns which would ultimately arm the ship. The new *Ford* (CVN-78) class carrier also uses electric power to a much greater extent than in previous ships, in its case partly so that in future it can be armed with electric lasers in place of the current close-in weapons.[2] In *Zumwalt* the rail gun is seen mainly as a shore bombardment weapon, but there is also interest in it as a defensive weapon against increasingly fast anti-ship missiles. To some extent the electromagnetic catapults on the carrier *Ford* are rail guns firing aircraft rather than projectiles. The US Army is already adopting a 60kW electric laser specifically to shoot down UAVs, and the Navy is interested in scaling it up to deal with incoming missiles. To some extent, too, electric weapons are seen as ideal counters to new hypersonic anti-ship missiles which the Russians claim they are about to field. It remains to be seen, of course, just how well either the rail gun or the electric laser performs in practice.

THE ATTRACTION OF ELECTRIC WEAPONS

The new electric weapons are attractive for a number of reasons. First, in theory there is virtually no limit on the number of shots, because the energy comes from the ship's electric system and not from a limited number of cartridges. In the case of a rail gun, the ship does have a finite stock of projectiles, but generally they are considerably smaller than conventional shells. They rely entirely on kinetic energy to do damage and hence have no dangerous explosive content. Indeed, it is sometimes suggested that electric weapons are inherently safer than chemical (powder) weapons because the ship does not need a magazine. That may be somewhat misleading. Each shot requires a powerful electric jolt, typically supplied by a charged-up bank of capacitors. When charged, the bank is a massive concentration of energy. Presumably a hit on a charged-up bank of capacitors would be devastating, very much like a hit on a magazine.

Both the light generated by a laser and the projectile from a rail gun are far faster than any conventional projectile. In theory that ought to simplify anti-aircraft fire control enormously. High speed and short reaction time (more for lasers than for rail guns) may become increasingly important as hypersonic anti-ship missiles enter service over the next decade or so. The faster the attacking missile, the less time the defence has to react. The faster the defending weapon, the greater the fraction of reaction time the defence can exploit, and the further from a ship it can engage an incoming weapon.

Proponents of directed-energy and hypervelocity weapons generally point to the extremely low cost per shot – about 2015 the US Navy's Office of Naval Research, which has funded American work in this area, claimed a cost per shot of US$1 for a 110kW laser.[3] However, the cost of the laser mechanism and of the pointing and beam-steering part of the system have generally been so high that previous attempts to weaponise lasers have failed as being grossly uneconomic. For example, the Israelis adopted a missile solution to the problem of rocket attacks although they and the United States had invested heavily in a variety of laser anti-rocket weapons. In their case, one reason why they abandoned the laser solution was that the laser had short range, so that large numbers might be needed to cover a wide area (that problem would not, however, apply to a naval close-in defensive weapon covering a single ship). Similarly, it is not altogether clear that the special projectiles used in rail guns will be much less expensive than conventional shells.

LASERS – DEVELOPMENT

The British Royal Navy deployed the first operational naval laser weapons, which it called 'dazzlers'. They were developed in a very secret crash programme during the Falklands War. The British were finding it very difficult to counter low-flying Argentine aircraft using conventional guns and missiles, which just could not react quickly enough. Lasers were built and deployed. However, apparently by the time they were available the war was over.

The new generation of US Navy combatants such as the destroyer *Zumwalt* (DDG-1000) and aircraft carrier *Gerald R. Ford* (CVN-78) are equipped with significant reserves of electrical power. One reason is to provide sufficient flexibility for them to be equipped with new energy-directed and high-velocity weapons that may reach maturity over their service lives. *(US Navy)*

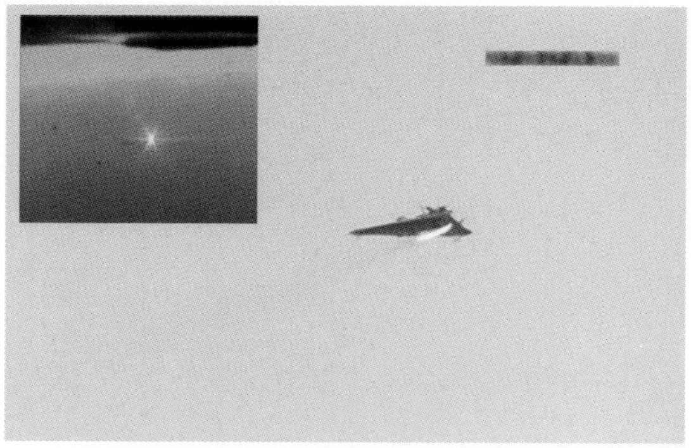

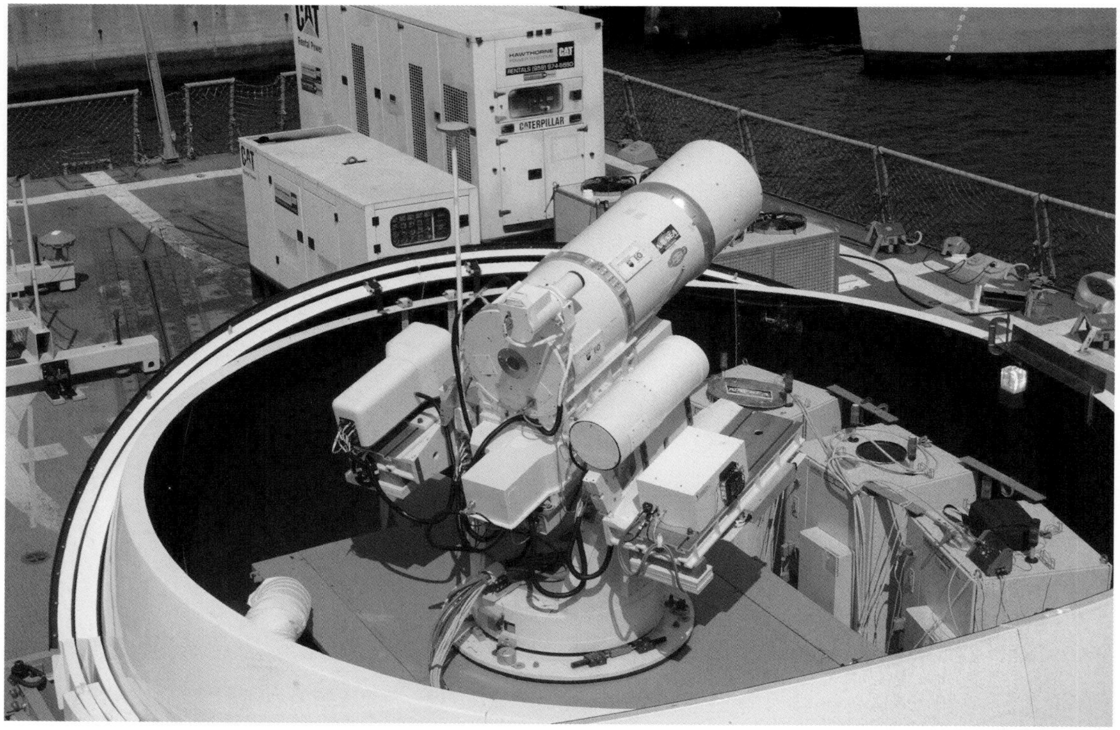

The Laser Weapon System (LaWS) demonstrator was temporarily installed aboard the DDG-51 type destroyer *Dewey* (DDG-105) in 2012. Constructed from commercial fibre solid-state lasers utilising combination methods developed at the Naval Research Laboratory. LaWS could be directed onto targets from the radar track obtained from the ship's Mk 15 Phalanx CIWS (or other targeting sources). The smaller, split image is a screen shot from a video showing the system being used successfully to down a remotely-controlled target aircraft. *(US Navy)*

They remained on board some ships as very secret weapons. To some extent the British programme was also inspired by reports (which were apparently false) that the Soviets were developing similar weapons, and indeed had deployed one on board the large cruiser *Kirov*.[4] The dazzler was intended not so much to blind the attacker as to confuse him and ruin his aim. It might well render his windscreen opaque. The dazzler also offered a less-lethal counter to small Iranian boats which British warships encountered during their Gulf patrols. A few years after these devices were unofficially reported, Britain acceded to a United Nations-sponsored ban on blinding lasers, and the dazzlers disappeared. Their existence was officially admitted only when a January 1983 letter from the Minister of Defence to Prime Minister Thatcher was released under the Thirty Year (declassification) Rule.

The problems which inspired these weapons remained. The small-boat problem in the Gulf is based on the fact that numerous potentially hostile craft often approach major naval units. They have to be fended off without causing them serious casualties, because any casualties would themselves become an excuse for more lethal attacks. Yet they can do real damage, as the attack on the US Navy destroyer *Cole* (DDG-67) in 2000 showed. Moreover, firing to discourage them may exhaust magazines just as they are needed. Indeed, their tactics may be intended mainly to exhaust stocks of defensive weapons. The new US Army 60kW electric laser is described specifically as a means of shooting down small unmanned aerial vehicles (UAVs) so that large air-defence missiles will not have to be wasted on them. This is generally presented in money terms: a big missile costs much more than a UAV. However, the more important point is that big missiles are available in only small numbers, and if they are all used up in response to UAVs, nothing is left to deal with more lethal threats when they appear.

This sort of problem explains the US Navy's pilot deployed laser project, using the Laser Weapon System (SEQ-3) or LaWS. This commenced on board the base ship *Ponce* (AFSB(I)-15) in the Persian Gulf in August 2014.[5] The system was declared operational in September 2014, and was reportedly so successful that the local command demanded that this experimental device be left in place after tests were complete. Laser power can be tuned up from the dazzle level to the attack level. In tests, the laser has shot down drones in about two seconds. It has also successfully destroyed boat engines, immobilising the targets.

Reportedly other countries – including China, France, Germany and Russia – have current research programmes into dazzling lasers. A German programme is developing a solid-state laser to counter rocket, artillery and mortar projectiles, with trials reported from 2012 onwards. Successful tests of a laser against airborne targets were announced by MBDA Deutschland in November 2016. The British Royal Navy also retains its interest in lasers, allocating £30m (US$37.5m) to the MBDA-led 'Dragonfire' consortium to develop a Laser Directed Energy Weapons (DEW) Capability Demonstrator towards the end of 2016.

LASERS – PROBLEMS, ALTERNATIVES AND SOLUTIONS

Laser anti-air defence is hardly a new idea, although it now seems a lot closer to reality than in the past. The attraction has always been grossly simplified fire control. Since the beam reaches its target essentially instantaneously, simply pointing it at an incoming missile would seem to be enough. In the 1970s, for example, there was a serious proposal to replace the 5in (127mm) gun on a *Knox* class frigate with a big laser.[6] Unfortunately, there was a problem. At the time, the only source of high laser power was chemicals, which unfortunately were quite toxic in liquid

Three images of the LaWS demonstrator on board the interim afloat forward staging base *Ponce* (AFSB(I)-15) during operational trials in the Persian Gulf in the autumn of 2014. The trials were reportedly so successful that the local command asked for the system to be retained for the remainder of the ship's deployment. *(US Navy)*

or gaseous form. This particular proposal died when it was pointed out that the exhaust from the only high-powered chemical laser then available would rapidly kill everyone on the ship's bridge.

At times during the 1980s models of 'nearly ready' laser anti-aircraft weapons were displayed at the annual Navy League show in Washington. Apparently nothing came of the idea; laser technology was not yet mature enough. Another problem at the time was the sheer mass of the laser director, which not only had to point at the target but also had to compensate for changes in the intervening air due to heating by the laser.

Roughly parallel to early naval laser projects was interest in beams of charged particles, which are normally generated by particle accelerators. Like a laser, they can be envisaged as a way of transforming a fraction of a ship's electric power into energy it can

The US Army's Mobile High-Energy Laser (MEHL) shown deployed on a Stryker combat vehicle. This is a relatively low-powered 5kW system capable of destroying small drones but the US Army has also tested a larger 60kW high-power system, designed and built by Lockheed Martin. This is twice the power achieved by the older 30kW LaWS. The US Navy is watching US Army developments with interest as it seems possible to build a modular version of this weapon. *(US Army)*

project against a distant target. How distant depends on how much energy is lost as the beam is projected through the air. For example, the same particles which are aimed at an incoming missile will also hit molecules in the air, using up energy by breaking them down. Also, as the particles in the air break down they interact with the beam, bending it and spreading it out. Particle beams were much discussed in the 1980s in connection with defence against incoming ballistic missiles, but they seem to have fallen out of favour.

There are also high-powered microwave (HPM) weapons, which may be able to disrupt or disable computers in much the way that an electromagnetic pulse (EMP) might, albeit at a much shorter range and on a much smaller scale. In theory an HPM weapon might be able to upset the guidance system of an incoming anti-ship missile. The US Air Force tested such a weapon carried on board a cruise missile in 2012 and, reportedly, the United Kingdom has also studied the idea as a cruise missile warhead. Russia is apparently interested in an HPM warhead for air-to-air missiles. The great drawback to any HPM weapon is that it may not be at all evident that the target has been disrupted. This is much the same problem as that of an ECM system. How can it be sure that the incoming cruise missile has lost its way, and is not simply manoeuvring evasively on its way to a target? That contrasts with a simple anti-missile weapon, which destroys the incoming threat in a very visible way.

Current laser systems solve the toxicity problem. Instead of energetic chemical reactions, they are powered by the ship's electric system. A solid-state device is excited electrically; a fraction (unfortunately small) of the electric power put into the solid-

The new US DDG-51 class destroyer *Rafael Peralta* pictured cutting through the early morning sea mist during trials off the Maine coast. Water vapour forming layers just above the sea can absorb energy from a laser beam and reduce its effectiveness. *(US Navy)*

The decommissioned *Spruance* class destroyer *Paul F. Foster* (EDD-964) now operates as a remotely-controlled self-defence test ship and has already been used in laser trials. It is intended to install a 150kW laser on the ship for further trials within the next twelve months. *(US Navy/ Northrop Grumman)*

state medium emerges as a coherent beam. That can be either a continuous beam (which heats the target) or a pulsed beam, which may in effect punch through the target. Electric power is turned into laser power by multiple solid-state devices working together. The Army's 60kW laser is interesting to the Navy because it may be possible to build a modular version.[7]

Lasers have presented other problems. Just above the sea, water vapour forms layers that absorb energy from a laser beam. Sand, dust, salt particles, smoke, and pollutants absorb and scatter the beam. Variations in the atmosphere refract (bend) the beam. Turbulence in the air can break up the beam, which is why lasers are often considered incapable of operating in all weathers. A laser's effect on an incoming missile depends on how much energy it can concentrate on the missile. Considerable work was done on ways of dealing with refraction, such as shooting a precursor beam through the refracting layers and using its behaviour to adjust the main beam, or trying to burn through the layers. As the laser beam passes through the air, it heats it, and that changes the refractive properties of the air and causes the beam to spread ('thermal blooming').

Another problem, which affects other exotic weapons, was to ensure that the laser, which might be visualised as a kind of death ray, would destroy a big missile. It probably would not simply set off the missile's warhead, and it could not just melt the whole missile airframe. Most likely it would destroy or damage a limited part of the airframe, for example tearing up the radome, or slicing off part of a fin or wing. How much damage would be enough? The answer might be, distressingly large. Many jet aircraft have continued to fly after suffering damage which might be expected to destroy their aerodynamics and bring them down.

Existing anti-missile weapons generally destroy much of the incoming missile (if they are explosive) or are intended to impart so much kinetic energy to the missile warhead that it explodes, destroying the missile. The Phalanx CIWS works in this latter way. A laser does neither. It heats a spot on the missile. How much damage it does depends on how long it can be focused on that spot and on how much energy it imparts. Laser and other directed-energy weapons have long been proposed for defence against ballistic missiles; during the Cold War it was thought by some that the Soviet Union had effective directed-energy anti-missile weapons. It might be imagined that the same technology would surely be good enough against a relatively small anti-ship missile, much flimsier than the re-entry vehicle of an inter-continental ballistic missile (ICBM). However, a spot burned through the heat shield of an ICBM warhead would be fatal, because the heat of re-entry would do the job. Anti-ship missiles suffer no such stress.

The great unanswered question about lasers – at least publicly – is how well an enemy can harden its weapons against them. During the Cold War, the Soviets armoured some of their missile warheads; they alone had ample size and weight to do so. Although no one builds such massive weapons now, some of them are surely large enough to accommodate some hardening. Much may depend on how fast and how manoeuvrable an incoming missile is, because that in turn may determine how long the laser beam can dwell on it to destroy it. The possible operational advent of laser weapons seems to be coinciding with the advent of Russian hypervelocity missiles.

In January 2016, the US Navy stated that it would release a directed-energy weapon roadmap the following month, but it had not been released as of May 2017. However, enthusiasm for such weapons is high. At the January 2017 Surface Navy Association meeting Rear Admiral Ronald Boxall, Director of Surface Warfare, said that within a year the navy would be testing a 150kW laser on board its self-defence test ship, the former destroyer *Foster* (ex-DD 964). A drawing showed a pair of large shipping containers forward of the ship's nest of vertical missile launchers, with a stabilised laser in a beam director turret atop a mounting between the vertical launchers and the ship's bridge.[8] The laser is the solid-state fibre type already in use at a much lower power on board the base ship *Ponce* in the Persian Gulf.

RAIL GUNS – DEVELOPMENT

The other current exotic electric weapon is the rail gun. The idea dates back to about 1911, but major problems have blocked development. In effect the rail gun is a high-powered linear motor. Current flows down one rail and up the other, the two being connected by a conducting projectile or by a plasma generated behind it (for example, by vaporising metal foil). The two rails set up magnetic fields which in turn generate the forces driving the projectile. There are some major complications. One is that to achieve very high accelerations the currents running through the rails must be huge. A second is that the magnetic fields which create the force driving the projectile also act to tear the two rails apart. The greater the desired acceleration, the stronger all the forces involved are.

To date, the question has been whether a rail gun can remain intact through hundreds of firings. The sponsoring Office of Naval Research (ONR) claims that initially it could hold the gun together through twenty firings, but that as of 2016 it was sustaining as many as 400 (sceptics noted that they may not have been at full power). The ONR request for gun designs included a capacity of 650 shots. On the other hand, if it can be made to work, a rail gun can accelerate a projectile to velocities well beyond those available from a conventional gun (although there is some possibility that a light-gas gun could compete with it). As of 2017 the US Navy was funding two competitive gun designs, one by BAE Systems and the other by General Atomics (which also builds the electromagnetic catapults on the new *Ford* class carrier).

Rail guns apparently first gained prominence within the US Navy as a result of a study conducted in the 1990s by the Strategic Studies Group (SSG) at the Naval War College. The group saw a rail gun as an ideal weapon to support US Marines and other troops well inshore. Given its very high velocity and a lofted trajectory, a rail gun might reach as much as 200 miles inland. The study assumed that it could be aimed very precisely, and the high velocity would be combined with a very short time of flight. The projectile would arrive at the target at very high velocity and, hence, with high kinetic energy, which the study group assumed could be transferred directly into a target. In this way an inert projectile, properly aimed or guided, might offer a combination of precision (previously unavailable in ship-to-shore fire) and devastating effect. Later it was pointed out that the hypervelocity offered by a rail gun might be even more valuable as a means of destroying incoming missiles. The rail gun project envisaged a finned projectile incorporating GPS guidance. In stark contrast to cruise missiles, rail gun rounds could easily be transferred at sea. That ability to take fresh rounds on board would give a surface combatant the sort of protracted attack capacity otherwise enjoyed only by carriers.[9]

The US Navy is currently funding development of prototype rail guns by BAE Systems and General Atomics. This picture shows a profile view of the BAE Systems' weapon. *(BAE Systems)*

The effect of a full-energy shot from an electromagnetic rail gun caught on camera. *(BAE Systems)*

The rail gun programme began in 2005; it is to fire a projectile at a speed of Mach 5.9 to Mach 7.4 (4,500 to 5,600mph at sea level). By way of comparison, a typical high muzzle velocity might be around 500ft/sec, which is about 2,390mph. In April 2014 the navy announced that it planned to test an experimental rail gun on board a Joint High Speed Vessel (now re-designated as Expeditionary Fast Transports) in FY16 for at-sea tests. The catamaran was an attractive test platform because it has a large helicopter flight deck on which the gun can be mounted, and also because it generates considerable electrical power for its waterjets. As of the spring of 2017, however, no such test had been reported.

The projected rail gun projectile is quite small. The Defense Department's Strategic Concepts Organization (SCO), which seeks improved capabilities by better using existing technology, has proposed that the projectile could also be fired from conventional guns. It would be a sub-calibre projectile comparable in concept to sub-calibre tank shells. Its quoted velocity is Mach 3. Although that is hardly what the rail gun is supposed to attain, such a shell might be fast enough to be useful against incoming missiles. In theory it (like the rail gun) would not require very complex fire control, which might be shifted rapidly from target to target in the face of a mass raid.

The rail gun projectile can also be fired from a conventional gun as a saboted round. Like those used ashore as anti-tank projectiles, the saboted round can be fired at high velocity (it is relatively light) and, because it is dense and has little drag, it retains that velocity longer than a conventional shell.[10] In advocating this kind of hypervelocity projectile, the sponsoring Office of Naval Research pointed to the manoeuvrability built into the rail gun projectile and its guidance system.

Sceptics of the rail gun note that by September 2016 Defense Department emphasis seemed to shift from the rail gun to the saboted projectile fired by a conventional gun. The new argument was that providing conventional guns with hypervelocity guided projectiles would turn them into effective anti-missile weapons capable of engaging so many incoming weapons that they could defeat saturation attacks. This emphasis, incidentally, seemed also to shift from the Navy to the Army and its 155mm howitzers.

RAIL GUNS – PROBLEMS

There are, moreover, some problems inherent in both the rail gun as a shore bombardment weapon and in the rail gun and/or its hypervelocity projectile as an anti-missile weapon.

As with the laser, the first question in shore bombardment is how lethal the rail gun is likely to be. Its projectile may or may not carry any explosive, but it carries enormous kinetic energy when it comes

An artist's rendering of a prototype rail gun on the flight deck of an Expeditionary Fast Transport. It had been reported that such a trial would take place during FY2016 but this has not yet been reported. *(US Navy)*

Nations other than the US are testing new-era weapons such as lasers for both military and naval applications. This image shows the German High Energy Laser Effector developed by MBDA Deutschland, which carried out a series of successful tests against an aerial target at a facility on Germany's North Sea coast in October 2016. *(MBDA)*

down. The question is how much of that is transmitted to the target. Although the projectile is guided (presumably using GPS), the ability to manoeuvre it is likely to be limited. GPS itself is not accurate to less than a few feet, so a GPS-guided projectile will hit, at best, within a few feet of a target. It may do worse if guidance is applied mainly when the projectile is high above the atmosphere, where it does not create a disturbance around itself (probably including charged particles). There is very little experience of what happens when a really fast object hits the earth. The question is how much of the energy is transmitted to nearby objects, and how much goes into just burying the projectile.

Notes

1. All opinions are the author's own, and should not be attributed to the US Navy or to any other organisation for which he has worked. All data are taken from public sources.

2. The *Gerald R. Ford*'s propulsion is still steam-driven. However, much auxiliary equipment that used to be steam-driven on previous carriers is now electrical, including the electromagnet catapults. Her electric distribution system reportedly has two and a half times the capacity found on older ships.

3. Actual required laser power levels are classified. However, a 2014 Congressional Research Service report surveyed various claims. Factors other than power level include beam quality (BQ), a measure of beam spreading. A 2010 navy briefing indicated that effective anti-missile effect would begin at 500kW, which is far beyond current capabilities. Another such briefing stated that at 50–100kW with a low (good) BQ, a laser would be effective against UAVs, the RAM anti-aircraft missile, and man-portable anti-aircraft missiles. With an output of hundreds of kW a laser would be effective at greater ranges and against a crossing anti-ship missile (power above 100kW, BQ about 2). It would take a megawatt laser to deal with supersonic, highly-manoeuvrable anti-ship missiles. Speed and manoeuvrability would presumably limit the time during which the laser beam could dwell on the target. A 2010 industry briefing associated a power of 50kW with UAVs and small boats. It would take 300 kW to deal with a subsonic anti-ship missile, and 500 kW to deal with a manned aircraft. An earlier, 2007 report by the Defense Science Board associated tens of kW with surface threats at 1–2km (up to about one nautical mile) and 1–3MW with a battle group (area) defence mission at a range of 5–20km. See Ronald O'Rourke, *Navy Shipboard Lasers for Surface, Air, and Missile Defense: Background and Issues for Congress - R41526* (Washington DC: Congressional Research Service, 31 July 2014).

4. It now appears that this was not the case, although the Soviets did develop a dazzler for their tanks (it was too expensive to deploy in quantity). The mistake arose because of Soviet deployment of elaborate-looking electro-optical devices on board ships, the nature of which was by no means clear at the time. The 1980 Convention on Prohibitions or Restrictions on the Use of Certain Conventional Weapons Which May be Deemed to be Excessively Injurious or to Have Indiscriminate Effects was amended during a 1994–6 review which added Protocol IV on blinding laser weapons. The United States ratified this protocol on 23 December 2008. It bars laser weapons intended to cause permanent blindness. It can be argued that a 'dazzler' is not so intended. The US Defense Department pointed out specifically that the wording of the Protocol '... reflects a recognition of the inevitability of eye injury as the result of lawful battlefield laser use'.

5. This Laser Weapon System (SEQ-3, LaWS) is essentially six welding lasers ganged together in what must have been an urgent test programme. They were coupled to excellent surveillance optics, a feature which proved useful in itself. This combination was judged twenty-five percent efficient: it would take 400kW of shipboard power to produce a 100kW beam (the other 300kW would emerge as heat the weapon would have to dissipate). The key discovery – at the Pennsylvania State Electronic-Optic Center in 2004–5 – that made LaWS possible was that it was not necessary to combine the beams coherently. LaWS was tested successfully against UAVs – initially on shore but later at sea – in 2009–12. In June 2009 it successfully engaged five out of five threat-representative UAVs. Subsequently, in May 2010, it successfully engaged four at a range of about one nautical mile over water from San Nicholas Island. It also showed that it could destroy a RHIB boat at a range of about half a nautical mile, and that it could jam and disrupt electro-optical and IR sensors. Between July and September 2012 it engaged three out of three UAVs while on board the destroyer *Dewey* (DDG-105) off San Diego. In 2010–11 the Navy successfully tested the Maritime Laser Demonstrator (MLD) against a small boat. The *Ponce* deployment using LaWS was announced in April 2013. This laser is rated at 30kW (output in a test was 33kW), half the power output of the Army laser more recently purchased. At one time the Navy envisaged scaling LaWS up to 100kW by FY14, but that seems not to have been done. A follow-on technology demonstrator, to become operational in 2018, is to produce 100kW to 150kW. In March 2014 the US Navy expected to reach initial operating capability with a shipboard laser in FY20 (2019/20) or FY21 (2020/21). The threshold to engage anti-ship cruise missiles is reportedly 200 or 300kW.

The important question in air defence is probably that of accuracy. Current anti-aircraft projectiles, including missiles, are proximity-fuzed because guidance systems are not precise enough to guide them into the incoming target. For example, an incoming missile is affected by gusts of air, which move it a few feet off its intended course, and thus may ruin the fire-control solution generated by the defending weapon system. Radars which feed a missile fire-control system generate beams up to a few degrees wide. In the past, such constraints have been accepted because they still allow a defending missile or shell to get close enough to destroy the incoming enemy missile. That still leaves problems. For example, the blast from a defending missile warhead takes a finite time to get from missile to target. The faster the incoming missile, the less time there is for the blast to hit the incomer. It might well be argued that a hit-to-kill solution is increasingly unavoidable as incoming weapons become faster and faster. It is not clear, however, that a command-guided weapon like a hypervelocity shell will always solve that problem, since it is still plagued by errors inherent in the guiding radar and other sensors. Nor does it seem likely that a hypervelocity shell or projectile can carry enough in the way of onboard sensors to guide itself. On the other hand, if a hypervelocity shell could hit an incoming high-velocity anti-ship missile, surely that would be enough to destroy it. There would be no lethality question.

WHERE DO WE GO?

Over the past few years, rail guns and hypervelocity shells have received both enormous publicity and a great deal of scepticism. In 2017 it is still by no means clear whether the boosters or the critics are closer to being right. Lasers seem to be a lot closer to becoming operational; that LaWS actually is operational, and that the US Army is currently fielding a fibre SSL suggest strongly that lasers are emerging from the experimental stage.

A schematic of the types of target the British 'Dragonfire' laser system is intended to combat. The system faces the same challenges and limitations impacting similar US technology. *(MBDA)*

This account emphasises US Navy programmes because they are public and because they seem well advanced. It would be foolish to imagine that other major navies (and armies) have shown little or no interest, only that they have been much less public about it. All of those interested in these weapons face the same problems and the same physics.

6. This period was a kind of false dawn for the US laser programme, all sorts of laser weapons being proposed. For example, there was an apparently serious proposal to replace the gun on board a Phantom fighter with a laser. It was soon obvious that those involved were overly-optimistic, and also that there were relatively simple means of overcoming laser attacks. None of the proposed laser weapons materialised. Presumably the differences between what happened then and what is happening now include much higher power and much more stable beam direction, so that beams dwell much longer on vulnerable parts of their targets. Periodically after the 1970s there have been reports of laser successes, such as the destruction of TOW missiles by laser.

7. The current programme uses a solid-state laser (SSL). In the past, the navy was also interested in an alternative free-electron laser (FEL), but that seems to have been abandoned. The navy has experimented with both fibre SSLs and slab SSLs. Fibre lasers are widely used in industry, for example in car manufacturing. The LaWS is a combination of six such welding lasers. An SSL operates at a wavelength of 1.064 microns, close to the 'sweet spot' (wavelength of minimum absorption) of 1.045 microns. Work is currently underway on SSLs operating above the threshold (about 1.5 microns) at which laser radiation endangers human eyes. The Maritime Laser Demonstrator (MLD) was a slab laser. By 2014, MLD had demonstrated a power level of 105kW, about three times that of the SSL LaWS. At that time a slab SSL was considered about twenty to twenty-five percent efficient; beams are generally combined coherently. The 105kW version was first tested in 2009; it combined seven 15kW slabs. MLD poses greater problems of heat dissipation, and it uses more complex optics. Around 2014, it was claimed that slab lasers could be scaled up to an output of 300kW without requiring any technological breakthroughs. The alternative FEL was considered attractive because it was tunable and, hence, could match frequency (wavelength) 'sweet spots' in the atmosphere. The Office of Naval Research (ONR) developed a 14.7kW FEL and wanted to build a 100kW FEL during FY10-15 as a step towards an ultimate megawatt (MW) laser. However, this effort was curtailed because the SSL seemed to offer much quicker results. Apparently the fibre and slap SSLs are limited in ultimate output, hence must combine beams (which adds complexity) to attain high power. An FEL can be scaled up more easily, and it does not generate heat within its laser mechanism (it still needs cryogenics to handle its chemicals). In an FEL, a stream of electrons from an electron gun is accelerated and then sent through a transverse magnetic field which causes them to 'wiggle' and to release photons. As in other lasers, the photons bounce between mirrors to induce further coherent emission and thus to create a beam. Beam quality would be much better than for an SSL. As of 2014, concept designs for FELs showed structures as much as 100ft (30.48m) long. They would emit X-rays as well as light, hence would have to be shielded. The size of the FEL could be reduced by using superconductors, but that would add the complication of cryogenics.

8. The smaller of the two containers is the thermal storage module, which stores excess heat from the laser. The larger of the two is the energy storage module which powers the laser. The structure forward of the bridge houses the beam director electronics. It appears that the laser itself is on trunnions atop it.

9. At present it is nearly impossible to transfer missiles to a surface combatant at sea, whereas it is relatively easy to transfer them to carriers. If a surface combatant could somehow produce the same sort of close support as a carrier to a few hundred miles inland, the economics of sea power would shift dramatically. It is not clear to what extent the SSG envisaged a guided rail gun projectile; it seems to have envisaged a low-cost weapon which could be carried in very large numbers.

10. BAE's saboted round is 26in (66cm) long and weighs 40lbs (18.1kg); a conventional 5in shell weighs about 70lbs (31.7kg). The projectile itself is 24in (60.9cm) long and weighs 28lbs (12.7kg), including what BAE describes as a 15lb (6.8kg) payload. That is presumably mainly the guidance system. Note that the rail gun was intended to use a GPS guidance package, presumably set before firing. A hypervelocity round used to deal with incoming missiles would need continuous communication with the firing ship. It is not clear to what extent the disturbance generated by the moving projectile might interfere with such communication.

4.3 TECHNOLOGICAL REVIEW

Author:
Richard Scott

ROYAL NAVY GUIDED WEAPONS

Recent developments

Drill Sea Dart missiles are pictured on the launcher of the Royal Navy Type 42 destroyer *Edinburgh* in the lead-up to the ship conducting the last-ever firing of the long-serving anti-aircraft weapon in April 2012. The Royal Navy's Cold War-era missiles are being replaced by a new generation of more capable weapons. *(Crown Copyright 2012)*

The decade between 2010 and 2020 will witness the replacement of the overwhelming part of the Royal Navy's above-water guided weapon systems inventory. A generation of missile systems developed in the Cold War era – Sea Dart, Seawolf, Sea Skua and Harpoon – will be gone. With the notable exception of Harpoon, all will have been succeeded by new guided weapons developed in the United Kingdom, or in collaboration with European partners.

This recapitalisation, and the means by which it is being delivered, reflects a number of dynamics. These include the obsolescence and life expiry of legacy systems; the need to counter new, more stressing air and surface threats; a changed operational environment with greater emphasis on operations in littoral regions; and the need to ensure a precise and proportionate surface attack capability given prevailing rules of engagement. Another important influence is a new relationship with British industry intended to secure and sustain a 'sovereign' complex weapons enterprise, achieve long-term savings to meet challenging affordability targets, improve delivery, and grow export business.[1]

Yet at the same time, Navy Command Headquarters' Maritime Capability area has had to accept a number of gaps and risks as it attempts to balance front-line requirements with a severely-stretched budget. The absence of a funded replacement for the Harpoon Block 1C anti-ship missile is a clear case in point.

SEA VIPER

It was the events of the 1982 South Atlantic campaign, and the weaknesses exposed in the performance of the GWS 30 Sea Dart medium range missile system equipping the Royal Navy's Type 42 destroyers, which shaped the requirement for what became the GWS 45 Sea Viper anti-air guided weapon system. The Falklands (Malvinas) experience showed the limitations of Sea Dart – a system designed in the 1960s to counter Soviet high altitude bombers – against anti-ship missile threats flying just a few metres above the sea surface. The conflict also served to highlight a wider shift in the threat environment: the threat posed by multiple anti-ship sea-skimming missiles, offering reduced warning times and potentially approaching along different axes, demanded a new type of rapid reaction local area defence missile system able to provide an 'umbrella' of protection for ships in consort.

It was against this background that the concept of the Support Defence Missile System (SDMS) was evolved in the mid-1980s to underpin requirement for a Sea Dart replacement. The thinking behind SDMS was that advances in the threat demanded a new type of capability that could prosecute multiple simultaneous engagements, in a very short time period, against fast, manoeuvring targets approaching at very low level.

The nature of this evolving and increasingly 'stressing' maritime air threat conditioned SDMS requirements for a highly automated 'system of

This overhead picture of the Type 45 destroyer *Daring* taken in 2013 gives a good view of many of the components of the GWS 45 Sea Viper missile system. Up to forty-eight Aster 15 and Aster 30 missiles can be housed in the SYLVER VLS located forward of the bridge, with the twin-faced Sampson radar housed within the radome at the top of the forward mast providing detection, tracking and guidance functions. The S1850M long range radar aft provides enhanced surveillance capabilities. *(Crown Copyright 2013)*

An Aster 30 missile in flight. The Aster 15 and Aster 30 missiles share a similar 'dart' but have different-sized rocket boosters. *(MBDA)*

systems'. This combined a very high performance multifunction radar (MFR) for surveillance and multi-target tracking, a fast, accurate and exceptionally agile missile interceptor assuring a very high probability of a kill, and a shipboard command and control (C2) system to perform system management and threat evaluation and weapon allocation (TEWA). Missiles would be vertically launched to provide omnidirectional coverage, being targeted and guided via a midcourse uplink from the MFR before switching to an active seeker for the terminal phase of an engagement.

The technical characteristics enshrined in the original SDMS concept have today been translated into operational capability on the front line in the form of the GWS 45 Sea Viper system equipping the Royal Navy's six Type 45 destroyers. Sea Viper is one of two variants of the tri-national Principal Anti-Air Missile System (PAAMS) developed by MBDA to meet the naval air defence requirements of France, Italy and the UK.[2] In that regard, it shares a number of common subsystems with the Franco-Italian PAAMS(E) variant – qualified in 2007 and equipping the new Horizon frigates of these two nations – specifically the Aster 15 and Aster 30 active radar homing anti-air missiles, the SYLVER A50 vertical launcher system (VLS), and the S1850M Long Range Radar (LRR).

Sea Viper – previously known as PAAMS[S] – shares the same munitions, VLS and LRR as PAAMS(E). Where it differs is in the employment of the BAE Systems Radar Type 1045 Sampson E/F-band active array MFR and a British-developed C2 system to meet a more demanding performance requirement.

As one of the two major subsystems unique to Sea Viper, Sampson is acknowledged to be a key performance discriminator on the basis that its extended detection range and early track formation gains additional reaction time. Housed in a distinctive 'golf ball' radome, and rotating at 30rpm, the Sampson antenna uses back-to-back arrays each

An Aster missile launch from the Type 45 destroyer *Daring*. A total of eight test high-sea firings have been carried out to date and all have been successful. *(BAE Systems)*

populated by multiple gallium arsenide transmit/receive modules. Functionality includes long- and medium-range search; surface picture and high-speed horizon search; high-angle search and track; multiple target tracking; and multiple-channel fire control (including confirmation, track classification, mid-course guidance for Aster missiles via an integral uplink, and kill assessment). The Sea Viper C2 system performs picture management, TEWA, and engagement planning and control. It also provides the primary interface with the Type 45's broader combat management system.

Each Type 45 destroyer has six eight-cell SYVLER A50 VLS, allowing for a maximum ship outload of forty-eight Aster missiles. The Aster 15 and Aster 30 munitions share a common terminal 'dart' – incorporating the seeker, electronics, proximity fuze, autopilot, warhead and uplink receiver – but are differentiated by their range (conditioned by the size of their first-stage rocket booster): Aster 15 is effective out to ranges from 1.7km to 30km, while Aster 30 extends range out from 3km to 100km.

Aster has two major performance discriminators. First, its highly accurate active radar homing capability obviates the requirement for shipborne target illumination and maximises the probability of a 'hit-to-kill'. Second, the missile's aerodynamic design enables manoeuvres in excess of 50 G, while the novel PIF lateral thrust-control system (acting upon the missile centre of gravity) provides for an additional 12 G to minimise miss distance in the terminal phase.

Following launch and turnover, an Aster missile starts to receive missile uplink messages via the MFR, augmenting the missile's inertial measurement unit data with missile position information obtained from the ship's radar. This is designed to ensure that there is a very high probability of acquisition by the time the missile's active radar seeker is switched on, and that manoeuvres during the terminal phase can be dedicated to countering target manoeuvres rather than correcting for errors in its own missile heading.

Sea Viper qualification testing was performed in the Mediterranean using the trials barge *Longbow*, culminating in a successful salvo-firing in June 2010. A first high seas firing was performed from *Dauntless* in September 2010. A further seven high-seas firings – all successful – had been conducted as of May 2017.

To address the emerging threat from anti-ship ballistic missiles, MBDA UK and BAE Systems are continuing work to scope and de-risk a potential integrated air and missile defence (IAMD) upgrade for the Sea Viper system. The United Kingdom, through the Missile Defence Centre, has for more than a decade conducted science and technology activities intended to de-risk and demonstrate critical ballistic missile defence (BMD) technologies and techniques, including two successful live demonstrations of an experimental BMD radar based on the Sampson MFR.

Informed by the success of these activities, the 2015 Strategic Defence and Security Review confirmed that further work would be conducted to investigate '... the potential of the Type 45 destroyers to operate in a BMD role'. MBDA has already completed an eighteen-month project definition and risk reduction study to explore an upgrade of Sea Viper to meet a postulated Initial Anti-Tactical Ballistic Missile Capability that would provide local area defence against anti-ship ballistic missile threats.

While the study has defined an initial level of IAMD capability to counter anti-ship ballistic missiles, such as the much-publicised Chinese DF-21D, MBDA and BAE Systems have been jointly conducting internally-funded study activities to scope an evolutionary path to address the medium range ballistic missile threat. The so-called Full IAMD capability, postulated for the mid-to-late 2020s, would incorporate more substantial improvements including more extensive radar (MFR and LRR) upgrades, and the introduction of the Aster 30 Block 1NT missile. Block 1NT, now in full-scale development under Franco-Italian funding, incorporates a new seeker and offers a lower layer intercept capability.

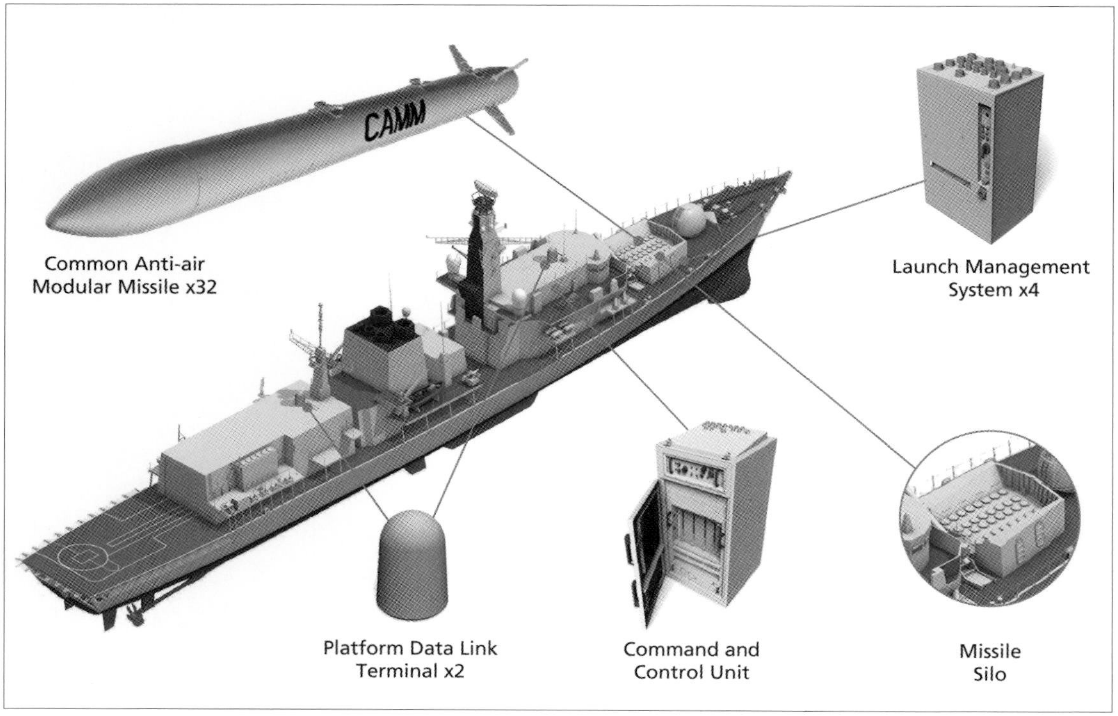

A schematic showing the principal alterations required to the Type 23 frigate to accommodate the new Sea Ceptor missile system. *(MBDA)*

SEA CEPTOR

The Royal Navy's Type 23 frigates were delivered from build with the GWS 26 Mod 1 Vertical Launch Seawolf point defence missile system. This system has latterly embodied the Seawolf Mid-Life Update package, which comprises a series of tracking, guidance and weapon management upgrades designed to sustain a capability to counter the anti-ship cruise missile threat through to a planned 2020 out-of-service date. The update intro-

The two principal components of the new Sea Ceptor system are the new MBDA CAMM and BAE System's Artisan radar. Target positional updates from Artisan are provided to missiles in flight by means of data links. The missiles themselves use active radar homing for the final stages of target engagement. *(MBDA, BAE Systems)*

duces an upgraded Type 911 tracker while also addressing obsolescence in the existing below-decks weapon management and guidance computers.

However, the Seawolf system is now betraying its age. It is inherently constrained in terms of the number of fire-control channels, is increasingly costly to maintain, and imposes a significant topweight penalty on the Type 23 platform (principally the two Type 911 trackers). Furthermore, the shelf life of the current missile warstock is firmly aligned to out-of -service date.

The replacement for Seawolf is the GWS 35 Sea Ceptor local area air-defence system, which represents the realisation of what was previously known as the Future Local Area Air Defence System-Maritime (FLAADS[M]). The FLAADS(M) project entered the Assessment Phase in July 2008 as one of six programmes taken forward under the Team Complex Weapons initiative; a follow-on FLAADS(L) variant, sharing the same effector, is being developed to meet the British Army's requirement to replace the Rapier ground-based air-defence system.

Sea Ceptor is founded on MBDA's new 'soft-launch' Common Anti-air Modular Missile (CAMM), which uses active radar homing (supported by mid-course guidance updates) to deliver an all-weather engagement capability out to ranges beyond 25km. This shifting of much of the guidance and homing burden onto the missile, while exploiting the track-while-scan functionality offered by a three-dimensional air surveillance radar (the RN is using Radar Type 997/Artisan) for target indication and intercept point prediction, means multiple CAMM missiles can 'launch on search' to prosecute multiple simultaneous air threats.

In its Type 23 guise, Sea Ceptor installation has been engineered to use existing GWS 26 Mod 1 infrastructure and interface points: CAMM missiles are fitted in the existing VL Seawolf silo (one canister per cell for a maximum of thirty-two missiles), while launch management system cabinets (providing power, ship and target reference data and initialisation impulses to the missile) occupy the same space, and re-use the same cabling, as the current missile firing units. Each Type 23 will also receive platform data link terminal equipments. These provide for two-way communications between the ship and the CAMM missile; target positional updates can be uplinked from the ship to the missile in-flight, while missile status information and diagnostics can be sent back to the ship. They are installed fore and one aft to ensure uninterrupted 360-degree coverage.

Argyll and *Westminster*, the first two Type 23s to be equipped with Sea Ceptor, have both now received the system and and completed harbour acceptance trials with *Montrose* also scheduled to be ready for sea imminently. *Argyll* is preparing to undertake first firing trials on the Hebrides range, with the system due to become operational before the end of 2017. All thirteen remaining Royal Navy ships are planned to have completed the upgrade by the end of 2020.

The new Type 26 Global Combat Ship will receive Sea Ceptor from build. The Type 26 embodiment will differ in that each ship will be able to carry up to forty-eight missiles, split in separate magazines forward and aft. In addition, each ship will have two main processor units, as opposed to a single unit in the Type 23 fit. Although only one main processor unit will be functionally utilised by the Sea Ceptor system at any one time, the presence of a second unit provides resilience/redundancy primarily in the event of battle damage, with the ability to quickly bring the second unit to a fully operative state and take over the same function as the primary unit as circumstances require.

Unlike Seawolf, which is a short-range point defence system, the extended range and multiple simultaneous engagement capability offered by Sea Ceptor will enable protection for both the host ship and high-value units in the local area. As such, its introduction to service will in part mitigate the reduced number of Type 45 destroyers available to the fleet.[3]

Sea Ceptor has achieved export success ahead of entering service with the Royal Navy. In 2014 MBDA was brought under contract by New Zealand to supply the system to meet the Local Area Air Defence requirement associated with its two-ship ANZAC Frigate Systems Upgrade (FSU) project. Later that year Brazil announced Sea Ceptor as its anti-air missile of choice for its *Tamandaré* corvette programme, although no formal contract has yet been signed.

The most recent success for Sea Ceptor came in

A view of the newly-refitted Type 23 frigate *Argyll* in Plymouth Sound in February 2017 following installation of the Sea Ceptor missile system. The most obvious change to her appearance is the deletion of the Type 911 tracking radars used by the old Seawolf from on top of her bridge and hangar. *(Babcock International Group)*

The two missiles being developed under the Future Anti-Surface Guided Weapon programme will be deployed on the Wildcat HMA.2 helicopter in Royal Navy service. This image shows both the Thales Martlet system housed in a five-cell launch pannier and the much larger MBDA Sea Venom. *(MBDA)*

early 2017 when the system was selected as part of the modernisation package for Chile's three ex-Royal Navy Type 23 frigates. As is the case with the FSU programme, the Chilean Type 23 upgrade is being led by Lockheed Martin Canada.

FUTURE ANTI-SURFACE GUIDED WEAPON

The retirement of the Lynx HMA.8 from RN service at the end of March 2017 also marked the out-of-service date for the Sea Skua anti-ship missile. Originally developed to give the Lynx a powerful punch against Soviet missile-armed fast attack craft and corvettes, Sea Skua was successfully used in combat during both the 1982 Falklands War and the 1991 Gulf conflict. However, changes in the operational environment and increases in target capability had, over time, reduced Sea Skua's effectiveness. An additional challenge was posed by stringent rules of engagement in areas of high background shipping density.

Sea Skua will be replaced from late 2020 by not one, but two new missiles developed under the Future Anti-Surface Guided Weapon (FASGW) programme. The Thales Martlet system will address the FASGW (Light) portion of the requirement whilst the MBDA Sea Venom/*Anti-Navire Léger* (ANL) will meet the FASGW (Heavy) component. In Royal Navy service, both are initially intended for deployment by the new AW159 Wildcat HMA.2 helicopter.

The decision to divide the FASGW requirement into two parts was the outcome of operational studies undertaken during the mid-2000s which identified a clear split in the FASGW target set. Whereas the FASGW (Heavy) requirement reflected the need for a Sea Skua replacement able to prosecute fast attack craft and corvettes from beyond the range of organic air-defence systems, the need for a complementary FASGW (Light) recognised the need for a lighter and more affordable precision guided weapon to deliver proportionate effect against smaller craft devoid of air defences.

Martlet is an exploitation of the laser beam-riding Lightweight Multirole Missile (LMM) developed by Thales UK. Building on the pedigree of the company's existing Starburst and Starstreak surface-to-air missiles, LMM introduces a number of new missile subsystems commensurate with its low-collateral, precision strike role. For example, while the fixed-fin Starstreak rolls in flight, the point accuracy sought for LMM requires that the missile fly 'nose stable', demanding a new control actuation system and fully controlled forward canard fins to impart skid-to-turn commands.

The LMM delivers a 3kg blast fragmentation/shaped charge warhead combining localised effect with good armour penetration. A laser proximity fuze, using simple low-cost gate technology set at the point of launch, is designed to ensure that the missile can successfully engage very-low metal, semi-solid targets, such as rigid inflatables. LMM uses a two-stage motor. The first stage is a very short in-tube burn to eject the missile from its launch tube with zero recoil; the second stage accelerates the LMM round to a speed of just over Mach 1.5.

In the FASGW (Light)/Martlet application, hermetically-sealed LMM missiles will be hosted on a five-cell launch pannier; the Wildcat will be capable of accepting up to four LMM panniers (two per weapon carrier port and starboard). Another key system engineering requirement associated with Martlet is the integration of an Active Laser Generation Unit (ALGU) inside the Wildcat's nose-mounted MX-15Di electro-optical/infrared turret. This will transmit a coded laser beam along which the LMM missile will fly.

Sea Venom/ANL, being developed by MBDA under a collaborative programme jointly funded by France, is a high-subsonic, drop-launch sea-skimming missile characterised by an imaging infrared (IIR) seeker, a two-way datalink allowing the infrared target image to be relayed back to the launch platform for operator-in-the-loop control, a 30kg semi-armour piercing blast/fragmentation warhead, and a maximum range in excess of 20km. While the 110kg missile will be capable of flying a fully autonomous 'fire and forget' profile, operator control will allow changes to the missile in free flight so as to enable, for example, in-flight re-targeting, aimpoint correction/refinement, and safe abort.

Sea Venom/ANL also brings the capability for salvo firing, with the ability to fire up to four missiles at a single target, or against up to four different targets. Another significant difference is the freedom of manoeuvre available to the helicopter post-release: while Sea Skua demanded that the launch aircraft must maintain position and continue to illuminate the target through to impact, Sea Venom/ANL offers the freedom to turn away from the target post-release.

The combination of infrared seeker and operator control has additionally endowed the Sea Venom/ANL with a coastal suppression capability unavailable to Sea Skua. This means that the missile can be employed to strike land targets such as radar sites and coastal missile batteries.

The Sea Venom/ANL boost/sustain propulsion arrangement adopts a fixed boost motor aft and a mid-body rocket sustainer with a downward canted ventral nozzle. MBDA argues that this configuration offers better drop stability, maintains the centre of gravity in flight, avoids the safety issues associated with jettisonable motors, and allows the data link antenna to be incorporated in the rear of the missile.

Leonardo Helicopters is leading FASGW (Light) and (Heavy) integration into the Wildcat HMA.2 under a contract signed in June 2014. Under this agreement, the company is taking responsibility for joint integration, trials and certification of the two missile systems. The FASGW fielding plan will see the two weapons introduced to Royal Navy service in parallel, with trials starting in 2018. Initial operating capability is planned for October 2020.

FUTURE CRUISE/ANTI-SHIP WEAPON

The GWS 60 Harpoon surface-to-surface guided weapon (SSGW) will be retired from RN service

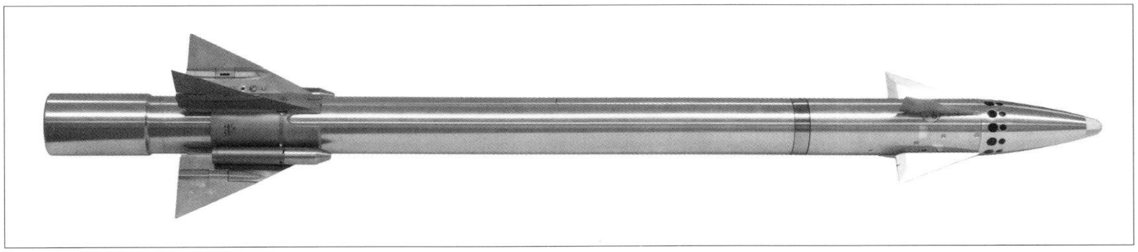

Martlet, the Royal Navy designation for the missile developed to meet the Future Anti-Surface Guided Weapon (Light) requirement, is an exploitation of Thales UK's laser beam-riding Lightweight Multirole Missile (LMM). Martlet is intended to counter smaller craft such as RHIBs and speedboats and is a relatively small missile weighing a total of c. 13kg, with a 3kg warhead. *(Copyright Thales)*

Although the Lynx HMA.8 has been retired from Royal Navy service, many variants of the Lynx remain in service around the world and MBDA sees potential demand for Sea Venom from these operators. This image shows a Sea Venom release from a Lynx Mk 8 in early 2017 as part of trials validating the missile's ability to be launched from these platforms. *(MBDA)*

An image of the Type 23 frigate *Montrose* operating in the Mediterranean in 2012. The Harpoon surface-to-surface missiles located directly ahead of the superstructure will be phased out of Royal Navy service by the end of 2018. Other changes will see the Type 997 Artisan replace the legacy Type 996 surveillance and target indication radar on her foremast, whilst the Type 911 tracking radars will no longer be needed following the replacement of her Seawolf missiles with the CAMM, which uses active radar homing. *(Crown Copyright 2012)*

at the end of 2018. Originally developed by McDonnell Douglas, now Boeing, the GWS 60 system was competitively procured in 1984 to meet the need, set out in Staff Requirement (Sea) 6548, for a second-generation SSGW to equip the Type 22 Batch 3 frigates and Type 23 frigates. The system currently equips all thirteen Type 23 frigates, plus three out of the six Type 45 destroyers; in the latter case, *Daring, Diamond* and *Duncan* are all fitted with equipment transferred from the now decommissioned Type 22 Batch 3 frigates.

The GWS 60 system uses the Harpoon Block 1C turbojet-powered sea-skimming missile. Cruising at a speed of Mach 0.9, and credited with a maximum range of about 130 km, the Block 1C missile flies out steering to a commanded heading (using inertial guidance) before then searching a designated area for its target with a J-band active radar seeker. Several flightpath waypoints can be programmed prior to launch.

Harpoon's retirement will leave RN warships without a heavyweight SSGW, opening up a gap in over-the-horizon anti-surface warfare capability. The decision to remove the system reflects both the age of the weapon system, and the limitations of the Harpoon Block 1C in crowded littoral environments. While Harpoon Block 1C still constitutes a potent all-weather, over-the-horizon capability in blue-water environments, it is acknowledged that current rules of engagement severely limit Harpoon's use in the crowded and cluttered littoral environment because its guidance and homing system does not deliver the required flightpath accuracy or target discrimination.

The United Kingdom retains a long-term requirement for a future offensive anti-surface capability. Early studies are being pursued with France for a Future Cruise/Anti-Ship Weapon (FC/ASW), but any production system is not intended to enter service until c. 2030. As well as Harpoon, FC/ASW is also intended to meet requirements for the replacement of the Exocet anti-ship missile, and the SCALP and Storm Shadow air-launched conventionally-armed cruise missiles.

A €100 million contract funding a three-year FC/ASW concept phase was awarded to MBDA in March 2017. Work is being split 50/50 in terms of both quantity and quality of content between the UK and France; according to MBDA, the activity will '... mature systems and technologies that will increase the survivability, range and lethality of anti-ship and deep strike missiles launched by both air and naval combat platforms'.

The aim of the concept phase, which is being equally funded by France and the United Kingdom, is to lay the ground work and inform both countries' decision making and requirements for a potential follow on assessment and demonstration phase of the next generation of cruise and anti-ship missiles, with a planned operational capability to be achieved by the end of the next decade.

Notes

1. The new relationship with industry is driven by a partnering arrangement known as Team Complex Weapons (Team CW) that originates from initial discussions in the mid-2000s. Its aims are to deliver the UK's Complex Weapons requirements in an affordable manner, while at the same time ensuring a viable industrial capacity.

2. MBDA is Europe's premier developer and manufacturer of missiles. Originally formed through the merger of various European missile businesses, it is currently owned by BAE Systems (37.5 per cent), Airbus (37.5 per cent) and Leonardo (25 per cent). MBDA is at the heart of the implementation of the Team CW approach following signature of a full Portfolio Management Agreement with the MOD in 2013.

3. The Royal Navy had originally planned a class of twelve Type 45 destroyers. This was later reduced to eight ships, and ultimately capped at six.

4.4 TECHNOLOGICAL REVIEW

Author:
Bruno Huriet

MODERN WARSHIP ACCOMMODATION

Some considerations

Descriptions and analysis of naval ships usually focus on their weapons, electronics and propulsion systems; by contrast crew accommodation is seldom detailed. This focus does not reflect the growing importance attached to living standards on board warships by navies across the world, as demonstrated by an ongoing evolution and cumulatively significant development in accommodation provision.

The importance currently given to accommodation is indicated by OPNAV Instruction 9640.1b, issued by the US Navy in 2012. It states, 'Habitability is one of several important factors included in the overall consideration of unit mission readiness. A warship cannot be designed around optimum habitability factors alone, but conversely, habitability factors cannot be progressively sacrificed to other readiness elements without eventual degradation of mission readiness. Maintaining the appropriate shipboard quality of life … supports positive morale and peak mission readiness.'

Up until the early 1980s, the gap between life afloat on a naval ship and life ashore was not considered a major issue; packing thirty or more sailors in a small compartment, bunks on three levels, with communal showers and toilets was considered normal. However, standards of accommodation on land were continuously improving – people were becoming used to more space, individual rooms and larger or – even – ensuite bathrooms. Eventually, there came a time when the difference between life ashore and life at sea grew so large that it had to be reduced. A number of additional factors – some specific to certain countries, some more common – also emerged to drive the greater attention now paid to habitability. These included:

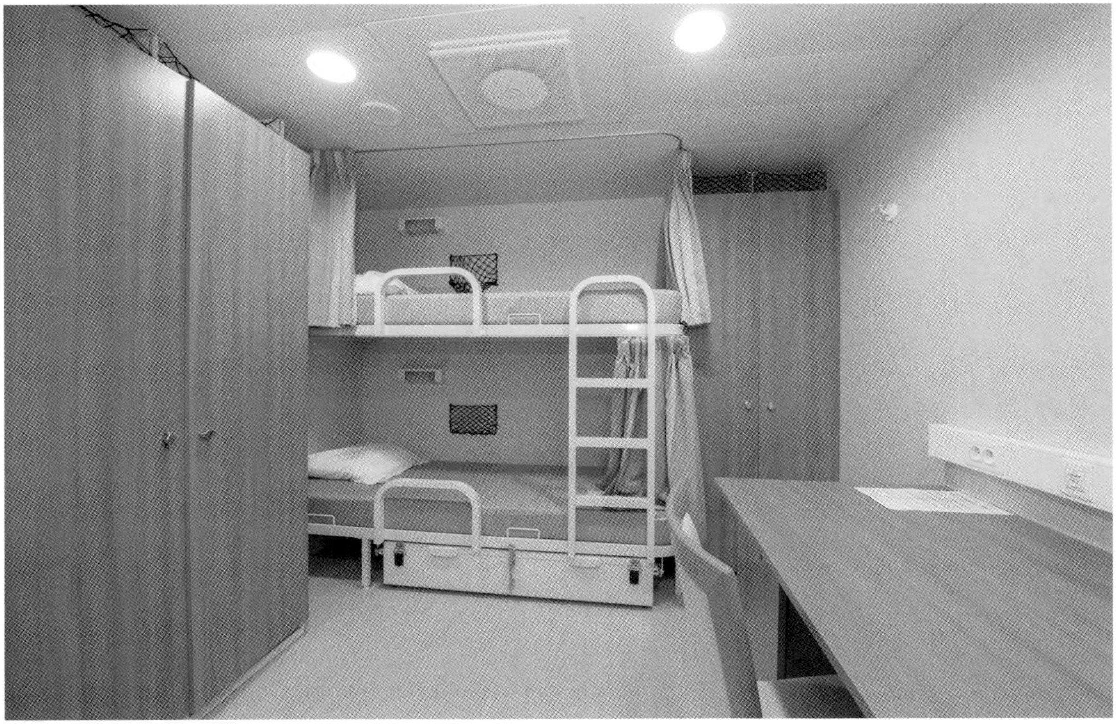

A two-berth cabin for junior officers onboard the French amphibious assault shup *Dixmude*, which was delivered in 2012. Constructed by the Chantiers de l'Atlantique facility at St-Nazaire famous for building many famous cruise liners, the quality of accommodation reflects the expectations of a new generation of sailor. *(V Orsini copyright Marine Nationale)*

- A move away from conscription to full volunteer manning.
- An increasing level of professionalism and expertise in crews as ships became more complex.
- The arrival of female sailors.
- Reductions in crew size due to financial constraints and technological development.
- Changes in the working environment as a significant part of crews came to be deployed behind computer screens.
- The development of a 'constantly connected' way of life ashore.

Crucially, most navies have been forced to take action to attract and retain the young, educated men and women who have to be willing to leave their home for several months at a time to serve afloat. Provision of decent accommodation whilst they do so has been a key part of this effort.

This chapter takes a more detailed look at the influences driving improvements in warship accommodation, explains how modern habitability standards are set and explains some of the key elements behind achieving effective living and working conditions for naval personnel

The French FASM-67 destroyer *Tourville* was commissioned in the 1970s and approaching the end of her career when this photograph was taken in 2010. The need to house a large crew of over 280, around half of which were conscripted ratings, had a major impact on accommodation standards. *(Bruno Huriet)*

FACTORS DRIVING CHANGES IN HABITABILITY STANDARDS

1. CREW SIZE AND COMPOSITION

The size and profile of warship crews has changed significantly in recent years. Most 'Western' navies are now totally professional. They typically deploy a reduced number of ships that have, however, become considerably more complex, incorporating technologies that require skilled operators. This has driven an increase in the cost of individual crew members, a significant factor in calculating overall lifecycle expense. Accordingly, the desirability of reducing crew size has become an important issue. This can be reflected in considering two classes of French surface combatant, both in the c. 6,000-ton category:

- ***Tourville* Class (FASM-67)**: Commissioned from 1974 onwards, these ships had an overall length of 153m and were driven by steam turbines. They incorporated a weapons outfit of two 100mm guns, six Exocet surface-to-surface missiles, Crotale surface-to-air missiles, anti-submarine torpedoes and a Lynx helicopter.
- ***Aquitaine* Class (FREMM)**: Commissioned from 2012 onwards, these ships have an overall length of 142m and incorporate CODLOG (diesel electric or gas turbine) propulsion. They are fitted with one 76mm gun, a VLS for Aster surface-to-air and SCALP land-attack cruise missiles, eight Exocet surface-to-surface missiles, anti-submarine torpedoes and a NH90 Caiman helicopter.

Table 4.4.1 shows changes in manning size and composition between these two generations of warship. Three significant facts are apparent:

- Overall crew size has been reduced by nearly two-thirds.
- The higher level of professional skills required is

Table 4.4.1: CHANGES IN WARSHIP CREW SIZE AND STRUCTURE

CLASS	***TOURVILLE* (FASM-67)**	***AQUITAINE* (FREMM)**
Year Commissioned:	1974	2012
Displacement:	6,100 tons	6,000 tons
Total Complement:	282	108
Of which Officers:	17 (6%)	22 (20%)
Of which Petty Officers:	122 (43%)	70 (65%)
Of which Ratings:	143 (51%)	16 (15%)
Of which Women:	0	8–10%

The modern FREMM type frigate *Languedoc*, pictured here around the time of her delivery in 2016, is approximately the same size as *Tourville* but requires fewer than half the personnel. However, a much higher proportion are officers or petty officers and all are volunteers. Fortunately, there is the space to provide a much higher level of accommodation. *(DCNS)*

demonstrated by a near-inversion of the rank pyramid – there has been a strong increase in the proportion of officers and petty officers amongst the crew, with only a modest number of ratings embarked.

- There has been a change from an all-male to a mixed crew.

The reduction in crew size is apparent in many other new naval programmes; for example the German Navy's F125 frigates have a core crew of around 120 compared with around double that found in the 1990s-era F123 *Brandenburg* class. Part of this reduction has been facilitated by technology; bridges are now far less crowded than in older ships as automated navigational aids supported by multi-function consoles enhance efficiency and situational awareness. It is also increasingly common to have relief crews; support personnel that allow high-value vessels to operate for long periods whilst part or even all the crew can rest or train.

It goes without saying that this trend to smaller crews backed by technology means that sailors need to be more self-reliant and multi-skilled than in the past. As mentioned in the introduction, the need to attract this new generation of sailors – more highly educated than their predecessors – is one of the drivers behind the improvement in habitability standards. The reduced number of personnel on board allows for more space per man but this, alone, is insufficient. Notably, Admiral Prazuck, the *Marine Nationale*'s Chief of Staff, recently stated, '… it remains the case that there is a real gap between the way the young generation lives and the requirements of life on board. What is most disturbing for the new recruits embarking is not having the possibility to be constantly connected to internet and social networks …'

Meanwhile, the growing number of women now serving in naval ships has also required changes. As the distribution of women and men is never fixed, living spaces have to be capable to accommodate any mix. The easiest solution is to have 'neutral' cabins, which include a bathroom, allowing them to accommodate either gender. One consequence of this new thinking is that urinals no longer have a place in the latest warship designs! In older ships, a complete berthing area with a separate washroom and toilet area has typically had to be dedicated to women; not always an easy setup to achieve.

2. CHANGES IN ATTITUDE

The changed nature of modern warship crews has been accompanied by changed attitudes that also have a considerable design impact.

Quite clearly, a skilled professional joining the navy does not look forward to execute chores such as cleaning communal toilets or large berthing areas. Such duties have to be kept to a minimum. This impacts the arrangement of compartments: it is one reason why some navies have one 'wet unit' (typically a prefabricated bathroom module with shower, sink and toilet) per cabin of four or six. Keeping one's 'own' sanitary area clean is expected to be less demotivating than cleaning rows of communal toilets. Accommodation specifications now often require materials and installations that make cleaning and maintenance as easy as possible.

The loading of provisions is another chore that has seen a considerable reduction due to design improvements. Once a time-consuming task undertaken manually by a long line of sailors passing boxes and crates from the pier to the ship's stores on the lowest decks, this has been made more efficient by changing the position of storerooms and installing modern handling systems.

Perhaps most significantly, privacy – not really considered in the ships of the 1970s – is now a key requirement.[1] Instead of having crowded berthing, smaller modules are specified. Another example of change is that access to the wet unit is typically directly from the cabin.

3. TECHNICAL REGULATIONS AND STANDARDS

In the Western world at least, there has been pressure for the military to become more aligned with broader civilian regulations: instead of living in a world apart, navies are now required to follow international standards or national work and safety regulations. Whilst there are quite often exemptions or adaptations, the overall tendency is evident. In particular, one or more of the following international conventions and standards are frequently mentioned as reference documents for new warship construction, even if acknowledging that all requirements can never be fulfilled on a naval ship:

- **ISO – International Organization for Standardization:** Has established several standards relevant to warship design, including those related to lighting, vibrations and electrical systems.
- **SOLAS – International Convention for the Safety of Life at Sea:** An international maritime treaty sponsored by the United Nations' International Maritime Organization (IMO) and signed by 162 countries, the convention covers areas such as ship construction, stability requirements, fire prevention, lifesaving equipment, the transport of dangerous goods and broader maritime safety. SOLAS – aimed at merchant ships – is not directly applicable to warships or naval auxiliaries but compliance is encouraged.[2]
- **ILO – International Labour Organization:** Another UN agency tasked with promotion of rights at work and enhancing social protection. Its Maritime Labour Convention defines minimal standards to be met by accommodation and recreational facilities on vessels engaged in trade but these standards are often referred to with respect to naval warships.

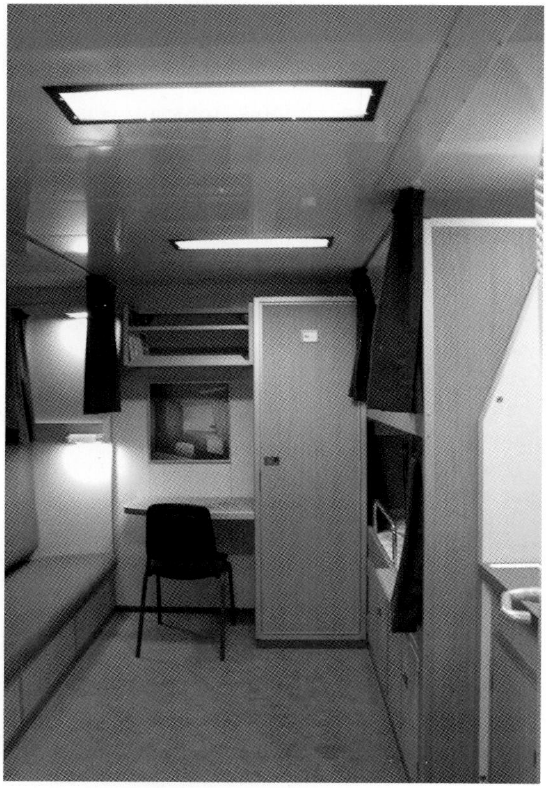

A mock-up cabin showroom produced by the German KAEFER group, a major naval outfitter. The materials used will typically be non-military products that have been tested in accordance with international requirements by one of the major classification societies and then specified for warship use. *(KAEFER)*

Public support for environmental protection has also impacted warship design: given merchant ships are prohibited from discharging garbage or oily water, naval ships – often engaged in environmental 'policing' missions – can hardly ignore the standards they are enforcing. Requirements in this area are set by another IMO convention – the International Convention Preventing Pollution of the Marine Environment (MARPOL) – that has had a particular impact on the design of naval tankers.[3] However, other standards – particularly relating to the treatment of garbage and sewage – have had a direct impact on habitability. Notably, the stowage of garbage on board in refrigerated stores and the installation of sewage-treatment plants have obviously required space and machinery not previously found in naval vessels.

The impact of international standards is often supplemented by national rules relating to specific areas of habitability. To give one example, the design and operation of food preparation and storage areas have often to fulfil the requirements of regulations used ashore or on civilian vessels. The US Navy refers to US Public Health Service (USPHS) rules applicable to cruise ships whilst the *Marine Nationale* is guided by regulations applicable to community kitchens.

GUIDELINES FOR HABITABILITY

The decades that have seen these evolving influences on warship habitability have seen a parallel increase in the importance of classification society rules (regulations) in setting technical standards for warship design. Classification societies are private companies offering two main services: certification and classification. Certification relates to the process of assessing or testing components; for example the fire-resistance of a panel or the strength of an anchor chain. Classification ensures that a ship's design, construction and through-life maintenance is carried out in accordance with rules covering areas such as structure, mechanical and electrical installations. Navies continue to control military aspects, such as weapon and combat systems.

The potential benefits of using a third-party classification society are multiple. Appointing a society provides an acquiring navy the assurance that design and construction are carried out in accordance with recognised, published requirements prepared by an entity independent from the building yard. International standards such as those set by MARPOL or SOLAS are more likely to be achieved. It allows navies and defence procurement agencies to work with experts who are constantly dealing with new technologies and rules. It facilitates the use of COTS (commercial-off-the-shelf) technology in areas where no specific military requirements exist. It provides access to a worldwide network of surveyors and other experts who can oversee newbuilding and inspect ships in service without a requirement for 'in house' teams of engineers.

Amongst the major classification societies involved in establishing warship design and construction rules are the American Bureau of Shipping (ABS), France's Bureau Veritas, Norway's DnV GL (formerly Der Norske Veritas), the UK's Lloyd's Register and Italy's RINA.

In general terms, classification society rules tend to be driven by international civilian standards, such as those established by SOLAS. A good example relates to requirements relating to fire protection, where rules relating to the resistance of materials and construction methods employed play a major role in avoiding the spread of fire. Typically a non-military product, tested in accordance with the SOLAS requirements on which all societies base their rules, will be specified for warship use. The fact that there will already be a large group of reference materials used internationally on merchant vessels and offshore platforms will assist competition and provide reassurance.

In spite of the underlying importance of mercantile standards, there have been moves to develop a specific Naval Ship Code, a goal-based standard setting minimum levels of safety for all naval vessels. Prepared by a working group, the International Naval Safety Association (INSA) first established in

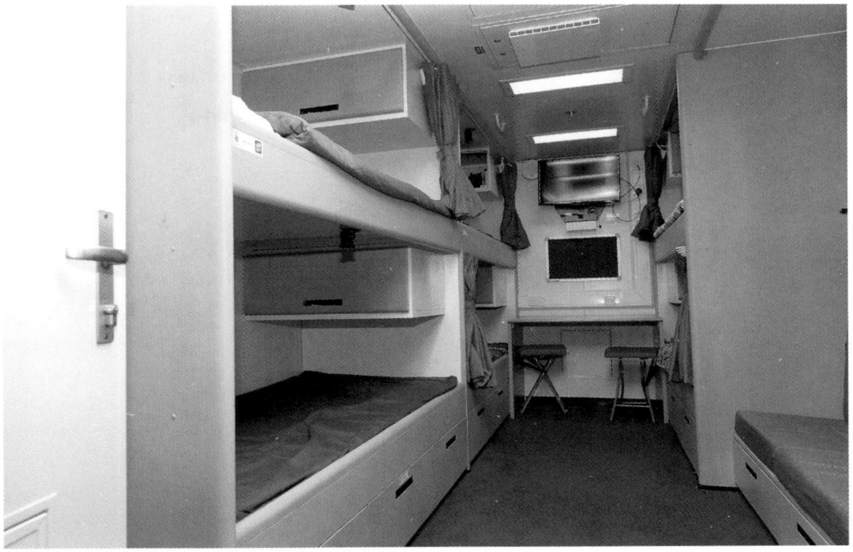

Pictures of berthing areas on the US Navy DDG-51 type destroyer *Mahan* (DDG-72) and the new British aircraft carrier *Queen Elizabeth*. Although there are standard NATO guidelines for minimum levels of surface combatant habitability, different navies have different practices; US Navy accommodation tends to be more austere. *(US Navy/Crown Copyright 2017)*

2008, the code is published as a NATO Standard, ANEP-77, of which version F has just been issued. There is heavy reliance on SOLAS requirements, thereby encouraging the harmonisation of mercantile and naval safety practice. INSA comprises all the main classification societies and a number of NATO and other navies.[4]

In spite of the increased importance of classification society rules, it is also important to appreciate that these do not usually determine how warship interiors are arranged, such as the amount of space per person or the size of the bunks. Most navies work on these requirements internally, taking advice from relevant experts such as those in ergonomics and obtaining feedback from warship crews. NATO has also played an important part here for Western fleets, publishing *Guidelines for shipboard habitability requirements for combatant surface ships*, under reference ANEP-24 in 1993. The document gives recommendations in areas such as berth spacing, the size of mess tables and the width of passageways. However, the standards set are now being significantly exceeded in many fleets; for example, it recommends crew berthing is limited to twelve sailors but current practice suggests four or six is the norm. Although some navies are happy to publish habitability guidelines widely, more limited distribution by other fleets limits meaningful comparison.[5]

BROADER FACTORS CONTRIBUTING TO WARSHIP HABITABILITY

Although guidelines on habitability have an important part to play in defining overall warship accommodation standards, it is important to appreciate that ensuring efficient and comfortable accommodation is about more than defining the size of berths or the number of seats in a mess room. Habitability is a key element of warship design that needs to be considered from the preliminary design stage onwards.

GENERAL ARRANGEMENT

Habitability considerations must therefore play a part from when the first draft layouts of a ship's compartments are produced. Living areas should not be crammed into whatever space remains after attention has been paid to armament, combat systems and propulsion! Instead, space needs to be reserved for living in that is as remote as possible from the machinery spaces (to limit noise and vibration), close to the ship's centre (to limit sea-induced movement) and to allow the hierarchical grouping of accommodation. Ratings, petty officers and commissioned officers will typically be allocated separate berthing areas, usually with two groups of cabins forward and aft for key personnel to ensure dispersion in case of combat damage or fire. Particular attention also needs to be paid to food storage, preparation and distribution, with the layout of galleys and mess rooms optimised for both efficiency and hygiene. Clearly compromises have to be made, often entailing long discussions between all parties involved!

INTERNAL CLIMATE

In order to resist nuclear biological and chemical (NBC) attacks, modern combat ships have no portholes and other openings are kept to a minimum. Moreover, some areas are filled with heat-generating electronic equipment. As a result, air-conditioning is an absolute essential. Temperature and humidity need be kept within set limits: for example US Navy requires sufficient heating to maintain at least 18°C in living areas and 24°C in sanitary facilities. Conversely, in hot climates, air-conditioning should maintain a temperature of not more than 25.5°C. Air renewal is also defined, at six volumes per hour. All this requires space for air-conditioning units and ducts, as well as consuming significant electrical power.

NOISE AND VIBRATION LEVELS

It is difficult to concentrate – or to rest – in a noisy environment. Although research has confirmed that sleep quality impacts concentration, this factor was not really considered until very recently. A major study about sleep was recently concluded by the respected Rand Corporation for the US Military; it suggested that insufficient sleep duration, poor sleep quality and fatigue were common across the services.[6] The consequences can be serious: lack of attention, poor-decision making and lack of awareness of dangerous situations are all potential risks. Whilst much of the solution relates to effective operational management, such as changing the sequence of watches on board ships, sleep quality can also be improved by decreasing the ambient noise level in relevant areas. Many current construction programmes now specify maximum sound levels; for example Italy's *Marina Militare* requires not more than 60dB(A) in berthing areas and 65dB(A) in the bridge or operations room. Vibrations can also have a negative effect on crew well-being and also have to be considered.

Of course noise and vibration levels have wider implications, particularly with reference to the greatly-enhanced focus on stealth that has been a major factor driving warship design. In order to achieve acceptable levels, a complete acoustic study will normally be prepared at the design stage. It will specify the type of noise-resilient supports required for the ship's machinery and the type of acoustic insulation to be installed. For example, bulkheads may incorporate special sound-dampening material, floating floors may be overlaid on decks (e.g. mineral wool will be laid on deck with special concrete on top or steel plates may be used with a visco-elastic layer) and wall panels separating berthing areas from corridors will have sound-attenuating properties. Similar considerations apply to vibrations; most notably machinery is likely to be installed on vibration-dampening material to avoid these being transmitted to surrounding areas.

Noise and vibration are so important now that some classification societies – Bureau Veritas for example – offer an additional notation when providing a classification to a ship. 'Comf noise & Comf Vib' comfort criteria establish detailed maximum values to be found in different areas of a ship.

LIGHTING

Living twenty-four hours a day in totally artificial lighting can also be problematic. It is therefore usual to see standards on light quality and illumination levels specified for a ship's various compartments. For example, 100 lux may be required anywhere in a cabin, but 300 lux at a desk or at a table seat.

The US Navy's LPD-17 type amphibious transport dock *Masa Verde* (LPD-19) pictured undergoing shock trials in 2008. Shock resistance requirements vary from navy to navy and are almost invariably classified; high shock resistance requirements will have an impact both on the materials used in accommodation and the way they are installed. *(US Navy)*

WEIGHT AND SHOCK-RESISTANCE

Although not making a direct contribution to habitability, requirements relating to weight and shock-resistance play a major part in the selection of the materials used in accommodation areas.

Weight is always an issue in designing naval ships. Naturally, areas such as weapons systems, machinery and even fuel capacity are accorded a high priority, as they define the mission capacity of the vessel. Every kilo saved on accommodation can be used to benefit fighting power, speed or range. Typically, a weight-allocation for accommodation and insulation is established during the early design phase; thereafter it is an ongoing engineering challenge to remain within the agreed figure. Lightweight materials will be used wherever possible; for example furniture could be made of aluminium rather than steel and newer, lighter products are replacing the rockwool previously used in fire insulation. Saving weight on insulation or accommodation is almost always cheaper than with respect to machinery or a weapons system.[7]

Requirements relating to shock-resistance are a very sensitive area and are almost invariably classified. However, it is evident that different navies can have quite different standards and – even within a given navy – standards can vary dependent on the type of naval vessel. Some navies will require all accommodation areas to incorporate resistance to a severe level of shock; others will focus resistance on key operational areas such as the operations room and main corridors. Whilst items such as wall panels and ceilings are not, of themselves, essential to a ship's mission it is important that they should not become loose and fall on personnel or prevent circulation within a vessel. Where shock-resistance requirements are high, installation is more complex as supports will need to be stronger and have some damping capacity. It should also be noted that the number of available products that have passed shock testing is quite limited.[8]

WARSHIP ACCOMMODATION TODAY

The various guidelines, standards and broader considerations outlined above have all had an influence on how the crew of a modern warship carry out their daily routine. Invariably, personnel need to

A four-berth cabin in the French amphibious assault ship *Dixmude*. There are a maximum of two levels of bunks and walls and ceiling panels are lined to give better acoustics, enhance comfort and ease cleaning. Standards across European navies are broadly consistent with these. *(V Orsini copyright Marine Nationale)*

An en suite facility on board the French amphibious assault ship *Dixmude*. Mixed crewing means that urinals are a thing of the past! *(V Orsini copyright Marine Nationale)*

sleep, wash, eat and move around, as well as to have some leisure time and – occasionally – receive medical treatment. The accommodation that supports these key functions is considered in more detail below:

BERTHING AREAS

As has already been indicated, berthing-area design is the product of several compromises. Space, cost, safety considerations, privacy and the needs of hierarchy have all to be taken into account. It is therefore interesting that most European navies, at least, have settled on quite similar berthing standards, with maximum numbers of berths per cabin ranging from four to eight. Similarly, it is now rare to see more than two levels of bunks installed. This allows sailors to use their bunk to read or watch a movie; something which was not possible with bunks on three levels.

Berthing areas for officers are better still; chief petty officers and commissioned officers are typically allocated single or double cabins, with a desk provided for performing the office work typically required of division heads and seating for visitors. The commanding and executive officers will always have single cabins, usually with a separate office able to seat two to four people for meetings. On most ships, the commanding officer's cabin is located to starboard, following an old mercantile tradition. It is often now the only cabin equipped with a porthole and some are also fitted with a bath; helpful for treating the victims of hypothermia in an emergency situation.

Whilst berthing is therefore generally more spacious than in the past, cabins remain cramped compared with room sizes ashore. On A French FREMM frigate, a ratings' cabin for four personnel is around 15m^2, a cabin for two junior officers is 12m^2, senior officers and department heads are located in single-berth cabins of 12m^2, whilst the commanding officer's cabin and adjoining office provides a total of 25m^2.

These berthing areas have linings – wall and ceiling panels – giving them a greater feeling of intimacy, better aesthetics and flat surfaces that are easier to clean than pipes and cables trays. Lining also contributes to acoustics, as wall and ceiling panels have sound-attenuation properties. Greater attention is also being placed on furniture, for example, the storage volume of lockers has increased to allow sailors access to more personal belongings.

Interestingly, US Navy practice shows significant differences with that seen in Western Europe. For example, berthing areas of up to forty enlisted sailors (compared with up to 150 on earlier ships) are still considered acceptable. Berths are still arranged in 'racks' of three, although they are grouped into cubicles to achieve greater privacy. In addition – with the exception of senior officers' cabins – the practice is not to use any lining, with ducts, pipes, cabling and the ship's structure all visible in berthing areas. The rationale is the priority given to damage control; the ease of access to structure and cabling is considered to facilitate the detection and repair of any problems.

WASHING

As already mentioned, communal washing spaces with rows of sinks, toilets and showers are now largely a thing of the past: the current standard is to have an en suite bathroom serving each cabin or, at least, several toilets and bathrooms serving a limited number of cabins. This avoids long walks to and from sleeping areas. The bathroom will be fully-equipped with all necessary sanitary fittings, a cabinet and mirror, towel racks (heated for some navies) and – of course – grab rails to use in heavy seas. The use of prefabricated modules is quite common, as it saves construction time: the wet unit is delivered to the yard in a fully-completed state, including piping and electricity. It simply requires fastening to the deck and connection to the ship's fluid and electricity systems.[9]

In order to cope with MARPOL requirements –

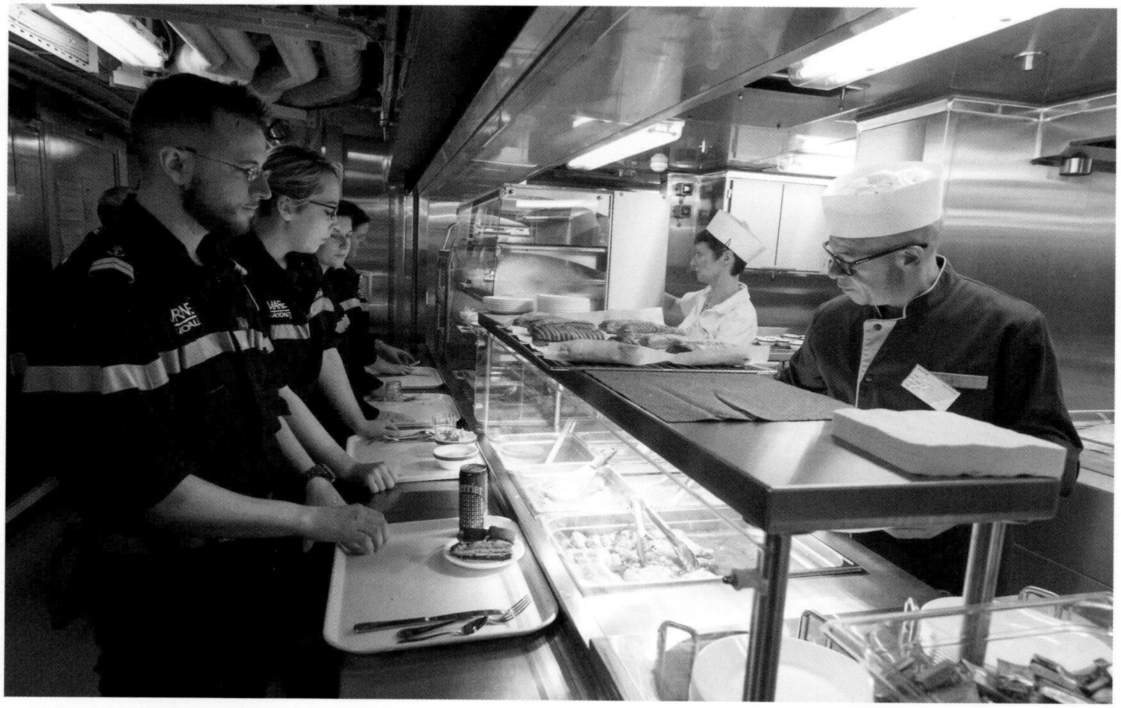

Naval food has a high reputation and this requires careful attention to providing an effective layout. As in civilian facilities, there is generous use of stainless steel to aid hygiene. This is the serving facility on the FREMM *Languedoc*. *(A Groyer copyright Marine Nationale)*

and to limit the size of piping – vacuum toilet drainage is frequently found on naval ships. The system includes a sewage treatment plant.

EATING AREAS

It is often said that food is great in the navy – this requires a catering team of the right size and an equipment layout that meets the same functional and hygiene standards as that achieved on shore. The flow of raw food, ready-to-serve food, soiled dishes and garbage has to be carefully delineated to avoid cross-contamination, no easy task given space is constrained. Galleys are totally clad in stainless steel bulkheads and ceilings with no visible piping or cablework to assist with cleaning. Equipment is generally similar to modern facilities ashore. One notable trend is an increase in the size of cold stores, as fresh food is preferred.

Dining facilities tend to be separated by hierarchical groups to allow various ranks to unwind and speak without constraint. Many navies arrange cafeteria-style service for ratings and petty officers; more senior officers are served in separate wardrooms.[10] As for berthing, there is a tendency for an increased allocation in space per crew member.

MEDICAL TREATMENT

The provision of medical treatment facilities can be an important consideration dependent on crew size and intended vessel function and deployment, even in peacetime conditions. Of course, in case of actual combat, such facilities become even more important. On a frigate, it is usual to find a treatment room complete with operating table for minor surgery, an X-ray machine, laboratory equipment; an intensive care room and an isolation room with dedicated bathroom. A secondary treatment area is usually provided in another part of the ship to ensure dispersion of medical equipment. This area, for example a recreation room, will have lockers storing medical kits to facilitate its conversion into an emergency station.

As in many other areas relating to habitability, medical equipment and procedures follows civilian standards and regulations; as an example, medical waste is to be treated separately.

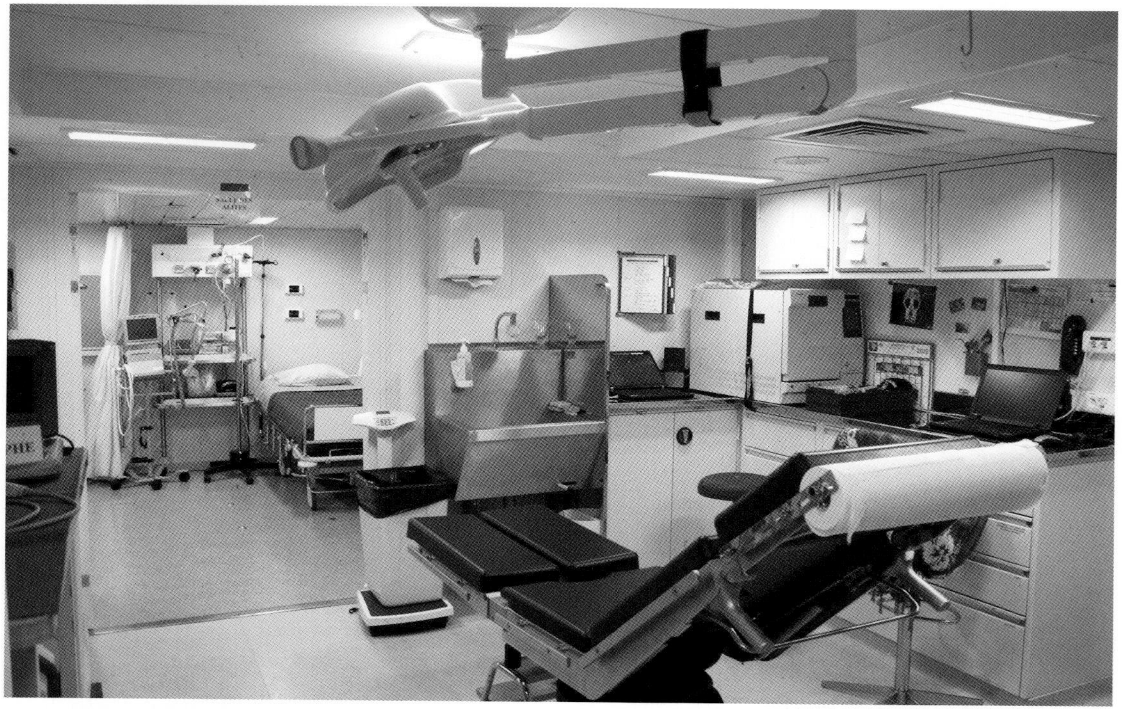

The medical facility onboard the French air-defence destroyer *Chevalier Paul*. Most warships incorporate some form of medical facility and those on ships of frigate-size and above can be quite extensive. They are often located next to the flight deck to aid the reception and/or evacuation of casualties. *(Bruno Huriet)*

RECREATION

One potential downside in the move towards small cabins is that there is no longer a lounge area like those sometimes found in the lobbies of larger berthing areas. Instead, separate recreational spaces are provided, separated by hierarchy for the same reasons as apply to the subdivision of dining facilities. These areas are usually equipped with a television, home cinema and a music system and, often, a bar. Here again, national considerations have an impact, some forces – such as the US Navy – are absolutely 'dry', some allow a certain quota of alcohol a day, some allow wine with meals (as for France's *Marine Nationale*). In larger ships – typically those of frigate-size and above – there is often a fitness room equipped with a range of exercise machines.

One negative consequence of today's connected society is the tendency for many sailors to retreat to their bunk and watch a movie instead of interacting with others by playing cards or watching a film in a group. This self-isolation and a potential resultant loss of 'crew spirit' is a growing concern.

CONCLUSION

It is clear that warship accommodation has altered significantly and positively in recent years. Changing materials, the widespread adoption of mercantile technical standards and the impact of broad social change on accommodation requirements have all contributed to much improved habitability. Most significant, perhaps, has been recognition that change has been essential to attract the skilled professionals needed to crew today's fleets.

Imagine, you are 24 years old, you are a hydraulics specialist and you have joined the navy as a petty officer. Ashore, you had a room with your own bathroom. You arrive on board your first ship. You get the bottom bunk in a stack of three, the one your buddies have to step on to reach their own sleeping space, in a large berthing space for fifteen. You have to walk through passageways to reach a communal bathroom with rows of washbasins. You have to store all your belongings in a ridiculously small cabinet. Won't it be a huge shock and a kind of disappointment? Instead, you find you are in a small cabin of four, with just one buddy above. You can sit in your bed, you have enough storage space, and your bathroom is shared between just four roommates and attached to your cabin. Won't you consider life on board differently?

Notes

1. This change is reflected in defence ministry standards related to accommodation. For example, the UK Ministry of Defence Standard 02-107 first issued in the year 2000 stated that '... the layout and co-location of accommodation spaces shall provide for privacy and personal territorial areas to allow individuals to regulate interaction with other people...' See *Ministry of Defence Standard 02-107: Requirements for Accommodation in UK Surface Warships & Submarines* (London: Ministry of Defence, 2000), p.13.

2. More information about SOLAS can be found by searching the IMO's website at www.imo.org. Although naval vessels and other government vessels in non-commercial service are exempt from SOLAS provisions, as close compliance as possible is encouraged. This is often reflected in national policy, for example a letter of understanding between the UK Ministry of Defence and the Maritime & Coastguard Agency confirms British naval auxiliaries will comply as closely as possible with SOLAS' requirements.

3. MARPOL regulations are also having an impact in other areas not related to habitability, including propulsion emissions. For example, it is intended that the British Royal Navy's Type 26 frigates will be fitted with a catalytic converter to lower pollution.

4. More information on INSA and the Naval Ship Code can be found at http://www.navalshipcode.org.

5. One up-to-date naval resource readily available on the internet is the US Navy's *Shipboard Habitability Design Criteria and Practices Manual (Surface Ships) for New Ship Designs and Modernization T9640-AC-DSP-010/HAB* (Washington DC: NAVSEA, 2016). It can be found at: http://habitability.net/WebData/T9640-AC-DSP-010_HAB.pdf. Reading this document gives a good idea of the way navies currently define habitability standards.

6. See Wendy M Troxel, Regina A Shih, Eric Perdersen, Lily Geyer, Michael P Fisher, Beth Ann Griffin, Ann C Haas, Jeremy R Kurz and Paul S Stenberg, *Sleep in the Military: Promoting Healthy Sleep Among U.S. Servicemembers* (Santa Monica CA: RAND Corporation, 2015). An interesting overview of current problems in the US Navy was provided by Lance M Bacon in 'Navy experts call for more sleep for ship crews' in *Navy Times* – 20 June 2015 (Springfield VA: Sightline Media Group, 2015).

7. Looking more generally at materials, most classification societies require metal furniture to limit the quantity of combustible material on board. This can be steel but, because weight is an issue, aluminium is common. Aluminium honeycomb can also be specified for flat surfaces, such as tabletops. Wood aspect laminates are also often used to give wood's warm feeling. Meanwhile, linings can be mineral wool sandwich panels (high-density mineral wool glued between two thin metal plates) or honeycomb panels. The latter are lighter but more expensive. Decoration can be laminates or a decorative foil, whilst ceilings are frequently 300mm-wide galvanised and oven-painted metal strips, containing a mineral wool inlay for acoustics. Two different philosophies tend to apply to warship flooring. Some navies rely on vinyl tiles glued to an underlay of self-levelling elastic cement – all certified as low flame spread – for appropriate areas, using ceramic tiles for wet areas such as galleys and bathrooms. Others, such as the US Navy, use epoxy resin floors in all areas. The merits and problems of each system are debated but, for both, quality of installation is a crucial factor. Any mistake or carelessness can have important consequences as the floor will become loose, water will creep underneath and deck corrosion occur. There can also be sanitary issues, for example dirty water being trapped under the tiles of a galley.

8. For the US Navy, test procedures are defined in accordance with the MIL-S-901 standard, which is widely published. Some classification societies also publish rules for shock testing, for example by DNV-GL at: rules.dnvgl.com/docs/pdf/DNVGL/CG/2016-04/DNVGL-CG-0063.pdf

9. This method of construction is now extending to cabins, reflecting techniques that have been used in commercial yards for some time. Instead of building every cabin on board and bringing on board every item piece by piece, prefabricated cabins are built in a workshop. The resultant 'box', complete with walls, ceilings, and furniture, and with all cabling installed, is loaded through an opening and rolled into place with the use of special fittings. The cabins of Royal Navy's latest *Queen Elizabeth* class aircraft carrier have been installed this way. Working ashore brings huge installation-time savings as one can work all around the structure, have all necessary components to hand and benefit fully from series production. A cabin factory is quite similar to a car assembly line.

10. Again, there are variations between countries. Perhaps reflecting a national pre-occupation with food, Italy's *Marina Militare* only adopted a shared galley for all ranks on its new FREMM frigates. Conversely, the Royal Netherlands Navy has common messing facilities on its frigate-sized *Holland* class patrol vessels.

Contributors

Richard Beedall: Born in England, Richard is an IT Consultant with a long-standing interest in the Royal Navy and naval affairs in general. He served for fourteen years in the Royal Naval Reserve as a rating and officer, working with the US Navy and local naval forces in the Middle East and around the world. In 1999 he founded *Navy Matters*, one of the earliest naval websites on the Royal Navy. He has contributed to *Seaforth World Naval Review* since the initial 2010 edition, and has written extensively on naval developments for many other organisations and publications, including *AMI International*, *Naval Forces*, *Defence Management* and *Warships IFR*. He currently lives in Ireland with his wife and daughters.

David Hobbs: Commander David Hobbs MBE RN (retired) is an author and naval historian with an international reputation. He has written eighteen books, the latest of which is *The British Carrier Strike Fleet after 1945*, and has contributed to many more. He writes for several journals and magazines and in 2005 won the award for the Aerospace Journalist of the Year, Best Defence Submission, in Paris. He also won the essay prize awarded by the Navy League of Australia in 2008. He lectures on naval subjects worldwide, including on cruise ships, and has been on radio and television in several countries. He served in the Royal Navy for thirty-three years and retired with the rank of Commander. He is qualified as both a fixed and rotary wing pilot and his log book contains 2,300 hours with over 800 carrier deck landings, 150 of which were at night. For eight years he was the Curator of the Fleet Air Arm Museum at Yeovilton.

Bruno Huriet: Bruno Huriet is a French merchant navy officer, holding a dual bridge & engine certificate. After a few years on bulk carriers he worked ashore as ship superintendent on a major naval export contract. Subsequently – for a period now extending to more than fifteen years – he has been employed by KAEFER, an international insulation and outfitting company involved in both civilian and naval projects. Having worked on several naval projects in different countries, he is particularly interested in comparing the varying design solutions adopted by different navies.
A long-standing pictorial contributor to *World Naval Review*, this is his first written article for the series.

Mrityunjoy Mazumdar: Mr Mazumdar has been writing on naval matters since 1999. His words and pictures have appeared in many naval and aircraft publications including *Jane's Defence Weekly*, *Jane's Navy International*, *Naval Forces*, *Ships of the World*, the USNI's *Proceedings* and *Warship Technology* published by the Royal Institute of Naval Architects. He is also a regular contributor to the major naval annuals like *Combat Fleets of the World*, *Flotes des Combat*, *Jane's Fighting Ships*, *Seaforth World Naval Review* and *Weyers Flotten Taschenbuch*. Mr Mazumdar lives in Vallejo, California with his wife.

Norman Friedman: Norman Friedman is one of the best-known naval analysts and historians in the US and the author of over forty books. He has written on broad issues of modern military interest, including an award-winning history of the Cold War, whilst in the field of warship development his greatest sustained achievement is probably an eight-volume series on the design of different US warship types. A specialist in the intersection of technology and national strategy, his acclaimed *Network Centric-Warfare* was published in 2009 by the US Naval Institute Press. The holder of a PhD in theoretical physics from Columbia, Dr Friedman is a regular guest commentator on television and lectures widely on professional defence issues. He is a resident of New York.

Richard Scott: Richard Scott is a UK-based analyst and commentator who has specialised in coverage of naval operations and technology for over twenty-five years, with particular interests in the fields of naval aviation, guided weapons and electronic warfare. He has held a number of editorial position with Jane's, including the editorship of *Jane's Navy International* magazine, and is currently group Consultant Editor – Naval.
Mr Scott is also a regular contributor to several other periodicals including the *Journal of Electronic Defense*, *Unmanned Vehicles*, *Defence Helicopter*, and *Warship World*. This is his first contribution to *Seaforth World Naval Review*.

Guy Toremans: Guy Toremans is a Belgian-based, maritime freelance correspondent and a member of the Association of Belgian & Foreign Journalists, an association accredited by NATO and the UN. His reports, ship profiles and interviews are published in the English language naval magazines *Jane's Navy International*, *Naval Forces* and *Warships IFR*, as well as in the French *Marines & Forces Navales* and the Japanese *J-Ships*. Since 1990, he has regularly embarked on NATO, Asian, South African and Pacific-based warships, including aircraft carriers, destroyers, frigates, mine-countermeasures vessels and support ships.

Conrad Waters: A lawyer by training but a banker by profession, Conrad Waters was educated at Liverpool University prior to being called to the bar at Gray's Inn in 1989. His interest in maritime affairs was first stimulated by a long family history of officers in merchant navy service and he has been writing articles on historical and current naval affairs for over thirty years. This included six years producing the 'World Navies in Review' chapter of the influential annual *Warship* before assuming responsibility for *Seaforth World Naval Review* as founding editor. Now taking a break from his career in the City, Conrad is married to Susan and has three children: Emma, Alexander and Imogen. He lives in Haslemere, Surrey.

Front cover – main: *Baden-Württemberg (German Navy)*; top left to right: *Hobart (AWD Alliance)*; *Spruance* & MH-60R *(US Navy)*; Rail Gun on Expeditionary Fast Transport *(US Navy)*; *Il Chul Bong (Republic of Korea Armed Forces)*
Back cover: - main: Russian Navy Day Fleet Review Saint Petersburg 2016; inset: *Queen Elizabeth (both Conrad Waters)*